The Development of the Organic Network

'Once again Philip Conford has produced a book in which the clarity and elegance of the writing parallels the clarity and elegance of his thought. In a tour de force mercifully free of theoretical musings he examines the complexities, tensions and contradictions within the modern organic movement and in considering its varied dramatis personae reflects upon the broader philosophical issues. A wide-ranging and often controversial account, it is essential reading for all those concerned with Green issues, food quality and the modern history of British agriculture.'
– *Richard Moore-Colyer, Professor of Agrarian History at the University of Wales*

'An authoritative, comprehensive and wide-ranging account of the development of organic agriculture and horticulture during the second half of the twentieth century. The author identifies the key individuals and organizations responsible for transforming organic growing from an unappreciated fringe activity into the force it is today, detailing its philosophical, scientific and spiritual underpinnings. Conford explores the influence on, and cross-fertilisation with, the environmental movement, the wholefood tradition, and issues such as animal welfare, vegetarianism and food safety. Thoroughly enjoyable, utterly engrossing, and magisterial in scope, this book should be read by everyone who cares about how our food is produced.'
– *Alan Gear, former Director of the Organic Gardens, Ryton*

'Philip Conford is the great historian of the organic movement, and this book brings the story almost to the present day. Anyone hoping to understand the idealism, contradictions, and fundamental importance of the movement must read Conford's work.'
– *Eric Schlosser, journalist and author of Fast Food Nation*

'Philip Conford is without question our movement's most authoritative living historian [...] as a chapter of agricultural history in which I have been personally involved has unfolded, Philip has been there quietly observing and accurately recording the events.'
– *Patrick Holden, Director of the Sustainable Food Trust and former Director of the Soil Association*

The Development *of the* Organic Network

Linking People and Themes, 1945–95

PHILIP CONFORD

Floris Books

Published in 2011 by Floris Books

Text © 2011 Philip Conford
Philip Conford has asserted his moral right under the Copyright, Designs and Patent Act 1988 to be identified as the Author of this Work.

All rights reserved. No part of this publication may be reproduced without prior permission of Floris Books, 15 Harrison Gardens, Edinburgh www.florisbooks.co.uk

British Library CIP Data available
ISBN 978-086315-803-2
Printed in Great Britain
by CPI Antony Rowe, Chippenham and Eastbourne

Contents

Abbreviations	11
Acknowledgments	14
Foreword by Jonathon Porritt	17
Introduction	19
Prologue	28
Lord Portsmouth has two 'encounters'	28
The organic movement faces post-war Britain	32
1. The Context: Agricultural Efficiency and Industrial Food	40
The implications of agricultural efficiency	40
Efficiency: its meaning	42
Efficiency: its implications in practice	43
The results of efficiency	56
The post-war development of industrial food	58
Addition	60
Subtraction	63
Distance	66
Corporate manipulation	68
2. The Organic Alternative: Farming	71
Organic farming — what the term means	71
The Biodynamic Agricultural Association	77
Helen Murray; John Soper; Deryck Duffy; David Clement; Alan Brockman; John Davy; Patrick Holden; Nick and Ana Jones	80
Post-war initiatives	94
Soil and Health and the Albert Howard Foundation of Organic Husbandry	95
The Scottish Soil and Health Group	98
Newman Turner and *The Farmer*	99
The Soil Association	101
From 1946 to the early 1970s	102

From the early 1970s to the mid-1980s	107
From the mid-1980s to the mid-1990s	111
British Organic Farmers	113
Permaculture	116
Education and training	124
The organic husbandry courses at Ewell	124
Working Weekends on Organic Farms (WWOOF)	125
Dr Nic Lampkin and Aberystwyth	126
Some overseas connections	128
How the case for organic cultivation was made	130

3. The Organic Alternative: Gardeners and Growers — 136

Organic gardening in the 1940s and '50s	137
Lawrence Hills, the HDRA and the Gears	141
Other notable organic gardeners	147
Organic growers	150
John and Shirley Butler	150
West Wales growers	153
The Tolhursts and the Schofields	155
The Organic Growers Association and *New Farmer and Grower*	156

4. The Organic Alternative: Health and Nutrition — 163

The importance of health: eugenics, economics and the fall from wholeness	164
Positive health	166
The Peckham doctors and their long-term influence	169
The Pioneer Health Centre and its aftermath	169
Dr Kenneth Barlow and the Coventry initiative	173
Dr Peter Mansfield	175
The significance of Peckham	177
Sir Robert McCarrison and his influence	178
Dr Hugh Sinclair	179
The McCarrison Society	181
Two medical scientists: Trowell and Cleave	184
Other medical influences	188
Scharff, Breen, Badenoch, Yellowlees, the Lattos; Stanton Hicks	188
Dental scientists	193
Health and industrial food	196
Positive alternatives	198

The wholemeal staff of life	198
Vegetarianism	202
Herbalism	209
Animal health	213
Newman Turner and Juliette de Bairacli Levy	213
Reginald Hancock	217
The Youngs of Kite's Nest	219
5. Commerce and Consumers	**222**
Newman Turner and the Whole Food Society	223
The Soil Association edges towards consumerism	226
The Wholefood shop	227
The Sams brothers and *Seed* magazine	230
Seed magazine: hippiedom and entrepreneurialism	231
David Stickland and Organic Farmers and Growers	236
Organic standards	242
Graham Shepperd, standards inspector	243
Peter Segger, organic entrepreneur	244
Government and supermarket involvement	246
Responses to the new commercialism	247
Its critics	247
The growing influence of retail	252
Alternatives to the retail system	254
Co-operatives	254
Farmers' markets and box schemes	255
6. Ecology, Environmentalism and Self-Sufficiency	**258**
Organicist ecology before *Silent Spring*	258
The organic movement and the new environmentalism	261
Robert Waller and Michael Allaby	263
Kenneth Mellanby and Edward Goldsmith	264
E.F. Schumacher	266
Resurgence	269
Self-sufficiency	269
John Seymour	269
Practical Self-Sufficiency/Home Farm	270
John Seymour and environmentalism	272
Sedley Sweeny and smallholdings	273
Environmentalist journals	275

The Ecologist	275
Vole	280
The 1980s: *The Ecologist* and Jonathon Porritt	282
Satish Kumar, organic history and a Green Books manifesto	284
The Soil Association and 1990s environmentalism	286

7. The Role of Science — 289

Lord Taverne's brainstorm	289
Science in *Mother Earth*	291
The Haughley Experiment	295
Dr Reginald Milton and Professor Lindsay Robb	297
Evidence from Haughley	299
Douglas Campbell	300
The Pye Research Centre	301
Dr Norman Burman	303
Municipal Composting	305
The research of Dr Ken Gray and Dr A.J. Biddlestone	307
Dr V.I. Stewart and the Bryngwyn Project	310
Dr David Hodges	312
The International Institute of Biological Husbandry (IIBH) and *Biological Agriculture and Horticulture*	314
Lawrence Woodward and Elm Farm Research Centre	318
Dr Anthony Deavin	323
Michael Allaby's case for science	325

8. The Politics of the Organic Movement: An Overview — 327

Organic politics: a sensitive issue	327
Right and Left in the 1940s and '50s	329
The 1960s: a leftwards shift?	334
A call for 'eco-politics'	337
The politics of the Seventies Generation	340
The Ecology/Green Party emerges	342
Growing Concerns and the organic movement	343
Organic responses to Green Party wooing	345
Involvement in policy-making	345
A 'new realism'	345
The politics of consumerism	347
The small-scale alternative	348

9. Earth and Spirit	351
Christian influences on organicist thought	351
Mother Earth and Eve Balfour	353
The Council for the Church and Countryside	354
The stewardship philosophy of Philip Mairet	356
Two Christian environmentalists: E.F. Schumacher and John Seymour	359
Robert Waller's 'ecological humanism'	363
A brief note on Little Gidding and on Judaism	365
The esoteric Christianity of Rudolf Steiner	366
'New Age' spirituality	368
Findhorn	368
The esoteric tradition	369
Bedrock philosophy: limits and wholeness	372
Recognition of finitude	372
Wholeness, holism and Smuts	373
Aesthetic unity and mysticism	375
Edward Goldsmith and the natural order	376
The persistence of husbandry	378
Appendix A: Leading figures in the organic movement, 1945–1995	380
Appendix B: Groups, institutions, organizations and journals in the organic movement, 1945–1995	403
Appendix C: List of Recorded Interviews	424
Endnotes	430
Select Bibliography	445
Index	458

I dedicate this book to the memory of
HANS LOBSTEIN (1921–2010)
in gratitude.

Abbreviations

ADAS	Agricultural Development and Advisory Service
AHFOH	Albert Howard Foundation of Organic Husbandry
AHNS	*Albert Howard News-Sheet*
ARC	Agricultural Research Council
BAA	Biodynamic Agricultural Association
BAH	*Biological Agriculture and Horticulture*
BDA	British Dental Association
BDJ	*British Dental Journal*
BNP	British National Party
BOF	British Organic Farmers
BPP	British People's Party
BSE	Bovine Spongiform Encephalopathy ("mad cow disease")
BTEC	Business and Technology Education Council
CAP	Common Agricultural Policy
CCC	Council for the Church and Countryside
CND	Campaign for Nuclear Disarmament
COG	Cornish Organic Growers
CPRE	Council for the Preservation of Rural England
CSA	Community Supported Agriculture
DDT	Dichloro-diphenyl-trichloroethane (synthetic pesticide)
DNOC	Dinitro-ortho-cresol (insecticide/weedkiller)
DoE	Department of the Environment
E	*The Ecologist*
EC	European Community
ECOP	Eastern Counties Organic Producers
EEC	European Economic Community
EFRC	Elm Farm Research Centre
ERCI	Economic Reform Club and Institute
EU	European Union
F	*The Farmer*

FJ	*Fertiliser, Feeding Stuffs and Farm Supplies Journal*
FW	*Farmers Weekly*
GATT	General Agreement on Trade and Tariffs
GC	*Growing Concerns*
GGA	Good Gardeners' Association
GM	Genetically modified
HDRA	Henry Doubleday Research Association
HF	*Home Farm*
HL	*Health and Life*
HS	*Health and the Soil*
ICI	Imperial Chemical Industries
IFOAM	International Federation of Organic Agriculture Movements
IIBH	International Institute of Biological Husbandry
JSA	*Journal of the Soil Association*
LE	*Living Earth*
LETS	Local Exchange Trading Systems
MAFF	Ministry of Agriculture, Fisheries and Food
MCC	Marylebone Cricket Club
MCPA	2,methyl-4-chlorophenoxypropionic acid (hormone herbicide)
ME	Myalgic encephalomyelitis
ME	*Mother Earth*
MEP	Member of the European Parliament
MERL	Museum of English Rural Life
MS	Multiple sclerosis
MSC	Manpower Services Commission
MWB	Metropolitan Water Board
NAAS	National Agricultural Advisory Service
NCB	National Coal Board
NE	*New Ecologist*
NEW	*New English Weekly*
NFG	*New Farmer and Grower*
NFU	National Farmers Union
NGG	National Gardens Guild
NH	*Nutrition and Health*
NPK	Nitrogen, phosphorus and potassium (as constituents of artificial fertilizers)

NUAW	National Union of Agricultural Workers
OAS	Organic Advisory Service
OECD	Organisation for Economic Co-operation and Development
OF	*Organic Farmer*
OFF	Organic Farm Foods
OFG	Organic Farmers and Growers
OGA	Organic Growers Association
OMC	Organic Marketing Company
OP	Organophosphorus compounds (synthetic insecticides)
PHC	Pioneer Health Centre
PN	*Permaculture Newsletter*
PRC	Pye Research Centre
PSS	*Practical Self-Sufficiency*
R	*Resurgence*
RRA	Rural Reconstruction Association
RSM	Royal Society of Medicine
S	*Seed*
SA	*The Soil Association* (journal)
SAQR	*Soil Association Quarterly Review*
SF	*Star and Furrow*
SH	*Soil and Health*
SOP	Somerset Organic Producers
Sp	*Span*
SPGB	Socialist Party of Great Britain
2,4-D	Dichlorophenoxyacetic acid (herbicide)
UKIP	United Kingdom Independence Party
UKROFS	United Kingdom Register of Organic Food Standards
USDA	United States Department of Agriculture
V	*Vole*
WF	*Whole Food*
WWOOF	Working Weekends on Organic Farms

Acknowledgments

My experience of researching for this book has led me to conclude that there is a great deal of generosity to be found among members of the organic movement. The many people I approached were almost invariably happy to spare me time in order to discuss their work for the movement, and in some cases they also provided me with documentary material which I have added to my ever-expanding archive of organic history. I am particularly grateful to Patrick Holden and his staff at the Soil Association for making me welcome there and providing me with space to work on the Association's journals and other material; to Lawrence Woodward, Pat Walters and Pam Tibbatts at Elm Farm for making my visits there so pleasant, and allowing me to make free use of their photocopier; to Sue Coppard of WWOOF for copying many back numbers of the WWOOF newsletter for me; to Dr Julia Wright and staff at Garden Organic for organizing a varied programme of research and social contacts for me when I spent a few days there, and to Dr Anna Ashmole, for letting me have a copy of her doctoral thesis.

Since being made voluntarily redundant from my job as an English teacher in 1998 in order to try to concentrate on researching the history of the organic movement, I have had no regular income and have been unable to raise funding for research through academic channels. I am therefore especially grateful to all those who have been willing to put money my way in order to support my work. Giles and Mary Heron provided an early token of their faith in my project, and it is possible that without their gift I would not have felt able to commit myself to it. Giles and Mary can claim a most impressive 'organic pedigree': Giles is the son of Tom Heron of the *New English Weekly*, and Mary, who worked for Eve Balfour in the early 1960s, can recall figures such as Rolf Gardiner and Richard St. Barbe Baker visiting her father in the 1930s. It is very appropriate that they should support work which emphasizes the continuity to be found in the organic movement's history, and I recommend Giles' book *Farming with Mary* (Heron 2009) to anyone interested in an account of the exhausting and rewarding efforts involved in running an organic hill farm. For financial support, I am also most grateful to the Biodynamic Agricultural Association, Gregory Sams, Don Gaidano of Horizon Organic, and Rachel Rowlands. Craig Sams arranged for the Soil Association to support my work with a substantial sum, and I am very grateful to him for this and for the fact that he made it clear, in writing, at my request, that my acceptance of the money did not in any way tie me to a party or institutional line. As *The Development of the Organic Network* is in certain respects a reproach to the Soil Association for neglecting its own heritage, this was particularly liberal of him.

Above all, though, this book would not have been possible without the financial support of Hans Lobstein, and for this reason I dedicate the book to him. Sadly, he did not live to see its publication. Hans first became involved with the organic movement about seventy years ago, when as a science student at London University he met George Scott Williamson and Innes Pearse, and worked for a time on Eve Balfour's farm at Haughley. He actively promoted the organic cause thereafter and, at nearly ninety, was still writing frequently to newspapers, magazines, politicians and bureaucrats on matters agricultural and nutritional. He was the sort of 'benevolent private individual' that Tom Hodgkinson (2007, p. 244) recommends artists to seek out: a modern-day patron, in fact, who did not wish to influence the content of this book, but offered constant encouragement. I was very fortunate to find so generous a supporter.

My thanks are due to everyone who spared time for me to interview them, and particularly to those who provided me with food and with accommodation when I visited. Appendix C lists all the recorded interviews: it is my intention that these will be lodged with the Museum of English Rural Life (MERL) at Reading University. Many of them already have been. Some of the interviewees also provided further information in conversations which were not recorded. The following people have also shared information and memories with me, either in interviews or in letters: Fred Birks; David Clement; Norman Foot; Jo Foster; Martin Rudy Haase; Maddy and Tim Harland; Meg Haver; Tessa Hosking; Doreen Huffman; Earl Kitchener; Cherry Lavell; Constance Leigh; Brian Leslie; Ali Mathieson; Frank Moore; Judy Muskett; Henry Nicholls; Lord O'Hagan; Howard Paton; Gerald Pearson; Jim H.S. Phillips; Mrs Olive Rose; Sedley Sweeny; H. Walmsley; Brian Walton; Julian Wood; Mrs Vera Wright; Dr Walter Yellowlees. I have enjoyed a trans-Atlantic correspondence with Mrs Helen Zipperlen (formerly Helen Murray), who has written me long letters packed with memories of her years working for the Soil Association at Haughley and in Scotland, and of the biodynamic movement, its leading personalities and its philosophy.

I have benefited from having valuable archival material entrusted to me. Some of it has already been passed on to MERL, and all of the rest will be lodged there in due course. This material came from John Hosking (son of J. Everard Hosking of the Kinship in Husbandry), Riccardo Ling, Joanna Ray (daughter of Dr Kenneth Barlow), Dr Victor Stewart, Robert Waller and John Wheals. Mick Stuart, son of Cdr R.L. Stuart, passed on to me a treasure-trove of his father's papers and acted as guide on the island of Islay. I am most grateful to Dr David Hodges for organizing and passing on to me his archive of papers relating to his many activities on behalf of the organic movement.

Ron Frith of Dunstable generously passed on to me a full set of *Vole* magazine and many back issues of *The Ecologist, Resurgence,* and *Practical Self-Sufficiency/Home Farm,* while Susan Maguire of Selsey provided me with a full run of several years' issues of *Mother Earth.* Vivian Griffiths arranged for me to borrow many issues of *Star and Furrow.* I was privileged to be able to work on the archives of Mary Langman following her death in 2004, and to rescue from obscurity certain invaluable historical documents.

I also thank those who, though not involved in the organic movement, have helped, encouraged or supported in various ways during what has been at times a struggle: Betty Boorman; Dr Jeremy Burchardt; Professor Richard Moore-Colyer and Dr Mike Tyldesley.

The staff of Select Office Services, Chichester, and in particular Tizzie Chandler, have been endlessly patient and cheerful in backing up innumerable versions of documents for me and printing off the various drafts of the typescript.

Once again, my thanks are due to Christian Maclean and my editor Christopher Moore at Floris Books, who have been willing to gamble on a second volume of organic history. In conjunction with my earlier book *The Origins of the Organic Movement,* the present study forms a diptych of the movement in Britain, from the mid-1920s to the mid-1990s. I am indeed fortunate to have found publishers for whom commercial considerations are not the sole purpose of their business, and am very grateful to them for enabling me to publish the results of several years of research.

Foreword

I was delighted to be asked by Philip Conford to write this short Foreword for *The Development of the Organic Network, 1945–95*, especially as he had quite properly spotted back in 1984 that I was someone 'who showed little awareness of organic history'! In the intervening 26 years I gradually acquired more and more 'awareness', but was only too conscious of the fact that there was a great deal more to learn. I now feel that all the gaps have been filled in — courtesy of Philip's impeccable research and authoritative interpretation.

My world (of sustainable development) and the world of organic farming overlap to a considerable degree. We share many 'inspirational founders' — including Richard St. Barbe Baker, Rachel Carson (whose 1962 critique of intensive chemical farming in the US is widely seen as providing the foundation of the modern environment movement), Fritz Schumacher, John Seymour and many others.

And we share what Philip Conford himself describes as 'a bedrock philosophy', anchored in two fundamental principles: the recognition of *limits* in the natural world; and the need for developing systems-based, holistic ways of looking at our role in the world today. These principles informed the work of many of the founders of the organic movement, including the redoubtable Lady Eve Balfour: 'nature always produces order. It is Man who causes chaos by his persistent attempt to resist or ignore natural laws'.

The history of the organic movement is one of fierce debate, personal and organizational rivalries and passionate political advocacy. Philip Conford steers us through those contested waters with great skill, not by adopting some studied, pseudo-neutrality, but by providing context, contrast and editorial insight. The debate about 'commercialism', for instance, runs deep and wide in the organic movement, and will in effect never be resolved. But it helps to understand why such fiercely contrasting views have their own logic and their own integrity.

At the same time, it helps to understand why concerns about health and nutrition have always provided the heartland of many people's commitment to organic farming, even as a self-perpetuating line-up of 'mainstream scientists' and the combined weight of the modern food industry have sought to disparage the principle and conscientious scientific foundations which underpin so much of the Organic Movement's work.

These battles with every aspect of orthodoxy and 'establishment thinking' provide an inspiring testament to the work of countless champions over the decades. Philip Conford quotes from Walter Yellowlees back in the 1970s as he tried to capture the mood of those early organic campaigners feeling 'like a faltering David, facing a monstrous Goliath, whose breast plate is professional indifference and scepticism, whose sword is blinkered technology, and whose helmet is mind-bending advertisements for worthless food'.

So how will history judge these pioneers? Beyond my own unstinting admiration for their cumulative endeavours, and my intuitive sense that today's 'mainstream' feels a lot more 'organic' than yesterday's mainstream, I have no idea. But I have no doubt that *The Development of the Organic Network, 1945–95* will provide a source of information and enlightened analysis that will better inform any subsequent judgments.

Jonathon Porritt
Founder Director, Forum for the Future
November 2010

Introduction

In January 2009, at a conference for organic producers held at Harper Adams College in Shropshire, a prominent and long-term member of the organic movement took issue with me over an article I had recently published in an academic journal.[1] During the course of the discussion, she gave it as her opinion that my forthcoming book — the present volume — would be seen as the definitive history of the British organic movement: a prospect which she evidently regarded with some trepidation. I therefore want to start by stressing that *The Development of the Organic Network* should not be considered the definitive history of the British organic movement, that it is not intended to be so, and that it is doubtful, given the organic movement's very complex and disparate nature, whether any book could ever legitimately claim such a title. If one ever does, then it will have to be based on a superhuman amount of research.

What I will claim for this book, though, is that it is the first both to give a picture of the organic movement as a whole, and to look in considerable, though not complete, detail at how it developed in Britain from 1945 onwards. Like its predecessor, *The Origins of the Organic Movement* (2001), it is a pioneering work, mapping out the territory and describing many of its features. But there is a great deal more work to be undertaken, so *The Development of the Organic Network* should be regarded as a starting-point, not as a 'definitive', conclusive statement. If it comes to be seen as the latter, this will indicate that, regrettably, organic activists have not recorded their own experiences, and that agricultural historians have not chosen to research this particular area. In general, the latter show a surprising reluctance to address the transformation of agriculture which has occurred during the past eighty years or so: a much more far-reaching agricultural revolution than that of the eighteenth century.[2] Ideally, then, I would hope that this book serves as a stimulus to further research, and that those who disagree with its views, or feel that certain important aspects of the movement have been neglected or insufficiently emphasized will respond by adducing evidence in favour of their own perspective. This, after all, is one of the main ways by which the work of history proceeds.

One reviewer of *The Origins of the Organic Movement,* Matthew Reed, pronounced that 'Conford has demonstrated a propensity to under-theorise'. If that is indeed my propensity, then I am content to call as witness in its favour the eminent historian the late Tony Judt, whose view was that 'the more Theory intrudes, the farther History recedes'. More poetically, there is the warning offered by the Victorian Lord Houghton to those aspiring to the clear heights of Reason, that 'distance leaves a haze/ On all that lies below'. One major purpose of this book is to dispel some of the haze that has begun to settle over half a century of organic history.[3]

In fact, I do have a theory about how we can find out what lies below the haze; though not a theory as Reed would understand it, in the sense of applying abstract sociological concepts to complex historical material. To pursue the mountaineering metaphor, we retrace our steps from the heights in order to slog around the foothills and across the plain. This is straightforward, unglamorous empiricism: seeing what is actually there. *The Development of the Organic Network* therefore concentrates on many of the figures who played an important part in organic history between the 1940s and the 1990s, and looks at the publications which promoted organicist theory and practice, drawing attention to the topics they dealt with and identifying issues which were debated — perhaps repeatedly, over many years, as in the case of vegetarianism — in their pages. This is an inductive process, considering the people, examining the recurring themes, and hoping that when sufficient evidence has been amassed some conclusions can be drawn. It seems to me that this more cautious approach should in time produce a greater degree of enlightenment than the *a priori* application of, say, Social Movement Theory.[4]

For it is inescapable that what one finds beneath the haze is multifarious and complex, replete with tensions and contradictions: so much so that LAWRENCE WOODWARD, one of the most powerful organic voices of the past thirty years, doubts whether one should even talk about an 'organic movement' at all. After studying fifty years of organic history, I can sympathize with this view. There have been internal divisions in major areas of organic concern, as the following chapters will demonstrate. Perhaps the best example, apparently trivial but in fact profoundly symbolic, is the difference of opinion between Woodward (farmer and scientist) and CRAIG SAMS (entrepreneur) over the value of an organic Mars Bar, if one were to be developed (see p. 253). During the course of the book I often refer, as a convenient shorthand, to the SEVENTIES GENERATION of organic activists: a very influential group of people who became involved in the organic movement during the 1970s and went on to form new associations, produce new journals, promote organic methods in the media and take them into the world of government policy. But the term Seventies Generation does not in any way imply a uniformity of outlook: their disagreements over philosophy, aims and tactics were many, and led in some cases to long-lasting antagonisms and lingering resentments. If a simple phrase like Seventies Generation, with considerable basis in fact, can apply to such a range of differences, there is little point in attempting to employ wider-ranging abstract categories.

Despite Lawrence Woodward's doubts about the existence of an 'organic movement', I use the term because it seems to me that philosophical differences, disagreements over tactics, factionalism and personality clashes are commonly to be found in movements, and that the term 'movement' does not necessarily imply a tightly-knit group of disciplined people all holding the same ideas. Rather, it can denote a range of people, organizations and publications opposed to an existing state of affairs (in the case of the organic movement, the dominance of industrial agriculture and technological food production) and working to curb or replace it. How this is to be achieved will be the subject of much debate, and it may be difficult to discern exactly what it is that different groups, let alone individuals, have in common beyond their dislike of what they oppose. Among members of the organic movement one can find every shade of political opinion; a variety of religious faiths, as well as rejection of the 'spiritual'; a desire to change the system from the inside and determination not to compromise with the system; belief that the case for organic cultivation can be made on purely scientific grounds and that the case is essentially ethical: and so on. Examination of what supporters of the organic movement have actually said or written — as opposed to what they 'ought', or might be expected, to think — brings home just how complex or confused the movement has been. At the end of the book, I suggest two ideas — limits, and wholeness — which might be regarded as the implicit bedrock of the organic philosophy; but I offer these more as a hypothesis for further investigation than as a definitive conclusion. They seem to emerge as constants from the plethora of ideas which one finds among the thousands of pages of organicist writings.

As far as the book's title is concerned, use of the word 'network' rather than 'movement' reflects my interest in the connections between the leading personalities of the organic movement, the organizations in which they were active, and the recurring themes, issues and debates of the half-century which the book covers. The book is therefore a work of synthesis, concentrating more on what people and things have in common than on seeking to differentiate between them. Such an approach implies an emphasis on historical continuity in the organic movement: a continuity whose existence some members of the Seventies Generation warmly dispute (as I can testify from personal experience), though others recognize it. This is a contentious area of organic history, but I have found little during my research to convince me that the developments of the 1970s and '80s mark a complete disjunction from what had happened earlier. On the contrary, there is much to support the view that the newcomers soon found themselves involved in what was already established, and that when they went on to found fresh initiatives of their own, they owed a good deal to senior mentors. No doubt this will be disputed — an evidence-based argument for the opposing position is long overdue — and so the debate will proceed.

One consequence of trying to create a disjunction between the developments of the 1970s and '80s and what went before is that the efforts, ideas and achievements of the earlier organic activists are downplayed or neglected, becoming at best merely the ineffectual precursors of their successors' more triumphant arrival in the mainstream. Thus do important figures get erased from organic history. According to the Seventies Generation, their elders were stuck in a time-warp; according to an article on Craig Sams in the *Independent on Sunday*'s magazine supplement, the SOIL ASSOCIATION was just a 'sleepy' charity until Sams transformed it.[5] Such an opinion denigrates the hard work of people whose efforts do not deserve to be dismissed in such cavalier fashion: people who kept the organic flame alive at a time when there was considerably less support for organic ideas than there has been during the past twenty years. While it is beyond dispute that the Seventies Generation brought remarkable energy and imagination to the organic cause, we need to see the movement as a whole and pay attention to its roots: surely an appropriately 'organic' response to the organic movement's history.

Like *The Origins of the Organic Movement,* then, the present book is an exercise in historical retrieval. In many cases, those who are being retrieved are still with us, a fact which should remind us of how close we still are to the period which the book examines. That so many important figures have been consigned to the shadows is a result of the increasingly intense spotlight which over the past decade or so has been directed at 'Beautiful People' (to use the phrase of writer and organic farmer Patrick Noble), as if the movement's success should be measured solely in terms of the number of organic products shifted from supermarket shelves. While fully accepting that the sale of organic products (particularly food, and the less processed the better) is important to consumers and vital to producers, I feel that something risks being lost in this attempt to transform consumer capitalism from the inside: that it indicates, ironically, a lack of the 'wholeness' which was once at the heart of the organic philosophy. (Fortunately, counter-balancing trends are developing, like Community-Supported Agriculture and the Transition Towns movement, which recognize the fragility of the existing retail system.) The current prominence of organic entrepreneurs, celebs and consumerism is perhaps the one real disjunction between past and present that exists in the movement, since in the early days one of its main aims was to reduce the influence of 'the unproductive middle-man' to the benefit of producers and consumers alike. In this regard, the organic movement has largely failed, despite various attempts over the decades to establish viable co-operatives, and the recent growth of farmers' markets and direct marketing schemes. Their time is still to come.

Although *The Development of the Organic Network* looks at the organic movement's attitude to commercial ventures, these take their place as just one strand in the network, not as the movement's be-all and end-all. Any history of the movement as it has developed since the mid-1990s would have to devote considerably more space to the commercial dimension; but it is too soon for such a history to be written, as we have yet to see where the liaison of the organic movement (strictly speaking, influential sections of the movement) with the commercial establishment and the social élite will lead it.

My own interests lie less with entrepreneurs and celebs than with peasants and philosophers: that is, with those who produce and with those who have tried to articulate a theoretical basis for the organicist view of the world. These are not of course absolute distinctions: a dedicated grower, such as PETER SEGGER, for instance, can also display a strongly entrepreneurial spirit. But when you study the history of the organic movement, you come across a host of people who were 'peasants' as opposed to 'traders': that is, they were dedicated to their own plot of ground — whether literal earth or an area of expertise like herbalism, nutrition or the science of composting — and worked steadily at it over many years. They were not 'sexy' — in the dismal contemporary sense of the word which indicates the likelihood of arousing media interest — but they tirelessly, and often at cost to themselves, promoted the organic message, worked always to increase the fund of available knowledge and skill, and took the trouble to leave some record of their endeavours.

The philosophers are fewer and farther between, but to compare the 'editorials' in the Soil Association magazine *LIVING EARTH* today with those which JORIAN JENKS and ROBERT WALLER once wrote for *MOTHER EARTH* is to be made aware of the extent to which the Association has jettisoned its intellectual heritage. Linda Theophilus, who as Linda Girling worked for the Association at Haughley forty years ago with Robert Waller and MICHAEL ALLABY, has recalled that Waller would spend a lot of time in his office thinking and drafting ideas. Whatever view one takes of the results — they had both their detractors and their admirers — it was at least recognized by many in the organic movement that belief in the organic approach implied an attitude towards nature and to humanity's relationship with nature that required systematic elucidation. What were the principles underlying the agri-business approach to the natural world, and why were they, in the organicists' view, mistaken? The answers lay in Europe's intellectual history, and thinkers such as PHILIP MAIRET, E.F. SCHUMACHER, Robert Waller and EDWARD GOLDSMITH attempted to uncover them, analyse their flaws, and argue for an alternative philosophy. In recent years, the organic movement has been so keen to advance its cause through persuading shoppers to buy organic products that the philosophical case for an organicist view of the world has largely disappeared from the movement's publications. But this was once considered an important element of the movement's strategy. Perhaps it is the organic movement itself which currently demonstrates 'a propensity to under-theorise'.

These broader philosophical issues bring us to the controversial areas of politics and religion. That the organic movement's roots were closely entwined with a 'radical Right' political outlook is incontestable, however unwelcome to most present-day organic activists the fact may be; this is perhaps one reason why the Seventies Generation was so keen to distinguish itself from what had happened earlier in organic history. Organic politics in the fifty years covered by this book were much more fluid, with people of all shades of political opinion playing their part in the movement. I have attempted an overview of the movement's political dimension in Chapter 8; such is the disparate nature of the subject-matter that very little in the way of conclusions can be drawn — except perhaps that support for the organic philosophy implies a sympathy for the regional and small-scale. But this is another area which requires more analysis — as long as that analysis is based on what views people actually held and what policies pro-organic bodies in fact advocated and supported. Again, I favour the inductive approach to such complex material.

The final chapter, on the religious/spiritual dimension to the organic movement, is likely to be as unwelcome to some readers of this book as the sections on politics were to some readers of *The Origins of the Organic Movement*. Since we live in a period of particularly aggressive secularism, it may well be tactically inappropriate to draw attention to the presence of some form of religious, or spiritual, or esoteric belief in organicist philosophy.[6] But since this book is a history rather than a manifesto, it recognizes that such belief has been not merely present, but highly influential. For many of its leading figures, the organic movement has had a spiritual purpose, of re-connecting people with God — or the nature spirits, or a greater spiritual power — through humility in the face of nature and recognition of dependence on a natural order. This is in contrast to what it sees as the arrogance and hubris of the technological mind, constantly seeking to change, dominate and defeat nature. However, Chapter 9 may also cause some surprise by the attention it pays, as did my earlier book, to the importance of Christian thinkers and activists in the organic movement. Once more, this is a question of looking and seeing what is actually there, and if I lay emphasis upon it, this is because it has either lain undiscovered, or been forgotten or ignored. Here is another project for researchers to undertake: why should the organic movement have attracted so many adherents of the Christian faith? Of course, the organic movement has also attracted many who have not been adherents, or who have been hostile towards it; and this raises the question of what the irreducible principles of the organic philosophy might be. I end the book by making a couple of suggestions, but, like so much else here, they should be regarded as spurs to further investigation.

And there is still so much to investigate. Although I have stretched my publishers' patience with the length of this book, I am all too aware that, even so, it cannot do justice to the wealth of material which exists in the books and journals of the organic movement, or to the experience and effort which organic farmers, growers, doctors, scientists, marketers and thinkers devoted to their cause. This is not a matter solely for academics: contemporary organic activists can only gain from becoming aware of how much can be learned from the organic movement's history. To remain in ignorance of it is to opt for a kind of collective amnesia. At various points during the book I provide lists in order to indicate the range of issues which the various organicist publications covered; these journals impress by their range, their depth, and the dedication which they embody. They were very informative, not just about matters organic, but about the wider context of national agriculture and nutrition.

I regret, then, that I have not been able to devote as much space as I would have liked to some of the interesting people and themes which feature in the history of the organic movement between 1945 and 1995. In particular, it seems to me that the history of BIODYNAMIC CULTIVATION in Britain deserves a whole book in its own right. There is no reason why a conventional agricultural historian should not undertake this task: one does not have to accept the principles of biodynamic methods in order to write about them, any more than one needs to be a Communist in order to write about Soviet Russia. Biodynamic cultivation is a major topic, and I have been unable to do more in this volume than indicate some ways in which the biodynamic movement was related to the organic mainstream. I also regret that I have been unable to include all the stories and memories which interviewees so generously shared with me. The organic movement counts among its supporters many unsung heroes, still with us, whose commitment has been expressed patiently over many years. The farmers Arthur Darlington, Simon Harris, GILES and MARY HERON, Riccardo Ling and Tony Reid come strongly to mind, as does the smallholder Dennis Nightingale-Smith, whose Organic Living Association newsletter maintains a stance of uncompromising integrity: simultaneously on the fringes of the organic movement and at its heart.

All these stories have been recorded and will be preserved; they either already have been, or in due course will be, lodged with the Museum of English Rural Life (MERL) at Reading University, where an 'organic archive' exists in embryo. So much material has already been lost — particularly during the Soil Association's move from Haughley to Bristol in the mid-1980s — that it is vital to ensure that any existing archival material or personal records find their way into the public domain. The more this happens, the greater chance there is that something approaching a 'definitive' history of the organic movement might one day be written.

I hope that there prove to be some members of the organic movement, some writers, and some academics, who will continue the work which this book has begun.

Author's note
Throughout the text, names of people in CAPITALS indicate an entry in Appendix A. Names of groups, organizations, institutions or journals in CAPITALS indicate an entry in Appendix B.

Prologue

Lord Portsmouth has two 'encounters'

What was going to happen to British agriculture and the countryside once the Second World War was over? Would farmers be 'betrayed', as they had been after the previous war, or would they be rewarded for their supreme efforts and granted long-term security, now that the nation recognized their literally vital importance for health and survival? And, if the latter were the case, what form would post-war national agriculture take? The war years had brought a dramatic increase in the use of machinery, chemical fertilizers and pesticides, while techniques of artificial insemination were now being developed. Influential agriculturalists such as Sir Daniel Hall and C.S. Orwin had produced wartime books advocating large-scale, specialized, mechanized methods as essential to a viable and productive post-war farming industry. They proposed a model of efficiency based on the measurement of output per man. But if this was adopted, it would presumably mean a decline in the rural population, and then what would happen to the social fabric and culture of the countryside?

It was with such questions in mind that, in the autumn of 1945, the Dorset landowner ROLF GARDINER organized a debate in central London, sponsored by the COUNCIL FOR THE CHURCH AND COUNTRYSIDE (CCC) and billed as an 'Encounter' between Agri-Culture and Agri-Industry. The CCC was a pro-organic initiative in which several leading organicists were active, and this particular occasion was presided over by the Rt. Hon. Lord Justice Sir Leslie Scott, chairman of the Committee on Land Utilization in Rural Areas, which had produced its report in 1943. Scott was not an entirely impartial chairman for the 'Encounter': his sympathies tended to the organic side of the argument, and his Committee's recommendations 'were interpreted widely as a charter for conservation and continuity (or restoration) of tradition'.[1]

The speakers for Agri-Culture were the EARL OF PORTSMOUTH, whose 1938 book *Famine in England* (written when he was still Viscount Lymington) had decisively influenced EVE BALFOUR, and who, with Gardiner, had established another organicist body during the war, the KINSHIP IN HUSBANDRY; another Kinship member, J. Everard Hosking, farmer, flax merchant and Chairman of the Council of the National Institute of Agricultural Botany, and Gardiner himself. Those whom they encountered were G.K. Knowles, General Secretary of the National Farmers Union (NFU); the Union's Technical Officer, G.D. Stevenson; its Public Relations Officer, W.A. Hill; H.W. Tomlinson of the Wiltshire War Agricultural Executive Committee, and F. Rollinson of the National Union of Agricultural Workers (NUAW).

The Editorial Note to the booklet summarizing the proceedings, which was published the following year, was written by Jorian Jenks, who would soon be appointed the Soil Association's first Editorial Secretary. Interestingly, he did not use the word 'organic', preferring to describe Gardiner and his allies as 'protagonists of the Husbandry school of thought': and not just of husbandry, but of 'true' husbandry, whose spiritual and cultural aspects must not be subordinated to technical and economic development for the sake of 'short-term material considerations'.[2] Jenks assured his readers that both parties to the Encounter desired the 'rehabilitation' of agriculture: but what did this word mean? On the face of it, British agriculture was in a far healthier state than it had been ten years previously. The husbandry school, though, pondered on the implications of the agricultural changes which war had so rapidly encouraged: in particular, the tendency of the machine to overpower natural methods. Or, to express it another way, for nature to be conceived mechanistically rather than biologically.

Two months earlier, Portsmouth had been a member of the British Mission to America to Study Farm Buildings in North America, and in California had been granted a vision of the mechanized, industrial future: Mexicans toiling at vast conveyor-belts; imported fertility; machinery obtained through interest-bearing loans; tasteless produce; large herds of dairy cows (though at 300–400 not large by twenty-first-century standards) milked and fed mechanically, and the few remaining labourers worked harder than ever in conditions not far removed from slavery. In some of the dairies there presided 'a white-robed scientist, a high priest among the test-tubes, ever alert to suppress disease': a 'laboratory hermit', as SIR ALBERT HOWARD had dubbed such figures. Portsmouth had run up against men of similar outlook in Britain, too, when he had pleaded with leading agricultural scientists for wide-ranging studies comparing natural with chemical manuring and had been told: 'Give us something we can measure, and we will work with you'. Nature's complexity was being dealt with through test-tube experiments and statistical analysis of tiny plots. Portsmouth revealed himself as an early exponent of the Precautionary Principle when he concluded by suggesting that 'the burden of proof lies with the proud innovators of modern technical progress that what they are inaugurating will not harm unborn generations'.[3]

For Hosking also, the criterion was that mechanization should benefit the human being who operated the machine; yet its effect in general was, and was intended, to reduce the demand for skilled workers. In America, the resulting de-population was a cause of the disintegration of rural communities. The *reductio ad absurdum* of the industrialist's concept of efficiency, Hosking pointed out, would be 'with *one man only* farming his mechanized paradise' and that worker would in all likelihood be bored senseless by the repetitive tasks he was required to perform. Rolf Gardiner saw machine-minding as 'more fatiguing than rhythmical manual toil', requiring physically uncomfortable or warping positions. Men were becoming adjuncts of tools rather than the tools being the servants of men; the industrial revolution was invading the fields, in defiance of biological realities. And with these developments disappeared a sense of wholeness, of using mind and body in harmony, of strengthening the spirit through working with nature. It was imperative, Gardiner felt, that the many returning servicemen who wanted some form of agricultural living should be given the opportunity to become smallholders, foresters, craftsmen, rather than specialists remote from physical life, serving a departmental machine. 'Vital practice' was far preferable to abstract agricultural education and the depressive effects of Science. (Evidently D.H. Lawrence's philosophy remained an influence on Gardiner.)[4]

The 'Officials' who responded to the three 'Husbandmen' met their caution with optimism. G.K. Knowles said that agriculturalists recognized the threat posed by machinery and would keep progress on 'sound natural lines'; nothing comparable to American factory farming existed in Britain. At the same time, fear must not be allowed to hold back 'progress'. G.D. Stevenson was similarly sanguine about the prospects for rural employment, foreseeing an increase in land workers and an equilibrium between man and machine; nor did he believe that farming systems would have to adapt themselves to machinery. F. Rollinson of the NUAW openly espoused efficiency defined as output per man, and ambiguously spoke of 'labour-saving machinery': did he mean that it reduced physical drudgery, or that it reduced a farmer's need to employ workers? The former, apparently; he proceeded to envisage a rural society of people enjoying all the facilities available to the urban population for developing their minds and spirits. As for 'efficient' methods having any detrimental effects on people, animals or the soil: Rollinson simply discounted such possibilities. He was that bugbear of the early organicists: a Planner. Whereas Gardiner and Portsmouth wanted to re-vitalize rural life 'from the ground up', through local initiatives, associations and estates rooted in regional structures, Rollinson anticipated the benefits which the emerging era of the planned economy would bring, with landowners and farmers subject to central control. As Jenks pointed out in his Editorial Note, the Officials were 'men of undoubted goodwill and religious sincerity', and they made a number of telling points in their responses to the Husbandmen, but with the benefit of sixty years' hindsight, one can see that their optimism was — as it so often is — misplaced and that the Husbandmen, whether or not one shares their dislike of agri-business, were the more perceptive about its likely effects on rural life and the countryside.[5]

Two years later, after the passing of Tom Williams' AGRICULTURE ACT (1947), Lord Portsmouth was involved in another 'encounter' with a proponent of the agri-business approach, when he and H.D. Walston (father of the contemporary scourge of organic farming, Cambridgeshire barley baron Oliver Walston) gave broadcast talks on the theme of *Rural England: The Way Ahead*. These talks present in microcosm the incompatible philosophies of agriculture which have clashed for more than seventy years. Walston argued for efficiency defined as output per man, to be achieved by 'a rational division of labour and by specialisation'; there should in fact be 'an ever greater attempt to raise output per man rather than output per acre'. In terms all-too-familiar to us today, Walston asserted that a drift of manpower from the land was 'a natural process of evolution', and that England should buy its food in the cheapest markets; though, perhaps inconsistently, he also thought that England should grow a lot of food in order to reduce transport costs. He aimed the usual sneers at Portsmouth, condemning him for lacking 'common sense', for indulging in nostalgia, for refusing 'to face the facts of life as they are today', and for, supposedly, wanting to preserve agriculture as a mid-nineteenth-century museum-piece. For Portsmouth, to ignore the health of the soil while concentrating on maximum productivity was simply short-termism, an attitude based on a desire to conquer, rather than work with, nature. England could not be secure, still less happy and healthy, unless she grew far more food and 'by good husbandry, maintain[ed] and even increase[d] the fertility of her soil'. Far from being an inevitable evolutionary process, the land's de-population indicated the dominance of 'wrong material and mechanistic standards', while mechanization meant the loss of variety of skilled jobs. Agriculture and horticulture should be the basis of rural society, with local areas as self-sufficient as possible.[6]

In these two 'encounters', it seems to me that one finds raised, explicitly or implicitly, all the issues which have exercised the organic movement for the past sixty years. The advocates of agri-industry — or, to use the more common term, agri-business — have had things almost entirely their own way during this period, which, given the provisions of the 1947 Agriculture Act, was always likely to be the case. The following chapter will examine in more detail how the principles and practice of agri-business developed in the post-war decades, and the remainder of the book will look at the organic movement's response to that development.

The organic movement faces post-war Britain

In the mid-to-late 1940s, there existed various organizations and publications which in different ways and from differing perspectives promoted the organic cause. Some of these dated from the 1920s. EDGAR SAXON had taken over from the Tolstoyan publisher Charles W. Daniel the journal *The Healthy Life* in 1920, re-constituting it as *HEALTH AND LIFE* in 1934. It advocated nature cure healing methods, 'honest' food produced organically and, for a period in the 1930s, Social Credit financial reform doctrines, which if adopted — so it was argued — would help create a healthier, more balanced, national economy. In 1922, the forester RICHARD ST. BARBE BAKER had established the MEN OF THE TREES movement, with which leading organicists became involved: among them, Sir Albert Howard, Lord Portsmouth and Philip Mairet. Its journal *Trees* took an environmentalist approach to issues of vegetation and soil erosion. Whereas Baker was influenced by his experiences in Canada and Africa, one of the chief organicist initiatives, in contrast, grew from urban Peckham, in South London. This was the PIONEER HEALTH CENTRE, which began in 1926 as a family health club and burgeoned in 1935 as a purpose-built, modernistic social laboratory for observing the conditions which make for individual and communal health. It closed during the war and re-opened afterwards. 1926 was also significant for the founding of the Chandos Group of Christian thinkers who in the mid-1930s took control of the *NEW ENGLISH WEEKLY (NEW)*. Founded in 1932 to promote the cause of Social Credit, from around 1937 the *NEW* increasingly turned itself into a comprehensive forum for organicist ideas. The third significant initiative to emerge in 1926 was the RURAL RECONSTRUCTION ASSOCIATION (RRA), a pressure group whose aim was to educate politicians about the economic problems faced by rural society; to that end, it undertook concentrated research and publicized the results. By the end of the war it had become in effect a front organization of the organic movement, its journal *RURAL ECONOMY* being edited by Jorian Jenks, who from 1946 was also editor of the Soil Association journal *MOTHER EARTH*. The RRA further demonstrated the organic movement's links with the monetary reform movement through working in conjunction with the Economic Reform Club and Institute.

Two other bodies were formed in the 1920s. The aim of the NATIONAL GARDENS GUILD (1927) was 'to unite all classes in restoring living beauty through gardening', and its journal *The Guild Gardener* taught methods of organic husbandry. The Anthroposophical Agricultural Foundation was formed the following year to promote the development and practice of RUDOLF STEINER's agricultural ideas — biodynamic, as they came to be called — outlined in his lectures of 1924. The leading exponent of the biodynamic approach,

EHRENFRIED PFEIFFER, was well known to the organic pioneers and highly respected by them, while another Steinerian, LAURENCE EASTERBROOK, was agricultural correspondent of the liberal newspaper the *News Chronicle*.

I have traced the coalescence of the organic movement during the 1930s in my earlier book *The Origins of the Organic Movement*. Of the pro-organic bodies or publications active when the war finished, three came into being during that decade.

Christendom, the quarterly journal of the Anglo-Catholic Christendom Group, was founded in 1931; the *New English Weekly,* already referred to, in 1932, and *Peoples Post,* newspaper of the DUKE OF BEDFORD's national-socialist fringe movement the BRITISH PEOPLE'S PARTY, in 1939. It was also during this decade that FABER AND FABER, thanks to their agriculture and horticulture editor RICHARD DE LA MARE, began to produce what would become a long series of classics in the canon of organicist literature. Eve Balfour's HAUGHLEY EXPERIMENT in Suffolk, which would play such havoc with the Soil Association's finances in years to come, was conceived shortly before the war.

The war years themselves produced the two organicist bodies we have already met, both initiatives of Rolf Gardiner: the Kinship in Husbandry group and the Council for the Church and Countryside. DR LIONEL PICTON began publishing his *News-Letter on Compost,* which Sir Albert Howard took over in 1946, changing its name to *SOIL AND HEALTH.* And two particularly influential books appeared: Howard's *An Agricultural Testament* (1940) and Eve Balfour's *The Living Soil* (1943).

Despite the commonly held view that the world of mainstream agriculture completely ignored the organic pioneers, it in fact paid a great deal of attention to them. Fisons and ICI during the war and the post-war years launched large-scale advertising campaigns against 'the humus school', whose practices, if widely adopted, would have rendered their products unnecessary; and the trade papers *Farmers Weekly* and *The Fertiliser, Feeding Stuffs and Farm Supplies Journal* gave considerable space to discussion of the merits and fallacies of the organic case. Eve Balfour's book, which was into an eighth edition by 1948, aroused such interest that in 1946, along with the Wiltshire farmer FRIEND SYKES and GEORGE SCOTT WILLIAMSON, co-founder of the Pioneer Health Centre, she formed the Soil Association, whose purpose was to investigate the relationship between soil fertility and the health of plants, animals and human beings.

Buoyed up by the wartime importance accorded to farming, by the widespread desire for a post-war society more healthy and just than that of the inter-war years, by the advances made in nutritional science, and by the considerable public interest in organic ideas evident during the past three or four years, the Soil Association's founders must have felt optimistic about the possibilities of advancing the organic cause, and the organization attracted as members almost all the leading exponents of that cause. The most notable exceptions were the agronomist SIR GEORGE STAPLEDON and, ironically, the Association's chief begetter Sir Albert Howard. Stapledon was both a mentor to the Kinship in Husbandry, particularly Gardiner, and a scientist of international renown highly regarded in the world of mainstream agriculture. His *New English Weekly* review of *The Living Soil* was sympathetic but not uncritical, and he preferred to retain a scientist's detachment towards the Soil Association. Howard was more obviously dissatisfied with the Association, withdrawing at a very early stage on account of its approach to the Haughley Experiment. He believed — correctly, as it transpired — that any attempt to make the Experiment sufficiently rigorous to withstand scientific scrutiny would be hugely expensive, and he objected to laymen having ultimate control over the proposed research.

Howard therefore ran his own journal, *Soil and Health*. After his death in October 1947 his friends created the ALBERT HOWARD FOUNDATION OF ORGANIC HUSBANDRY to commemorate his pioneering work. This was 'a fighting organisation'[7] dedicated to promoting the Rule of Return (of wastes to the soil) and opposing artificial fertilizers. In contrast to the Haughley Experiment, the Foundation envisaged a network of demonstration farms, market gardens and allotments, both in Britain and overseas. One of these centres was Goosegreen Farm near Bridgwater in Somerset, run by FRANK NEWMAN TURNER. Goosegreen was the location of the INSTITUTE OF ORGANIC HUSBANDRY, the world's first centre for instruction and advice on organic methods, and for natural (that is, herbal and homoeopathic) treatment of animal disease. Turner, a journalist who wrote regularly for *Farmers Weekly,* set up his own journal in 1946; this was *THE FARMER,* which ran for ten years and contained a wealth of valuable material on organic husbandry, horticulture, nutrition and animal health. He also established the WHOLE FOOD SOCIETY, which appears to have been the earliest instance of an organic marketing co-operative, bringing together organic growers with consumers wanting organic produce. Many years before the growth of box schemes in the 1990s, the Whole Food Society was organizing direct deliveries from farms to homes.

One other cottage publishing industry requires mention: tucked away in the Home Counties, near East Grinstead, was the journalist L.B. POWELL, who produced Country Living Books, a series of collections of writings by leading personalities in the world of organic husbandry: H.J. MASSINGHAM, Ronald Duncan, Lord Portsmouth, Laurence Easterbrook, John Middleton Murry and the broadcaster Ralph Wightman were among the contributors, and there was much advice on the sorts of topics on which JOHN SEYMOUR would later become an authority: goat-keeping, flour-milling, growing vegetables and other aspects of self-sufficiency. Powell went on to contribute to the Soil Association journal and *THE ECOLOGIST* in the 1970s, but is now one of many figures in organic history whose commitment has been forgotten.

Looking back at the emergent organic movement in Britain, as it existed in the immediate aftermath of the war, one can see that it was about much more than just the relationship of soil fertility to health. J.S. Blackburn's symposium *Organic Husbandry* featured on its back cover the symbol of the Wheel of Life, a self-contained cycle involving soil, plant, man and animals, and wastes being returned to the soil, but the organic philosophy involved more than observance of the Rule of Return. It held implications for the nation's economy, for its politics, for its culture and for its religion. As I have argued in *The Origins of the Organic Movement,* the concept of a God-given Natural Order, interpreted through Christian theology, was integral to the early organic philosophy, and the involvement of many organicists in the Council for the Church and Countryside was no coincidence. The Kinship in Husbandry's desire to re-invigorate the life of the countryside, redressing the imbalance between urban and rural, was largely based on a belief that urban conditions separated people from awareness of God, an awareness which could be strengthened, or created, through contact with the natural world. This was clearly a very ambitious project, in effect amounting to an attempt — however unrealistic — to alter the direction that English society and culture had taken since the early nineteenth century. It is implicit and often explicit in most of the work and writing of the organic movement at that time.

When, in the summer of 2000, BBC Radio 4 broadcast a two-part series on the organic movement's history, the organic veteran DR WALTER YELLOWLEES wrote to the Soil Association with characteristic vigour to complain of the emphasis placed on politics: one of the programmes had looked at the movement's early links with the radical Right and Social Credit. But the organic movement after the war was indeed political, albeit not linked to one particular party. Various noble supporters had spoken in the House of Lords debates on food and farming during the war; the influential National Liberal LORD TEVIOT was President of the Soil Association from 1946 to 1950; Jorian Jenks remained an active Mosleyite; there was a strong dislike expressed in organic writings for State Socialism, and a particular fear that the land might be nationalized (though the Labour peer LORD DOUGLAS OF BARLOCH helped establish the Soil Association as a legal entity, and both Portsmouth and Jenks appear to have thought well of the Labour government's Minister of Agriculture Tom Williams). The proposed National Health Service was mistrusted, and there were still, as already mentioned, close links with the financial reform movement through the Rural Reconstruction Association, itself a political pressure group.[8] Biodynamics might have appeared non-political, but in the background were Rudolf Steiner's ideas on the Threefold State. Opposed on the one hand to State Socialism and on the other to Big Business, finance capitalism and free trade, the organic movement was in effect seeking a 'third way', favouring the regional and small-scale over the centralized, the co-operative over the profit-driven, and the sustainable economy over the trade-dependent. It would be disingenuous to pretend that these are 'non-political' aims. Of course, it would have been possible to support the organic movement — to join the Whole Food Society or the National Gardens Guild or the Soil Association — without being concerned about the movement's more far-reaching agenda, and no doubt this was the case for many people. But the organic philosophy is holistic, and you cannot, logically, ignore its wider implications.

Another noteworthy feature of the organic movement at this time was that it could call on the authority of some leading figures in the fields of agriculture and health. Howard, Stapledon, Portsmouth, Easterbrook and Lord Bledisloe were all noted agriculturalists; Richard St. Barbe Baker and SIR ROBERT McCARRISON were internationally renowned as, respectively, a forester and a nutritionist. Sir Norman Bennett was perhaps Britain's outstanding dental scientist. They were voices crying in the wilderness, but their status at least ensured that the organic message could be taken into the heart of the mainstream establishment. The movement was also able to propagate its ideas through the respected publishers Faber and Faber, thanks to their editor Richard de la Mare, who continued to publish organic books through to the 1970s, and through the books and journalism of, for instance, H.J. Massingham, Ronald Duncan and Adrian Bell.

In its favour, then, the organic movement during the first two or three years after the war had various influential supporters in the agricultural, medical and political establishments, as well as some prolific literary communicators. The organic pioneers' case for organic methods was receiving considerable attention, and a certain amount of sympathy, in the agricultural and national press. The Pioneer Health Centre was flourishing again after its wartime closure, and, in addition to the already-existing bodies established before the war there were fresh organizations emerging, of which the Soil Association appeared the most widely supported and prestigious. The movement had also permeated the Church of England, via the Council for the Church and Countryside. There existed a sense that society lay open to change and reconstruction now that the war was over, and the organic movement was able, for a short time at least, to share that sense of opportunity.

However, various disappointments and deaths followed this initial optimism. The *New English Weekly* folded in 1949; in 1950, a major blow, the Pioneer Health Centre was forced to close, unable to find funding after being excluded from support by the NHS. Howard died in 1947, and 1948 saw the deaths of Dr Picton, Sir Norman Bennett, the biologist M.C. Rayner and Montague Fordham, founder of the Rural Reconstruction Association. Massingham died in 1952 and Scott Williamson in 1953. The Kinship in Husbandry lasted little more than a decade, its final gesture being a memorial volume for Massingham. Stapledon had retired, and Portsmouth chose to spend more of his time in Kenya. Edgar Saxon entered his seventies, and *Health and Life* lost a good deal of its sparkle. The biodynamic movement remained active, but esoteric. And it is important to bear in mind that the same fairly small group of leading spirits tended to crop up in a variety of organizations: Jenks, for instance, in the Soil Association, the Rural Reconstruction Association and the Council for the Church and Countryside (CCC); Philip Mairet in the Kinship, the CCC, the Men of the Trees and at the *New English Weekly;* Rolf Gardiner in the CCC, the Soil Association, the Kinship ... and so on. The organic pioneers spread themselves thin, and, although some of them were wealthy, the movement as a whole could only dream of matching the resources available to their opponents.

In military terms, one might see the organic movement as a band of guerrilla fighters in territory which had been rapidly occupied by opposing forces. What lines of attack would they take against their more powerful enemy? Before we examine their tactics and arguments, which will take the greater part of this book, we can look at some general features of agriculture and of the food industry during the second half of the twentieth century.

1. The Context: Agricultural Efficiency and Industrial Food

The implications of agricultural efficiency

At the end of the previous chapter, I suggested that we might regard the post-war agricultural orthodoxy, as promulgated by legislation and directives, advisory services, trade journals, and books by eminent agriculturalists, as an invading power which had colonized British farming in a remarkably short time. This was largely the result of the war. How the fertilizer companies gained such a grip on government policy that by the 1940s use of their products was identified with good husbandry and to all intents and purposes made compulsory by the War Agricultural Executive Committees, is a topic which demands investigation, though the appointment of ICI's agricultural advisor Sir William Gavin as chief agricultural advisor to the Ministry of Agriculture and Fisheries early in the war was clearly a crucial moment in that process.[1] The marketing departments at ICI and Fisons must have worked hard during the 1930s, since at the beginning of that decade the general complaint among progressive agriculturalists was that the ingrained conservatism of farmers made them suspicious of new methods. By 1945, chemical fertilizers were being used on almost every farm in Britain.

Traditional farming methods, based on a biological conception of agricultural processes, were under concerted attack from a new approach which sought to apply instead the techniques of industrial production. As is the case with the bio-technology industry today, shortage of food was offered as the irresistible reason — and moral high-ground — for adopting the new methods. The Americans had suspended the Lend-Lease arrangement in August 1945, Britain's foreign exchange reserves were severely depleted, and worldwide food shortages meant that Britain's farmers were urged to even greater efforts than they had achieved during the war. The new Labour government's Minister of Agriculture, Tom Williams, recalled in his autobiography that the situation was more threatening than it had been in 1941, with droughts in South Africa and Australia, and the failure of the Indian monsoon, which resulted in the loss of four million tons of rice.² In 1947, the government announced a plan for raising the UK's agricultural output by a further twenty percent over the next five years, to fifty percent over pre-war production figures. In these daunting circumstances, it seemed clear that past methods were inadequate and irrelevant, and companies which stood to benefit financially from a modern, technological approach (but which were, no doubt, chiefly motivated by humanitarian concern for the plight of the starving masses) were energetic in promoting products for the new era of efficiency.

The farming which was being left behind was presented as ignorant, superstitious and fearful; in contrast, the visions offered of the age to come at times verged on the realms of science fiction (though these visions have since been reduced to the prosaic by reality). For instance, an advertisement for the Pegson-Marlow Pump, showed a farmer — if that is the right word — standing at a huge window, looking out over his fields and fiddling with a knob on a long panel filled with dials and temperature gauges. 'Farming from a control tower' had not arrived *'yet'* [my emphasis], but the Pegson-Marlow Pump was 'a step in that direction': a direction which, it was strongly implied, was the only correct one to take. If the fertilizer manufacturers and machinery firms looked to do well out of progressive farming, so did financial institutions. The Westminster Bank Ltd advertised its services beneath a picture of 'The Farm of the Future!' which featured a tractor, a milking-machine, metal silos and a prairie-like field stretching away into the distance. 'Much of our agricultural inheritance which was more picturesque than effective will have disappeared, to be replaced by modern structures designed to fit a purpose ... Farming practices, whose only merit lay in their antiquity, will be discarded ...' The Westminster Bank hoped to draw farmers' cash and notes into its bank accounts, but this was all in the name of 'efficiency, rather than tradition'.³

Efficiency was a 'hurrah-word', used by commercial vested interests in the propaganda war of agri-industry against established forms of husbandry. What exactly did it mean when used by serious agriculturalists and by policy-makers?

Efficiency: its meaning

In the House of Commons, on 15 November 1945, Tom Williams announced that the objective of the government's agricultural policy would be 'to promote a healthy and efficient agriculture capable of producing that part of the nation's food which is required from home sources at the lowest price consistent with the provision of adequate remuneration and decent living conditions for farmers and workers'. The agricultural industry could not afford just to be ticking over; it needed to follow the example of manufacturing industry and strive for 'maximum production'.[4]

Another socialist, F.W. Bateson, appeared to take the view that Britain might dispense altogether with its agriculture, whose existence could be justified only if it increased its productive efficiency. Writing the same year, 1946, Viscount Astor and B.S. Rowntree were adamant that farming was essential to national life, and certain it could be both prosperous and efficient. The Oxford economist A.W. Ashby argued that the nation's agriculture was already in fact considerably more efficient than it was given credit for, since the sociological evidence demonstrated that a declining proportion of the population was involved in the production of foodstuffs, this being accompanied by an increase in the security and variety of supply. Ashby insisted that agriculturalists 'must' always measure efficiency by output per unit of human labour, since the workers' standard of living depended on this; though other measures might be necessary in times of shortages or inadequate nutrition. Money values best represented the ratio of total inputs to total outputs, this ratio being the measure of total efficiency.[5]

A change of government in 1951 did not result in any change of approach. Lord Carrington, the new Parliamentary Secretary at the Ministry of Agriculture, required the county agricultural executive committees to take stern action against inefficient farmers. As defined by the politicians, efficiency appeared simply to mean producing greater amounts of food. The concept of efficiency achieved such dominance in agricultural discourse that Stephen Cheveley of ICI, giving the opening address to a conference on *Agriculture in the British Economy* in November 1956, referred to farmers who were 'sick and tired' of hearing about it: perhaps in part because the concept was so difficult to define. The Hampshire farmer and pundit John Cherrington disputed this last view, stating bluntly that 'The only criterion of efficiency in farming is profit'. Cheveley believed that the element of food quality must also be considered, but his own attempt to define efficiency — 'the power to produce the result intended' — was otherwise more than somewhat vague.[6]

A 1960 study by the Nuffield Foundation, *Principles for British Agricultural Policy,* offered a rather more sophisticated attempt at pinning down the elusive concept. It defined 'efficiency' as 'an increasing degree of economy in the use of human, physical, and technical resources and the achievement of any given level of output with a minimum expenditure of resources'. How to quantify it, was more difficult: the usual measures (output per man, per acre, or per unit of capital) might all be misleading, so the best approach was to take into account '*all* the resources used in relation to the output'. Complex though such calculations were, they indicated that in all the major types of farming, the larger the farm, the more efficiently resources were used. Subsequently, the efficiency of British agriculture, whose average farm sizes steadily increased in the post-war decades, was to become one of the reasons for resistance to entering the Common Market and having to subsidize, through the Common Agricultural Policy (CAP), the more inefficient farmers in Europe, and particularly in France. In the 1980s, one of the CAP's most persistent critics, SIR RICHARD BODY, MP, defined an efficient farmer as 'one who is able to earn a livelihood from agriculture ... without being given any subsidy by the government'. According to this criterion, British farming was not remotely efficient.[7]

Efficiency: its implications in practice

(i) Planning

Efficiency was one of the two main concerns of the 1947 Agriculture Act, and was linked to the other, which was stability, to be achieved through guaranteed markets and prices. Efficiency was in effect the price to be paid for stability. In return for what was soon to be criticized as a 'feather-bed' system, the farmers were required to demonstrate good husbandry, a phrase whose meaning was distinctly different from what the organicists understood by it. What the government meant by this term was the increased use of modern technology and the products of the burgeoning agri-industry companies. It encouraged this process through a combination of incentives (which took the form of subsidies for fertilizers, drainage grants and provision of new buildings), and threats (to be backed by the disciplinary powers of county agricultural committees, which could, as a last resort, include dispossession).

In 1946 the government established the National Agricultural Advisory Service (NAAS), which took on, rather ambiguously, both advisory and regulatory functions, responding to farmers' requests for assistance or grants but also initiating improvements and, on occasions, bringing wayward farmers into line. Farmers thus found themselves under considerable pressure to use chemical fertilizers. To the organicists, the efficient 'socialist' agriculture which F.W. Bateson and C.S. Orwin encouraged, looked like a sinister alliance of State control with Big Business. Such a state of affairs would have been unthinkable ten years earlier, but the war had, of necessity, concentrated the direction of agriculture in the government's hands. Efficiency therefore went hand-in-hand with planning. Agricultural policy since the war has been so scathingly criticised by organicists (most notably Sir Richard Body and GRAHAM HARVEY), and economic planning and protectionism are so reviled in the dominant ideology of neo-liberalism, that we need to remind ourselves of the good intentions of those who advocated this approach for the post-war era. Williams himself was concerned to 'prevent a repetition of the depressed countryside of the twenties and thirties', and believed that fluctuating prices made it impossible for British farmers to plan cultivation, breeding or indeed anything else. He attracted the support of Laurence Easterbrook, the noted agricultural journalist and advocate of organic methods, who, despite sympathizing with those who felt that the farming community was 'more planned against than planning', saw in government policy 'something broad in its conception, solid in its foundations'. In a posthumously published essay, Sir Daniel Hall, one of Britain's outstanding agriculturalists, argued that a planned, progressive farming industry was the only one in which agricultural workers would be prepared to remain, since its efficiency and profitability would ensure decent wages and hygienic living conditions. An optimist, Hall believed that the coming of industrial agriculture was inevitable and that it could be controlled in the interests of workers and of the population at large. British captains of industry were, he maintained, much more sensitive to the welfare of their workers than were the more 'primitive' agricultural entrepreneurs of the USA, and the spread of industrialism was the spread of economic advantage: of greater choice and a higher standard of living.[8]

Post-war planning in agriculture must therefore offer greater scope to the new production methods which had appeared during the war, and this in turn necessitated a wholesale reconstruction of the lay-out of British farms. Large-scale machinery required large fields to be effective in practice, and long periods of use to be economically justifiable. There were many small, grazing farms which in the national interest should give way to large, more productive enterprises through amalgamation. But such a process could not be allowed to happen slowly through the effects of competition: it should be actively managed by the State, which must acquire all agricultural land and proceed to adjust boundaries in order to rearrange estates into more cohesive entities. Nationalization of the land — the prospect of which appalled the organic school — was not an end in itself: it was a means of preparing the land for modernized farming.[9] This would have been top-down agricultural planning on the most ambitious scale possible, as opposed to the localist/regional 'from the ground up' approach which the organic school favoured.

(ii) Larger farms and fewer workers

Nationalization of the land never occurred, but the developments which Hall and Orwin had hoped for occurred to a considerable extent without it. Between 1950 and the mid-1980s the number of farms of less than twenty hectares fell from 158,000 to only 61,000, a reduction of 61 percent; the number of farms of between 21 and 40 hectares fell from 60,000 to 34,000, a reduction of 43 percent. During the same period, the number of farms of more than 200 hectares increased by 21 percent, and of more than 240 hectares by over 30 percent. By the late 1960s, 'holdings of more than 120 ha. accounted for more than one third of all agricultural holdings'. The agricultural historian John Martin explains that after 1967 more than 47,000 holdings were excluded from the statistics as insignificant, but the fact that they were considered so is itself an indicator of where national priorities lay. Another historian, David Grigg, has offered figures which confirm the picture: in 1944, only 4.1 percent of holdings were over 300 acres; by 1983 the figure was 13.7. There were regional variations in this pattern;

small farms were most vulnerable in the east and south-east of England and in Wales.[10]

The policy of amalgamation which Hall adumbrated was accomplished without significant government pressure. A Farm Amalgamation Scheme, which in effect bribed farmers to sell to their larger neighbours in return for a pension, attracted little response; instead, according to one authoritative observer, 'the individual farmer ... followed his own perceptive nose and discovered the benefits of farming on a larger scale'. In fact, when it was recognized that the reductions in levels of guaranteed prices introduced by the 1957 Agriculture Act would disadvantage small farmers, the government set up the Small Farmers Assistance Scheme the following year, allocating money for modernizing buildings and increasing productivity. Such gestures did nothing to divert the general trend towards larger units. Keith Dexter and Derek Barber, in their 1961 book *Farming for Profits,* which became standard reading for agriculture students, recommended reducing the number of small farms as a long-term solution for problems in the UK's agricultural industry. For individual small farmers they proposed either increased production or part-time jobs away from the farm. At the end of the 1960s, Barber assisted J.G.S. and Frances Donaldson with a major survey of British agriculture, *Farming in Britain Today,* which argued that, although smaller farms were environmentally beneficial, there were too many of them, and that this posed a threat to the balance of the industry as a whole; they were, in fact, a 'problem'. The minimum size of farm capable of providing a living was gradually increasing, and any measure which encouraged people to try to make a living on farms of less than fifty acres would be a waste of state money. The Donaldsons and Barber granted that small farms were far more productive per acre than large ones, but considered this irrelevant unless there were another war or the threat of food shortages for some other reason. The purely economic case for large farms was incontrovertible.[11]

A decade later, and after nearly ten years in the Common Market, the desired process was continuing apace. Sir Richard Body expressed the situation succinctly: '[T]he small farmer is being driven out, and the large farmer is becoming larger'. Small farmers had been made unprofitable by the way the agricultural support system operated, with many grants and tax allowances which benefited the larger farmer more than the smaller. The Common Agricultural Policy (CAP) pushed farmers into arable cultivation and penalized livestock farmers, thereby boosting the incomes of the larger farmer and diminishing those of the smaller farmers, which in turn caused the amalgamation of thousands of units and the spread of 'prairie' agriculture. By the 1990s, the UK had by far the largest average holding size of any of the twelve European Union (EU) states: 67 hectares, as opposed to the EU-wide average of 14. David Grigg has pointed out that farms have grown larger even beyond the point at which economies of scale are achieved, but that these increases in size have occurred in part because farmers *have believed* that they were gaining economies of scale — such has been the effect of the ideology. There are two other main reasons for the growth in farm size: to increase gross income, and to acquire land which will serve as an investment.[12]

At the end of the twentieth century, the agricultural economist Sean Rickard felt that the process of size increase must be pushed still further; he recommended the removal of all compensatory payments to farmers as means of producing 'a hollowed out [*sic*] industry with larger, more efficient farms increasing their share of output', and the removal of price support to speed up the disappearance of small farms.[13] The agricultural industry, as Rickard saw it, depended for its efficiency on the continued decline in the number of farms and of those employed on them. J.E. Hosking, at the 'Encounter' between agri-culture and agri-industry, had truly foreseen the logic of measuring efficiency according to output per man.

Looking back over the twentieth century, Rickard must nonetheless have taken satisfaction in what he saw. The reduction in the number of farms and of farm workers is a major feature of British agriculture during that period, as many writers testify. In 1931, eight percent of the UK workforce was involved in agriculture; by the mid-1960s only four percent; by 1996, it was only two percent. In the first quarter-century after the war, more than half the agricultural workforce was lost, at an average of 20,000 people a year. In 1948, there were 563,000 farmworkers; by 1979 only 133,000.[14] From an economist's point of view, this is an admirable 'hollowing out' for the sake of efficiency; from the point of view of anyone concerned with British society and culture, it is a worrying de-population of the countryside. The decline has continued precipitously during the last ten years for a variety of reasons, but the most significant reason for the disappearance of farm workers since 1945 has been the advance of mechanization.

(iii) Mechanization

As we have seen, Sir Daniel Hall advocated large units in order to facilitate the pace and scale of mechanization; he would not have been disappointed by the advances in mechanization during the fifty years from the end of the war.

First came the triumph of the tractor over the horse. In 1945 there were more than 436,000 working horses in England and Wales; twenty years later there were fewer than 20,000, and most of those were used more for display or pleasure than because they were commercially advantageous. In 1942 there were 102,000 tractors; by 1950, 295,000, and by 1981 over 473,000. Even after the saturation of primary demand had been completed, replacement and additional demand kept sales going: in the late 1970s, the average age of tractors on large arable farms was less than three years. By 1986, there were 532,000 tractors in use: more than one per full-time farmer or farmworker. Tractors became ever more powerful: between 1956 and 1980 the number of tractors increased by 15 percent, but the average horsepower rose from 14 to 45. With their hydraulic systems and more flexible suspension, tractors became much more than just haulage machines; farmers could buy ploughs, drills, sprayers, hedge-cutters, hoists and other devices to fit them. The increase in use of combine-harvesters was similarly dramatic: in 1942 there were fewer than one thousand in use; by 1965 there were 58,000. Their number has not significantly increased since then, but, as with tractors, their power has done so considerably.[15]

Mechanization has made a lasting impact on the rural landscape. Marion Shoard began her influential polemic *The Theft of the Countryside* (1980) by identifying mechanization as a primary cause of the removal of landscape features: machines need a clear sweep in which to turn, and farmers want their expensive machinery 'driven as quickly as possible round unencumbered fields', to consume as few man hours as possible. Earth-moving equipment has obliterated many so-called 'obstructions' in preparing the way for machines to plough and harvest on a mass-production scale, while the machines themselves require new buildings to house them, and new roads to travel on.

When Shoard wrote her book, farmers were able to provide these facilities for themselves up to a considerable size, without planning permission. Another important factor was the system of minimum price levels guaranteed under the CAP, which, in Shoard's words, made it 'profitable for farmers to plough up almost any kind of uncultivated land ... to increase their output'.[16]

(iv) Factory farming

Mechanization proceeded not outdoors only, but indoors too, as factory methods were applied ever more widely to living creatures. Within twenty years of Tom Williams' announcement of an 'efficient' agricultural policy, the conditions in which animals and birds were kept had become a matter of widespread concern, powerfully expressed by Soil Association member RUTH HARRISON in her book *Animal Machines* (1964).

An intensive animal production system exhibited five features, according to the journal *Farmer and Stockbreeder* in 1961: rapid turnover; high-density stocking; a high degree of mechanization; a low labour requirement, and the efficient conversion of food into saleable products. To express it more philosophically, a mechanistic model was being applied to biological entities. Intensive systems were not a post-war innovation: battery cages for hens had been pioneered in Britain by a Lancashire farmer, Mr Winward, in the mid-1920s. Winward had built his own cages, but from 1930 cages were being manufactured commercially. The real advance in use of batteries came, as one would expect, once the policy of efficiency took hold after the war. In 1950, only eight percent of laying hens were in batteries; twelve percent were in litter or barns, and the remaining eighty percent were free-range. By 1965, the year of the Brambell Report on the welfare of animals in intensive husbandry systems, only ten percent were free-range, and more than half were in battery cages. After a further fifteen years, 95 percent were in batteries and only one percent was free-range. Here was truly a revolution in agricultural methods, in just one generation. Battery hens were unquestionably more 'efficient': they produced more eggs; they required less labour to look after them and they achieved economies of scale. The 1950s also saw the start of chicken's growing popularity as an alternative to red meat. Fast-growing hybrid hens in a controlled environment ('broiler house') could now be mass-produced. By 1967 there were 3,700 broiler producers, with an average flock size of 9,800 birds; twenty years later there were only 2,000 units operating, with an average flock size of 33,000 birds. Turkey producers likewise broke with natural patterns: the addition of antibiotics to feed removed the danger of parasitic disease, while selective breeding and artificial lighting moved the egg production cycle from late spring to summer, enabling younger turkeys to be sold at Christmas. One other aspect of industrial efficiency in poultry breeding needs attention: the development of vertical integration — that is, the control of the production chain, from the raw material to the end product, by the same company.[17]

Pigs, too, were turned into 'animal machines'. The pig industry had severely contracted during the war, and the government was keen to expand it again. Advances in breeding techniques gradually made sows more productive: between 1950 and 1985 the average number of weaners that sows produced annually increased from 12.8 to 20.6; piglets were weaned in considerably less time, and the feed conversion ratio was substantially reduced, thanks to the development of more efficient — and unnatural — housing conditions. Among these was the sweat house system, which aroused considerable opposition. Pig keeping had become a factory-based industry, concerned with economies of scale. As with poultry keeping, the average number in herds increased dramatically: in 1957 it was 34; by 1990, 470.[18]

Poultry and pigs are the most extreme examples of the application of efficient, industrial methods to livestock production, but intensification applied also to dairy, beef and sheep. The same general principles were involved: the desire for maximum production at lowest cost, achieved through breeding programmes (including artificial insemination), larger herds and flocks, and reduction of the labour force through mechanization. Traditional breeds of dairy cow and beef cattle more or less disappeared. Sheep farmers have generally been the most conservative among livestock specialists, as sheep are of all farm animals the least responsive to intensive methods; but even here, breeds have been 'rationalized' (which means that variety has been sacrificed), antibiotics have been introduced to deal with disease, flocks have become larger and stocking rates have increased.

(v) Use of fertilizers and other chemicals

The concept of efficiency, then, involved breaking with tradition and moving away from biological methods to a mechanistic, technological approach. Where treatment of the soil was concerned, the major change was the widespread abandonment of the rotation system in favour of monocultures whose productivity was sustained, not by the use of biological wastes, but by applying chemical fertilizers, or 'artificials'.

The progressive agriculturalists, prophets of agri-business, rejected traditional mixed farming as inefficient in comparison with specialization. In 1946, the year that the Soil Association was established, Astor and Rowntree launched an assault on mixed farming, condemning it as 'an obsolete survival from the nineteenth century' which could no longer pay its way, and claiming that pleas for its survival were self-interest on the part of farmers who wanted state subsidies. Expressions of concern about the state of Britain's soils (the organic school was undoubtedly the target here) were merely 'bluff for the non-technical public', that is, for those who did not understand that a mixed farming system was *not* essential to soil fertility. Certainly it was one method, but not the only: ley farming, deep ploughing, and artificial fertilizers were at least as effective.[19]

The fertilizer companies had done well out of the war, and were determined that the organic movement would not stand in the way of their continued prosperity in peace time. They launched persistent advertising campaigns in which the views of the organicists were either treated with patronising respect as representing a limited truth, or were implied to be hopelessly primitive. Fisons, for instance, attempted to appeal to the traditionalist with an advertisement headed 'As sure as God's in Gloucestershire', featuring an engraving of idyllic rusticity that could happily have taken its place in one of H.J. Massingham's topographical books for B.T. Batsford. The words beneath the picture claimed, with remarkable impudence, that Fisons was a feature of country life more permanent than traditional sayings, which 'change with time. But "Fisons" means the same thing always — and to everybody, everywhere'. The advertisement went on to refer to the company's twenty factories, naming in particular one of the main production units at Avonmouth; but, unsurprisingly, did not provide a picture of it. ICI, though not averse to presenting images of traditional landscapes, was more aggressively modern, contrasting the science of its products with the fertility myths of Hopi Indians and Ancient Greeks and Egyptians, as if to imply that any opponents of chemical fertilizers were irredeemably backward. More direct opposition to the school of organic husbandry was evinced in an advertisement titled 'The Constant Factor', which pooh-poohed any suggestion by 'alarmists' that modern farming methods were in danger of turning England into a dust-bowl.[20]

Tom Williams was at pains to reassure the vested interests of the fertilizer companies that they would continue to flourish. At a dinner of the Fertiliser Manufacturers' Association in July 1946, Williams told his audience that the government's desire for a high level of food production necessitated a continuing high demand for fertilizers, and that 'the government would view with considerable anxiety any marked tendency for the present level of fertiliser consumption to decline'. A direct connection between increased fertilizer use and increased food production was axiomatic, and the organic movement's doubts about this juggernaut were dismissed out of hand. The *Fertiliser, Feeding Stuffs and Farm Supplies Journal* stated uncompromisingly in 1950: 'Many plausible reasons are given for withholding "artificials", but *none* of them has *any* foundation in fact' [my emphasis].[21]

So the 'NPK mentality', as the organicists termed it — the belief that soil fertility could be maintained purely by applications of nitrogen, phosphorus and potassium — bore down all opposition. All this was made possible by government subsidies and the information campaigns run by the NAAS. According to Quentin Seddon the situation 'was such that some thought the NAAS a division of ICI, others that ICI was part of the government'. One other event at the start of the post-war period is worth noting in this context. In 1941, the Agricultural Research Council (ARC) had established the Unit of Soil Enzyme Chemistry in order to examine the biological basis of soil fertility, particularly at the micro level; this was disbanded in 1947, its initiatives never pursued. Writing for the ARC in 1981, G.W. Cooke recorded that, since 1941, the use of potassium had increased nearly six-fold and of phosphates by two and a half times. Most dramatic, though, was the increase in use of nitrogen fertilizers. At the end of the war, about 100,000 tonnes were used; by 1985 the figure was close to 1,600,000. Fifty years after the war, the philosophy of 'efficiency' was still being articulated by large-scale farmers. During the 1995 Oxford Farming

Conference, one of them, Vincent Lewis, spoke out against proposals to control the input of fertilizers and other agro-chemicals, on the grounds that this would penalize efficient farmers.[22]

Along with the increased use of chemical fertilizers ran an increased reliance on various forms of chemical sprays: herbicides, pesticides and fungicides. During the war, the discovery of the so-called 'hormone' herbicides, MCPA and 2,4-D, which were effective and considered relatively safe — certainly more so than the sulphuric acid and DNOC sprays which preceded them — meant that chemicals began to replace mechanical cultivation. The development of the boom sprayer, from 1947 onwards, ensured widespread use of hormone weedkillers on cereal crops. Initially targeted only at cereals, they became more generally applied in the post-war decades. By 1975, the amount of pesticides applied had increased twelve-fold. Significant advances were also made with insecticides, through the introduction of DDT and lindane, organochlorines which were to be followed by others such as aldrin and dieldrin. It was not until after the war, though, that the surge in pesticide use would occur. 'In 1950, 15 chemical ingredients were used to produce 352 products; by 1975, 200 to give over 800 products,' says David Grigg.[23]

Again thanks to government subsidies, the manufacturers of these products fared very comfortably. In 1945, their sales were worth around £24m; by the early 1950s they had reached well over £40m. There followed a long plateau which lasted until the mid-1960s, after which there came another remarkable surge. In 1973, the amount farmers spent on pesticides was nearly £38m; by 1983, it was £298m. According to John Martin, expenditure on pesticides finally peaked in the mid-1980s, but during that decade, despite an overall reduction in the use of pesticides, the area treated with them increased by 9 percent.[24]

In 1995, the agricultural journalist and *Archers* script editor Graham Harvey witnessed a display of crop production achieved by using high-input chemical controls: winter wheat free of any weeds or blemishes, thanks to weed grass herbicide, three applications of insecticide, four of fungicides and three of chemical growth regulator. In fifty years, the traditional pattern of farming had been largely wiped out: arable farmers no longer needed crop rotations, animals and human labour to eradicate weeds and control diseases. To Sir Kenneth Blaxter, writing the same year, chemical controls were an unmixed blessing, enabling farmers to 'plan their businesses with greater certainty, confident that, apart from exceptional natural disasters, they can budget for the production of expected yields at economic prices and attempt to run a business, not a lottery on the fickleness of the weather'.[25] But, as both Harvey and Sir Richard Body have forcefully argued, farmers could apply high levels of chemicals only because government subsidies enabled them to do so. Take those away and low-input methods, properly experimented with, might prove equally profitable.

(vi) Dependency, complexity and insecurity

Organic cultivation is often defined negatively, as an approach which refrains from using artificial fertilizers and other forms of chemical input. But hardly any of the organic pioneers — Sir Albert Howard may have been as exception — rejected the use of artificials outright, accepting that they could give an early boost to productivity on severely depleted soil. The point was, that in a well-run system which returned wastes to the soil, artificials would simply be unnecessary. The build-up of humus, though a slower process than the rapid results which artificials could achieve, would ensure a long-term fertility free from dependence on external inputs. It would also, the organic movement has argued, avoid the dangers of reducing the earthworm population, polluting water-courses and damaging the health of the animals and human beings who eat the chemically produced crops.

Agri-business systems, in contrast, have led to farmers depending increasingly on external inputs and on a complex technological and economic infrastructure. Whereas horses could be bred on the farm indefinitely, tractors and all their accompanying appliances have to be supplied, at substantial cost, by industry. Fertilizers, pesticides and the rest of the chemical armoury are supplied by large industrial concerns, and the whole system itself depends on finite resources, most notably oil. At the end of the twenty-first century's first decade, the British are starting to become aware of their potentially parlous situation in the face of oil shortages, nearly forty years after E.F. Schumacher drew attention to the vulnerability of technological agriculture.

This dependence on industry also raises questions about the much-vaunted efficiency of agri-business. Since it is measured by output per worker, it is arguable that the number of workers involved should include not just those employed on farms, but those employed in the factories that produce the tractors, combine-harvesters, ploughs, fertilizers, pesticides and all the other products which have become integral to modern agriculture. This number should presumably also include the botanists engaged in plant-breeding and, if GM crops become a feature of British agriculture, those in the bio-tech industry and universities who have worked on their development. Once these people are included in the equation, the degree of efficiency achieved will substantially diminish. Even such a defender of agro-technology as the animal nutritionist Sir Kenneth Blaxter has grudgingly admitted the justice of this argument: 'Extolling the virtues of the farming industry, on the basis of increased production per person, needs to be tempered somewhat in recognition of the additional manpower employed outside the farm in maintaining its productivity'.[26] The degree of efficiency similarly diminishes when assessed in terms of the amount of energy used to produce crops. And by the time one adds in the fuel and manpower used for transporting all the various inputs to the farms, the whole system looks considerably less impressive, even when judged on its own terms, and far more vulnerable.

In another sense, the agri-business approach is the opposite of complex. Farmers simply apply the amounts of chemicals that the companies tell them to; monoculture replaces variety and rotation; landscapes in arable counties become more featureless; there is a drive towards uniformity. But it is complex in its dependence on a long and increasingly unsustainable supply-chain and in its need to find ever more sophisticated solutions to the problems that its techniques engender. To take one example: keeping thousands of hens together in batteries increases the threat of infection, so the administering of antibiotics is adopted to deal with the problem; but, as time goes on, the antibiotics begin to lose their effect. As Blaxter says: 'Today the success of the industry is dependent on disease control'.[27] A further technological fix is therefore required: perhaps some form of genetic engineering will provide the solution? Thus the system moves ever further from a preventive approach and depends on the complex and costly research of scientists remote from the farm.

Underlying the system is what Blaxter has termed 'technological optimism', a quality which, according to his disciple Noel Robertson, 'it is *necessary* [my emphasis] to subscribe to' and to support financially, in order 'to fund and organise our scientific effort to conquer the problems of world starvation and, on a quite different scale, to give a new thrust and excitement to the British farming industry'.[28] This stirring plea for more taxpayers' money provides a good example of the outlook that Schumacher used to deride as 'A breakthrough a day keeps the crisis at bay'. Published exactly fifty years after the end of the war, and after a half-century during which agri-business had had things almost entirely its own way, the passage exhibits various features to which the organic movement would take exception. There is the authoritarian insistence that permits no alternative to technological solutions for problems in farming; the aggressive desire to 'conquer' problems; the macho enthusiasm (reminiscent of a sort of agricultural Jeremy Clarkson) for 'thrust' and 'excitement', and the implication that those who are sceptical about technological agriculture cannot really care about world starvation. If technology brings problems in its wake, then more funding will enable more sophisticated technological research to solve those problems; and if they in turn bring further problems ... so the complexities will multiply.

Blaxter was quite sure, though, that farmers were happy with '[t]his process in which increased production leads to the ready acceptance of new methods and techniques', and which had been called 'the technological treadmill'. This term (of which Sir Richard Body used the variant 'the chemical treadmill') implied 'servitude, but most farmers did not regard it as such. Farmers take pride in the accomplishment of higher yields and in other indices of good husbandry ... Nevertheless, the treadmill effect certainly contributed to the avidity with which farmers embarked on technological change'. While the use of the word 'husbandry' in this context is guaranteed to raise organic hackles, Blaxter does seem to imply that agri-business has given farmers little option but to career along on the technological juggernaut. Elsewhere in his book he rejects the call of 'some *enthusiasts* [my emphasis] to reverse the trends and return to simpler methods of farming'.[29] (One might reasonably ask whether Blaxter is not an 'enthusiast' for technological farming.) Complexity, apparently, is essential to overall productivity.

The results of efficiency

Sir Richard Body, one of the most trenchant critics of agri-business (or 'aggro-business', as he sometimes referred to it), entitled the first of his attacks on post-war British farming *Agriculture: The Triumph and the Shame.* The triumph consisted in the enormous increase in productivity since the 1940s, which Body described as 'verg[ing] on the miraculous'. In his book *Red or Green for Farmers?* published forty years after the 1947 Agriculture Act, Body itemized many features of this near-miracle: output of wheat, up from just under 2 million tons a year to nearly 14 million tonnes; of barley, from just under 2 million tons to more than 10 million tonnes; of milk from 1,653 million gallons to 3,470 million; of eggs, from 451 million dozen to 1,021 million dozen; of pork, from 15,000 tons to 760,000 tonnes, and of sugar, from 593,000 tons to 762,000 tonnes.[30]

Despite their different perspectives, Body and Sir Kenneth Blaxter agreed that this 'triumph' could not have been achieved without what the latter described as 'the unprecedented scale on which successive governments have financed the industry'. But whereas to Blaxter this was an admirable means of maintaining a prosperous and stable agriculture, for Body it should have been a source of shame.[31] Under the CAP, 'mountains' of surplus food were produced at tax-payers' expense; the Third World poor were disadvantaged; the British landscape lost its diversity and bio-diversity; animals were subjected to overcrowding and drugged into bogus health; farmers were driven out of the industry, and the food produced by agri-business techniques was potentially harmful to its consumers: such were the arguments adduced against UK farming by Body and, later, by Graham Harvey, two of the best-informed polemicists for the cause of sustainable agriculture.

Another consequence of the drive for efficiency was the increasing unpopularity of farmers with the public at large. As we saw earlier, the 1947 Agriculture Act granted farmers economic security (subject to certain requirements) as recognition of the essential contribution they had made to the war effort; there was a determination that no 'betrayal' should occur again, as it had after the First World War. Within a few years, though, this security was being criticized by the MP Stanley Evans as 'feather-bedding', and already the seeds of resentment were being sown.[32] By the end of our fifty-year period, farmers were the object of widespread mistrust, disliked primarily for their intensive use of polluting chemicals, their enormous machines, their cavalier attitude to countryside features, their treatment of birds and animals, their substantial financial handouts and their obliteration of wild creatures and plants.

How far such complaints were justified, and how far the public was ungrateful to the people who provided much of the food which they were happy to take for granted, is beside the point: many of the public would in any case have added to their complaints a suspicion about the quality of that food. The fact is that technological agriculture aroused considerable opposition and its excesses were a significant factor in the declining respect for, and interest in, the farming profession. So marginalized has agriculture become that Andrew Marr, in his bulky tome *A History of Modern Britain,* includes only one index reference to it, which concerns the agricultural depression of the decades before 1940. The hardships of the farming community during the foot-and-mouth epidemic of 2001 are not considered worth a mention, and the agricultural revolution which has occurred since 1945, and which is much more far-reaching in its effects than that of the eighteenth century, is not discussed at all, rendering it apparently of less significance than the antics of the Sex Pistols.[33]

Most of the fears voiced by the supporters of organic husbandry at the immediate post-war 'Encounter' with agri-industry have, it seems, been realized and surpassed. Since the organic pioneers proposed that agriculture be adopted as the basis of a preventive national health service, we should now consider how the food industry developed during the half-century following the Second World War.

The post-war development of industrial food

If the post-war history of agriculture meant disappointment for the organic movement, so did the development of the food industry and its putative effects on the nation's health. Concern about the British diet's quality and affordability had been widespread during the 1930s. Much of the population suffered from poor health and physique but, thanks to the work of nutritionists such as Sir John Drummond and Sir John Boyd Orr and to the carefully calculated rationing system, the standard of health and fitness improved as the war years passed. Particularly important for this advance, in the view of the organicists and various nutritionists, was the increased extraction rate of flour for the national loaf, which ensured that people were consuming more of the valuable constituents of wheat.

Given the organic movement's general mistrust of the State, there is a certain irony in the fact that the standard of national health improved as a result of strict State regulation and that the post-war loosening of restrictions facilitated a profit-driven trend towards a diet which the organicists regarded with antipathy. This loosening did not arrive rapidly: in some respects, the food situation deteriorated in the two or three years following the end of the war, with the introduction of bread rationing (not found necessary during the war itself) and the dire weather of early 1947, which resulted in considerable loss of crops. Rationing was not entirely removed until 1954, by which time the British had long been avid for relief from austerity. Professor Peter Hennessy sees food as such a powerful political issue in British life during this period that it played a significant part in changing the government in 1951.[34]

The prophet of the age of greater dietary choice, appearing just at the point when austerity was starting to make way for the affluence which dawned during the 1950s, was Magnus Pyke, who in later years was to develop an alternative career for himself as a comic turn on national television. His book *Townsman's Food* (1952) was an analysis of, and exercise in propaganda for, the food processing industry, lent authority by his status as a scientist. DR. G.E. BREEN of the Soil Association, reviewing it in *Mother Earth,* described it as 'terrifying', and, from his point of view, one can see why. In it, Pyke displayed the optimism which is so typical a feature of the progressive approach to food and farming. Like his equivalents in the world of agriculture, he was quick to indulge in jibes at the expense of those who 'feel romantic regret for the simple foods of earlier centuries'; and he implied that anyone not wishing to enjoy a diet of canned meat, fish and fruit was downright unreasonable. While admitting, in a casual aside, that not all of the 'preservatives, emulsifiers, colouring-agents, flavours, flour "improvers" [Pyke's use of inverted commas is interesting], anti-staling agents, sweetening agents, anti-oxidants and antiseptics' were harmless, he reassured readers that 'enlightened technologists' now realized their products needed to be 'wholesome as well as attractive'. Tapping in to the widespread dissatisfaction with the post-war diet, Pyke argued that although the nation's 'siege' diet had been nutritionally satisfactory, it could be accepted only under stringent compulsion. Since the justification for such conditions no longer existed, a complex, technical food industry, producing processed and sophisticated foodstuffs, was essential in order to compensate for the limitations of a mixed-farm diet. Its purpose was to assemble large quantities of food, make it palatable, and keep the population 'in tolerable [sic] health'.[35]

Tim Lang and Michael Heasman have termed this approach 'the productionist paradigm', its origins lying in the need to increase national self-sufficiency and in the food processing industry's capacity to preserve, store and distribute food on a large scale, using industrial techniques. It has, in their view, aimed at throughput rather than quality and has emphasized cheapness, attractive appearance, convenience and homogeneity while displaying few anxieties about the likely effects of highly processed foods on consumer health.[36]

The entire concept of 'food technology' was alien to the organic approach, and as the processing methods which Pyke described took hold of the food industry during the 1950s, the likelihood of a wholefood or organic diet for the British people — never very great — dwindled to the point of gross improbability. We shall look, with necessary brevity, at some dominant features of the food industry as it changed in the half-century from the end of the Second World War. It will come as no surprise that there are similarities to be discerned between developments in food technology and those in agriculture. In its early years, the organic movement was as much concerned with 'whole' food (also referred to as 'honest' food) as it was with food grown in humus-rich soil. The food industry since the war has concentrated on processing, with substances being added to food, or removed from it, and some natural elements being replaced by artificial substitutes. As in agriculture, the tendency has been to move away from biological processes and the land to chemical processes and the factory or laboratory. Production and distribution have become increasingly centralized in the hands of large-scale corporations, and consumers have become increasingly ignorant of the source of their food and of how to prepare food which has not already been processed. And, also like agriculture, the whole system requires for its smooth functioning a complex logistical structure dependent on oil.

Addition

Some of the additions made to food result from the industrialization of agriculture, as described earlier. Herbicides, pesticides and nitrogenous fertilizers have been liberally applied to crops, grassland and fruit, thereby making their way into the food chain with results that have been vigorously debated. Broiler houses necessitated the use of antibiotics to prevent the rapid spread of disease in cramped conditions. Owing to an accidental discovery in the early 1950s they also came to be used, in small doses, as additions to the rations of chicks, calves and piglets, to speed up their growth. The Agricultural Research Council carried out large-scale experiments in adding penicillin and aureomycin to the rations of fattening pigs, substantially increasing their weight gain and improving the efficiency of their food conversion rate.[37] The drive to breed bigger ewes and increase output of fat lamb led to the covering of grassland with nitrogen fertilizers; the drive to intensify beef production led to the implanting of the synthetic oestrogen hexoestral in bullocks, to make them fatten more quickly. In some European countries, including France and West Germany, and even in the United States, use of oestrogens was controlled by the early 1960s, but British authorities were unwilling to impose restrictions. Another major addition to the treatment of animals is the use of concentrates: industrially-produced feeding stuffs with a high food value relative to their volume. Grazing pasture has become inadequate to the needs of high-yielding Holstein cows, and the grass — which has itself been deprived of its mineral richness by over-use of nitrogen fertilizers — must be supplemented with artificial feeds and mineral blocks.

Even before any food processing took place, then, fertilizers, growth hormones, antibiotics and concentrates were being added to those materials — crops, grass, animals and birds — which formed the basis of British food. But many other substances, bearing little or no relation to biological processes, were developed for what Magnus Pyke called 'the busy world of modern food technology'. Food historian Barbara Griggs has noted that chemicals added to food were once known as 'adulterants' but subsequently re-christened 'additives', a term which 'sounds so much more innocuous'. (The contemporary food writer Michael Pollan resists this change, uncompromisingly asserting that 'Adulteration has been repositioned as food science'.) In fact, though, the term 'additives' itself became a term of implied condemnation, and the substances thus named were already objects of suspicion when Pyke wrote his book. In the previous year, 1951, a conference on the topic of 'Problems Arising from the Use of Chemicals in Food' had been held in London, in the opening address of which Dr G.R. Lynch, a world authority on forensic medicine, had stated that not one of the long list of chemicals used in food technology was 'of the slightest value to the nutrition of the human organism'. To Lynch, it seemed clear that they posed the risk of harm and that this risk should be examined; yet at the end of the 1960s *The Lancet* felt that the long-term effects of food additives remained a mystery.[38]

In the intervening years, the use of such substances had greatly increased. By 1960, around one thousand 'alien chemicals', to use Dr Franklin Bicknell's term, were being used in the British food industry; this number had doubled by the mid-1970s.[39] Some of these were 'natural', in the sense of being derived from naturally occurring food: these included sugars, starches, oils, gums and forms of alcohol. The others were synthetic, mostly derived from petroleum, coal tar and cellulose, their purpose being to improve taste, texture or appearance, to delay deterioration or to make production cheaper. There were the dyes used to enhance colour; preservatives; anti-oxidants to prevent fats from turning rancid; emulsifying agents, such as lecithin, to prevent separation; stabilizing agents; sweeteners (including sugar, saccharin, salt, monosodium glutamate, and citric acid); flavours; tenderizers; chelating agents and humectants. This is far from being a complete inventory.

In Britain, the labelling of such ingredients was lax. The Food and Drugs Amendment Act of 1954 paid greater attention to the issue of cleanliness than to controlling non-nutritive additions to food, with the result that policy allowed any additives unless specifically prohibited (as were some colours and preservatives). This contrasted with the approach of many countries, where nothing could be added to food unless specifically permitted. The history of additives in Britain, in the decade following the end of the war, is of stubborn resistance by the food industry to suggestions and scientific evidence that additives might prove harmful, and to attempts at introducing more detailed labelling.[40] When, in 1978, the European Economic Community issued a directive requiring additives to be listed on food labels by name or 'E' number as well as general category, Britain insisted on a derogation period, eventually submitting in 1984. Consumer suspicion about additives could, of course, only be increased by the fact that the British food industry was so unwilling to reveal which ones it was using. Fear of cancer and concern about the possible effects of additives on children's behaviour were two of the chief reasons for this mistrust.

One final addition to food processing is worth mentioning: the plastics in which the products of food technology have so often been wrapped.

Subtraction

The food industry has also corrupted the organic ideal of whole food by removing vital elements from natural products. In general usage, words such as 'pure' or 'refined' tend to indicate approval and carry a moral or aesthetic overtone; but for the organic movement they instead imply danger when used to describe manufactured foodstuffs. The nutritionist Professor John Yudkin exploited this divergence for effect in the apparent oxymoron of his book title, *Pure, White and Deadly*, about refined sugar. The other basic foodstuff which underwent refinement, thereby losing much of its nutritive value, was flour.

To take flour first, given the traditional symbolic importance of bread in our culture: the process of refining flour was transformed by the nineteenth-century invention of steel roller-mills, which enabled millers to separate wheat into its component parts and take commercial advantage of this separation. They could then produce a white flour which was more easily baked into a uniform loaf: a loaf which for reasons of social status and snobbery would have wider public appeal. White flour keeps longer, and holds more air and water; production of wholemeal bread requires more time and skill. Roller milling enables the nutritionally valuable elements of the wheat — its germ and bran, which are rich in fibre, essential fats, B vitamins and various minerals — to be sold as animal feed, or sold to patent medicine companies and then back to the public as health supplements. The nutritional inferiority of white bread was widely recognized before the Second World War, and in 1942 it was decided that the National Loaf should be made from flour with an 85 per cent extraction rate (that is, one hundred bags of wheat would make 85 bags of flour and 15 of offals, ensuring that the flour was higher quality than the pre-war flour of 70 percent, as it contained more of the wheat's nutritionally valuable constituents). When in 1945 a conference was held on the future of the post-war loaf, the scientists who attended agreed that the National Loaf had helped improve the nation's health and should remain brown, even though the milling combines argued that white flour could be enriched by the addition of synthetic thiamine.

It was a decade before the millers had their way. In 1955, the Ministry of Agriculture, Fisheries and Food (MAFF) established a Committee of Enquiry to weigh up the relative merits of brown, white and enriched white bread. The scientists who presented evidence were unanimous that brown bread should be protected; the millers claimed that brown bread (which, they said, was in reality grey) was very unpopular with the public, who might give up eating bread if not given the chance to eat white. The Committee opted to favour the milling interests, and a standard white loaf became available, 'fortified' with vitamins B1 and B3, with iron and with chalk. The loss of nutrients in white, 70 percent extraction, bread, compared with wholemeal, makes sobering reading: it includes 100 percent of Vitamin E, 85 percent of manganese, 80 percent of thiamine, 80 percent of fibre, 68 percent of folic acid, and 50 percent of potassium and iron.

Worse, from the organic movement's point of view, was to come just a few years later with the invention of the Chorleywood Process. This further transformed bread-making, with mechanical beating used to replace yeast as a raising agent, 7 percent more bread being produced from the same amount of (70 percent extraction) flour, and less time required to prepare the dough. The resulting 'flabby' loaf, to use Barbara Griggs' adjective, might with equal justice have been included in the previous section, as it contained up to 26 additives, among them bleaches, flour improvers, emulsifiers and anti-oxidants.[41]

Vertical integration was instituted in the milling industry as the millers took over the bakers, with consequent loss of consumer choice. Ruthless 'rationalization' of the market left it in the hands of just two companies: Associated British Foods and Rank Hovis McDougall. GEOFFREY CANNON provided a glimpse into the politics of this world when he wrote that in 1967 Joseph Rank established the British Nutrition Foundation (BNF), that its first Chairman was Sir Charles Dodds, the bio-chemist who had suggested fortifying white flour with thiamine, and that in the mid-1980s its President was Dr Elsie Widdowson, whose unrepresentative experiment with German orphans in the late 1940s had, the millers were eager to believe, suggested that white bread was probably as nutritious as brown.[42]

Refinement of sugar preceded the refinement of flour, being an offshoot of the Napoleonic Wars: in particular, of the British blockade of the West Indies, which denied the French their supply of cane sugar. The crystallization of sugar from beets was developed to the point where a pure, hard, white form of sugar could be cheaply produced, offering the advantages that it kept indefinitely, was uniform and stable, and travelled well. The resulting substance satisfied the acquired sweet tooth of Europeans but was no more than a concentrated carbohydrate with a sweet flavour, with all traces of B vitamins — necessary for the assimilation of the carbohydrate — and minerals — iron, calcium, copper and magnesium — eliminated.

Heavy use of refined sugar was one of the main features of British inter-war diet deplored by nutritionists, and its rationing was a factor in the improved health of the British during the Second World War; the ever-increasing amounts of sugar used in British food from the early 1950s onward were a cause of nutritional concern among some in the medical profession. In 1901, annual consumption of sugar per person was 93 lbs.; by the mid-1970s it was 126 lbs. White sugar was being eaten in sweets, chocolate, biscuits, cakes, soups, puddings, canned vegetables, condensed milk, ice cream, jams, yogurts, alcoholic drinks, soft drinks, and breakfast cereals — where the sugar content might be as high as forty or even fifty per cent. Within thirty years of the war's end, UK consumption of sugar per head was greater than in all but four countries, and those were Gibraltar, Greenland, Hawaii and Iceland. Refined sugar, an additive to a remarkably wide range of foodstuffs, has been regarded as a contributory factor to a range of 'civilized' diseases, including dental caries, cancer, ulcers and diabetes. Its harmful effects stem not just from the removal of its nutritionally valuable constituents, but from its tendency to impair the appetite for more natural foods of higher nutritional worth: an augmented subtraction.[43]

These various processes of manipulation through addition and subtraction have resulted in what the organic movement has not hesitated to deem 'dishonest' food. Franklin Bicknell, in his study of *Chemicals in Food* (1960), devoted a chapter to 'perverted foods': bread, margarine and cooking fats. James Lambert Mount considered that one of the most subtle dangers of refined foods was 'the deception of the palate by sweetening agents and flavours which encourage overconsumption of poor quality foods'. For Richard Mackarness, the deception lay in the tricks of packaging and display which persuaded gullible consumers (men were more vulnerable than women in this regard, he believed) to buy 'fabricated' food. By the mid-1980s, Geoffrey Cannon was arguing that many packets of vegetable and fruit products (such as soups, or Angel Delight) could be guaranteed to contain no vegetable or fruit. What has been achieved is a triumph, if that is the right word, of physiological psychology: 'the successful separation of our physiological needs from our senses, and the separation of those senses from one another'.[44]

But Magnus Pyke had anticipated and dismissed such criticisms, back in the early 1950s. He reassured his readers that manufacturers were aiming to provide '"honest" food', and that they should not be blamed for ministering to the innocent pleasure of a housewife who bought, for instance, a Swiss roll dyed chocolate-brown. An all-round increase in happiness would result from 'the use of harmless deception' in food production — just as it did from ladies using make-up. So there was no cause for concern. There never is.[45]

Distance

The food industry's development since the 1950s has been marked by an increase in various forms of distance. As the previous two sections have illustrated, one aspect of this change is the technological, laboratory- and factory-based nature of most contemporary British food: its remoteness from its biological sources and original wholeness; its separation — often total — from the natural world and re-location in the factory and laboratory.

Then there is the issue known as 'food miles': the physical distance that food products travel on their way to consumers. This has been a major cause of environmentalist concern in recent years, but is nothing new in organic thought. At the beginning of our period, in 1945, it had been identified by H.J. Massingham as an integral feature of industrialism, since industrialism 'depends upon transport, usually over vast distances', with a consequent detrimental effect on quality. He identified two reasons for this effect: in order to be carried, perishable goods had to be doctored; and products needed to have some contact with, or closeness to, their destination if their quality were to be assured and recognized. Thirty years later, in 1976, Michael Allaby of *The*

Ecologist was urging a reduction of the distances travelled by food for another reason: the need to economize on transport in a world where the spectre of oil shortages had been raised by the fuel crisis in the winter of 1973-74.[46]

Not only were such concerns disregarded, but, as supermarkets came to take a grip on retailing, the amount of food transported by road saw, in the period between 1984 and 2004, an increase of thirty percent; while, according to Tim Lang and Michael Heasman, during the 1990s there was a ninety per cent increase in the number of food and agricultural products traded by road. Supermarket dominance also meant that travel by car to buy food more than doubled between the mid-1970s and the early 1990s.[47] Airfreighted produce from overseas, ever more common in a globalized economy, has lengthened the distance between producer and consumer still further; national food security has been diminished as a result.

What might be termed 'cultural distance' is another factor. In her social history of British food habits since 1945, Christina Hardyment has suggested that the emerging gulf between shoppers and the sources of production served to underline the importance of money and 'shrewd shopping around' rather than 'the careful management of a seasonally varied larder'. As convenience foods became more widely available during the 1950s and '60s, the advertising industry increasingly portrayed housewives as glamorous appendages to their husbands' careers, a role apparently more worthwhile than that of a skilled and knowledgeable provider of healthy meals. Perhaps the earliest symbol of the new age of affluence was the Mother's Pride loaf, whose 'foamy white, feather-light slivers' contrasted with the coarse texture of the National Loaf.[48] A wide range of easily prepared foodstuffs followed.

There is little question that since the 1940s an entirely new food culture has emerged, one aspect of this process being the widespread loss of many traditional skills represented for instance by the proliferation of 'ready meals'. While in one sense the cornucopia of products offers a wealth of choice, it might also be argued that those who buy them are in effect allowing someone else to choose for them what they will be eating, given the number of additives in such products.[49] The range of choice is in any case not as broad as it may appear, since a dramatic dwindling of varieties of natural foods has occurred under the impact of food corporations and supermarkets: one thinks in particular of the restricted variety of types of apple available, in comparison with the enormous range which English orchards once offered. It is logistically simpler and therefore more profitable to deal with large numbers of fewer varieties. More generally, one might even argue that the emphasis on 'instant' preparation of food in the home (whether by heating, just adding water or micro-waving) and on 'fast food' outside the home marks a relapse into barbarism, if Lévi-Strauss' theory of 'the raw and the cooked' and the Old Testament tale of Jacob and Esau are correct in their suggestion that a culture can emerge only when people are prepared to wait patiently for results.

Whether or not the proliferation of celebrity chefs now represents a reversal of the trend, one can see that in the decades since the Second World War the British have become culturally distanced from food: from what their own culinary traditions have to offer; from the land, creatures and farmers which produce it; from awareness of how it is processed and distributed; and from the skills traditionally required to cook it — even to the extent that some urban apartments no longer feature a kitchen among their facilities. To what extent these changes have been the causes or the necessary results of new social patterns is one of those issues which sociologists and social historians might forever debate.

Politically speaking, the main change to occur during the post-war decades was the move from a high degree of state control to a free-enterprise permissiveness which, the organic movement persistently argued, resulted in a national decline in nutritional health in comparison with the war years. The same sinister alliance of the State and Big Business which the Kinship's husbandmen feared would blight agriculture was forged in the food industry, as Geoffrey Cannon demonstrated with a wealth of evidence in the mid-1980s. The profit motive was allowed to take priority over people's well-being, and it required considerable effort to persuade the State to intervene and try redress a little of the balance.

Corporate manipulation

The issue of genetically modified (GM) foods became a matter of organicist concern only at the very end of the period covered by this book, but needs to be mentioned because it involves so many of the features this chapter has identified as typical of agri-business and the food industry. Lang and Heasman see the drive towards GM crops as part of a new paradigm for food production, which they term the Life Sciences Integrated Paradigm.[50] It works towards integration of the food chain, is top-down in approach, operates on an industrial scale, has close links with government, and is laboratory-based and capital-intensive. Its attitude to the organic philosophy calls to mind that of the fertilizer companies back our starting-point in 1945. For the biotech industry, GM crops represent the latest exciting, progressive advance into a world of plenty. Optimism is obligatory, and GM crops are presented as a cure-all for the problems of world hunger, just as chemical fertilizers were in the 1940s. Opponents of GM are therefore not merely Luddites resisting the inevitable progress of science, but are, supposedly, keen to let other people starve rather than compromise their own principles.

****

The science has moved on, agri-business corporations and large-scale retailers have taken control of food production and distribution, and the philosophy of efficiency and manipulation continues to attract influential supporters in research, business and government. Despite facing such powerful forces, the organic movement has tenaciously expounded a contrary philosophy and experimented with methods of agriculture, horticulture and food production derived from it. During the past ten to fifteen years these have met with more widespread sympathy than ever before (and, one must add, with continued hostility from commercial and technological vested interests). Much of that support has been generated by anxiety about food quality and by assorted celebrities endorsing organically grown products. The resulting surge in sales of organic goods has been regarded by many as a vindication of the organic case, and by others, more sceptical, as a sign that celebrities have found a new bandwagon on which to jump.

But who built the bandwagon? How did it come to be there in the first place, and what were its component parts? The presence of the celebrities has served to obscure the dedicated efforts of those who maintained their commitment to the organic cause, at personal cost, during the decades when it was considered marginal at best and downright eccentric at worst. We can now proceed to examine, for the remainder of the book, the organicists' critique of the agri-business/food technology approach outlined in this chapter, and the philosophy and practice which, during the second half of the twentieth century, they developed, patiently and unglamorously, as an alternative to it.

2. The Organic Alternative: Farming

Organic farming — what the term means

'But wasn't *all* farming organic before chemical fertilizers and pesticides were invented?' This frequently asked question is based on the premise that organic methods of cultivation are merely negative, consisting in the avoidance of synthetic substances. The organic arable farmer BARRY WOOKEY described such a negative definition as 'simplistic', though he evidently regarded it as a starting-point for the understanding of what organic farming is. Nevertheless, it is a misconception. The Romans had no chemical fertilizers, but the large-scale estates (latifundia) of their empire, which ignored the rule of return and exhausted the fertility of North African soils, were far from 'organic'. Other cultures, particularly in the Far East, have provided examples of fertility maintenance which served to inspire the twentieth-century pioneers of the organic movement. This movement arose as a conscious and scientific reaction against the industrial-chemical approach to farming outlined in the previous chapter, and its roots in the period before 1945 are analysed in my earlier book *The Origins of the Organic Movement*.[1]

For simplicity's sake, we shall keep to the term 'organic' in this study, but it is worth noting that different terms have been used to denote methods of cultivation which seek to observe the rule of return. Invention of the phrase 'organic farming' has been attributed to Lord Northbourne, but Northbourne himself was adamant that this was not the case, writing to Ned Halley of the Rodale Press: 'I was certainly *not* the first to apply the word "organic" to farming or gardening. I have never known the ideas and practices under any other name'. In fact, the term 'organic husbandry' was in use at least two years before Northbourne's *Look to the Land* was published; in December 1937 and April 1938 Rolf Gardiner held at his Dorset estate meetings of local farmers to discuss that topic and adoption of its methods. Gardiner's near neighbour RALPH COWARD would insist that the term 'husbandry' was sufficient in itself, according to the dairy farmer Will Best. 'Humus farming' was another term used, while Frank Newman Turner titled his 1951 book *Fertility Farming*. Crops and food were sometimes referred to as 'compost-grown'. In the 1970s, DAVID STICKLAND and DR DAVID HODGES were instrumental in establishing the INTERNATIONAL INSTITUTE OF BIOLOGICAL HUSBANDRY, and in the 1980s Hodges and Dr Tony Scofield founded the academic journal *BIOLOGICAL AGRICULTURE AND HORTICULTURE*. Other terms which are used as possible synonyms for 'organic' include 'sustainable', 'alternative', 'traditional' and 'ecological'. This book will follow NIC LAMPKIN, in his textbook on organic farming, in rejecting the academic pedantry which might devote entire articles to defining the differences between all these terms, concentrating instead on what they have in common. (The one exception will be the biodynamic methods of Rudolf Steiner's followers, which are part of a distinct esoteric philosophical system.) The term 'organic' has been much the most widely used to describe the alternative approach to that imposed by agri-business. Over the past sixty years or more it has been defined by a variety of writers, and we can begin by taking a representative historical spread of their definitions.[2]

In 1951 Friend Sykes, the Wiltshire farmer, racehorse breeder and friend of Sir Albert Howard, declared that he had been farming organically for forty years because he believed that chemical sprays and fertilizers represented an 'unbalanced' attitude to farming problems. An organic farmer was someone who had faith in the power of the green plant to thrive, so long as the soil from which it grew was rich in humus. Such a soil was a 'living' soil, granting the power of disease-resistance to the plants grown in it and to the animals and humans fed on its products. Sykes followed Howard in believing that 'Nature, left to herself, frequently farms much better than the average farmer'.[3] One can see implied here the religious philosophy of belief in a God-given natural order which was such a central feature of the organic movement's early days: an organic farmer is someone who trusts in that order, seeking to understand its workings, and following the rule of return of wastes to the soil.

These issues were analysed more fully towards the end of the

1950s. Taking stock of the organic movement's progress, the Editorial Notes in the Soil Association journal *Mother Earth*, written by Jorian Jenks, found it necessary to wrestle with questions of meaning. The organic approach was gaining ground, but in what exactly did organic husbandry consist? The avoidance of synthetic chemicals was one aspect of it, certainly, but Jenks preferred to emphasize the '*positive and creative character*' of the organic approach, as '*a continuous practical demonstration of workable alternatives to increasing dependence on artificial aids and to all the hazards that their use involves*' [italics in original]. Jenks quoted SAM MAYALL, who two years earlier had defined an organic farmer as 'one who conserves and fosters the biological activities of his soil, relying on the indirect feeding of crops through soil life, feeding his livestock on crops so grown, and returning to the soil *all* so-called waste products, animal and vegetable.' But organic farming was not just a set of practices; those practices were derived from particular principles, and to be an organic husbandman one therefore needed to develop a mental attitude which understood the reasons for the practices: an attitude of mind which Sir Albert Howard had described as 'humility in the face of Nature' and which implied responsibility for the well-being of land, animals and those who consumed his produce.[4] An organic farmer was one who co-operated with natural processes. It is evident that such a definition contains a strong ethical and even religious component.

PROFESSOR R. LINDSAY ROBB who, like Howard, was an agricultural scientist of world-wide renown and shared the religious faith of the pioneers, defined organic farming as 'good husbandry designed for optimum yields of highest quality food free from anything detrimental to health and without impairing future productivity'. The organicist acceptance of limits and finitude is implied in the word 'optimum', but the key question this definition raises is what the phrase 'good husbandry' means; indeed, Robb considered the term more effective than 'organic farming'. Elsewhere in the article, Robb defined it by contrast with the philosophy of industrialism, which values quantity and speed, downplaying agriculture's biological dimension. Husbandry is simply 'care of crops and livestock and keeping the land in good heart'.[5]

Anne Vine and David Bateman found 'good husbandry' to be a major consideration in the early 1980s when they examined 'Some

Economic Aspects of Organic Farming in England and Wales'. Almost all the farmers canvassed listed it as a reason for adopting an organic system. Because of the variety of practices on the farms they examined, Vine and Bateman found it difficult to define organic farming, but concluded that its central idea was 'that of seeking actively to foster biological cycles and natural disease resistance mechanisms'.[6] The concept of balance was integral to such an approach: use of synthetic chemicals risked upsetting any complex biological pattern with its interacting processes and feed-back systems.

Also in the 1980s, the Soil Association's Agricultural Advisor R.W. Widdowson became aware of the need to write a book drawing together all aspects of sustainable farming; it appeared in 1987 under the title *Towards Holistic Agriculture: A Scientific Approach*. Eve Balfour wrote the foreword, demonstrating her own doubts about the term 'organic', which she considered 'misleading', preferring Widdowson's term 'holistic' as 'the clearest and most accurate description of what the ecological alternative to intensive chemical farming really is'. When Lady Eve herself doubts the validity of the term 'organic farming' the waters might appear muddy indeed, but since holism has always been integral to the organic philosophy we can take it, with her approval, that Widdowson's definition of holistic agriculture will capture the essence of the organic approach, particularly as he considered holistic agriculture a logical development of the ideas found in *The Living Soil*. His chief objection to 'organic' farming was its tendency to lay down standards which ignored variations in climate, topography and soil type, and it is clear that what he termed 'holistic' agriculture involved exactly the attention to interconnectedness which Balfour had outlined in her book, being 'concerned with working ecosystems'. Inputs of fertilizers and pesticides upset those systems, which are more than the sum of their parts.[7]

Three years after Widdowson's book appeared, Dr. Nic Lampkin of the University of Aberystwyth — who had reviewed it unfavourably in *NEW FARMER AND GROWER* — published his textbook on organic farming. In its first chapter he devoted four pages to answering the question 'What is organic farming?' No 'short, sharp, clear' definition was possible, he believed, owing to prejudices about the topic, the variety of names used in different countries, and the need for an intellectual grasp of the ideas which underpinned the practice.

However, Lampkin rejected the view that organic farming could be defined negatively, and he also pointed out that such farming requires neither a return to the agriculture of the early twentieth century, nor the adoption of a hippie lifestyle. Wary of the term 'holistic', he identified as central to the theory and practice of organic farming the simple recognition that 'within agriculture, as within nature — everything affects everything else': that, as Eve Balfour had suggested nearly fifty years earlier, the soil is part of a living system, inextricably linked to plant, animal and man. Lampkin quoted the principles and practices identified in the standards document of the INTERNATIONAL FEDERATION OF ORGANIC AGRICULTURE MOVEMENTS (IFOAM) as expressing the essence of organic farming. We may note among them particularly an emphasis on working with natural systems rather than seeking to dominate them; on encouraging biological cycles within the farming system; on maintaining bio-diversity; on use of renewable resources in local agricultural systems; on treating livestock in a way enabling them 'to perform all aspects of their innate behaviour'; and on considering 'the wider social and ecological impact of the farming system'. In this last instance, organic farming is seen as part of a more inclusive social organism, while in its use of organic matter and nutrients it aims to operate as near to a closed system as is feasible.[8]

For Lawrence Woodward of the ELM FARM RESEARCH CENTRE, writing early in the present century, the definition of organic offered by the United States Department of Agricuture (USDA) in the early 1980s remained the most accessible:

> Organic farming is a production system which avoids
> or largely excludes the use of synthetically compounded
> fertilisers, pesticides, growth regulators, and livestock feed
> additives. To the maximum extent feasible, organic systems
> rely on crop rotations, crop residues, animal manures,
> legumes, green manures, off-farm organic wastes, and aspects
> of biological pest control to maintain soil productivity and
> tilth, to supply plant nutrients, and to control insects, weeds,
> and other pests.... The concept of the soil as a living system
> ... that develops ... the activities of beneficial organisms ... is
> central to this definition.[9]

While arguing that there exist significant differences between the main strands of the organic movement, Woodward nevertheless identified 'an essential core of agreement' at the heart of it:

—the conception of the farm as a living organism, tending towards a closed system in respect of nutrient flows but responsive and adapted to its environment;
—the understanding of soil fertility in terms of a 'living soil', which has the capacity to influence and transmit health through the food chain to plants, animals and humans;
—the idea that these linkages constitute a whole system within which there is a dynamic yet to be understood;
—a commitment to science, and an insistence that whilst the above ideas challenge orthodox scientific thinking, they should be explored, developed, and will eventually be explained through appropriate scientific analysis.[10]

Woodward also attaches importance to the theory of holism (most fully developed by J.C. Smuts in *Holism and Evolution*), which he sees as a constant feature of the organic philosophy. He quotes from Smuts' 1929 article in the *Encyclopaedia Britannica,* leading one to infer from the passage quoted that organic farming involves an approach to the world which sees natural objects as wholes which are more than the sum of their parts and cannot be separated from their surroundings, which they influence and are in turn influenced by.

This is a long way from the 'simplistic' definition of organic farming which Barry Wookey offered, but it is right, in my view, to keep in mind that, whether it is recognized or not, a philosophical attitude towards the natural world underlies the basic practices of organic cultivation. Woodward's emphasis on holism also serves to remind us that the organic movement's origins can be traced, in part, to the biological philosophies of other thinkers prominent in the early twentieth century: Hans Driesch and, particularly, Sir Patrick Geddes. (Philip Mairet, Kinship member and editor of the *New English Weekly,* had worked with Geddes and was familiar with Driesch's work.)

David Lorimer uses the term 'agroecological' to describe the philosophy which the organic movement opposes to the agri-business model. As the name indicates, this philosophy gives primacy to ecological over purely economic values: it adopts a systems approach and values quality over quantity. It sees Nature as a Wheel of Life — a metaphor which can be traced back to the organic movement's early years — and seeks to work in harmony with this cycle. It is, to use Kirkpatrick Sale's term, 'human scale' rather than industrial scale. Since Lorimer is writing about the PRINCE OF WALES, it is odd that he does not emphasize the philosophy's spiritual dimension, but this is presumably implicit in the idea of working in harmony with Nature.[11]

Having undertaken this survey of how various organicists have defined organic farming, we can now look at some of the organizations and personalities which promoted it during the second half of the twentieth century.

The Biodynamic Agricultural Association

The BIODYNAMIC AGRICULTURAL ASSOCIATION (BAA) (originally the Anthroposophical Agricultural Association) was established in Britain in 1928, eighteen years before the founding of the Soil Association. In *The Origins of the Organic Movement* I gave an account of the relationship between the BAA and the mainstream organic movement during the 1930s and '40s, concluding: 'Without ignoring Steiner's considerable influence, it is highly probable that a movement for organic husbandry would have existed in Britain even if biodynamic methods had never been developed'. This now strikes me as an over-cautious judgement: Howard, not Steiner, is the key figure in the development of the British organic movement. The ecologist and compost-scientist Alwin Seifert, a good friend of Rolf Gardiner, described biodynamics as 'unquestionably the source and inspiration of almost all the up-to-date methods of rehabilitating soil, animal, plant and environment generally', but one has to question this claim where Britain is concerned. The organic movement's opponents, however, prefer to place the emphasis on Steiner, since the esoteric nature of his agricultural theories provides better ammunition for mockery of the movement's supposedly irrational philosophy. As an outstanding agricultural scientist, Howard is harder to attack on such grounds, but the more historically aware critics of the organic movement are now starting to turn their fire on him. Nevertheless, the biodynamic philosophy has exerted an influence on various important figures, and its relationship with the main stream of the organic movement is close and somewhat complex. It deserves a full-length, specialist study, but in the space available here I shall present some of the connections between the biodynamic movement and the wider organic movement as exemplified by certain representative individuals.[12] The biodynamic approach is sometimes referred to as 'organic plus', the corollary presumably being that the organic approach is 'biodynamic minus'. Michael Rust has recalled that Eve Balfour regarded biodynamic methods as the 'university' of cultivation, with organic methods as the 'kindergarten'. According to Vivian Griffiths, Robert Waller once said to the biodynamic farmer DAVID CLEMENT: 'We [i.e. the Soil Association] have the technical knowledge of organic husbandry; you have the poetry'. Charles Ellis considered that the biodynamic and organic approaches had in common 'an intuitive awareness of the necessity of working with Nature, rather than fighting against her.' Supporters of organic methods, though, confined themselves 'to the problem of producing healthy food from a healthy soil', which, although 'a useful contribution to the common good', was not enough for supporters of biodynamic farming. 'They want their activities to be related, as an integral part, to the whole of life'. One had to work not just with the earth, 'but with the forces streaming into our planet from the stars and outer space, and from the Earth outwards.' To organic farmers, Ellis recognized, such ideas would seem far-fetched, 'and outside the bounds of ascertainable knowledge', but biodynamic farmers believed that they were justified, at least to some extent, by the results they produced. In November 1974, a biodynamic enthusiast wrote to the Soil Association journal to suggest that biodynamics were far superior to organic methods, since only biodynamics could help bring new life forces to the 'living organism' of planet Earth. Organic methods merely maintained the Earth's energies at the same level, looking to the next few generations, while biodynamics looked 'to the far future'. David Stickland, the journal's editor, commented neutrally that the distinction 'would not be fully accepted by Soil Association members'.[13]

If the earlier generation of biodynamic cultivators regarded organic methods as in some way inferior, this did not prevent them from playing an active part in the Soil Association. During its first twenty years its Council and Advisory Panel of Experts included as members committed exponents of biodynamics such as MAYE BRUCE, Lady Cynthia Chance (at one time the BAA's Honorary Secretary), Lance Coates, Deryck Duffy and Laurence Easterbrook, along with others who experimented with biodynamic techniques or studied ANTHROPOSOPHY: Rolf Gardiner and AUBREY WESTLAKE, for instance. Steiner's disciple Ehrenfried Pfeiffer was another important link with the early Soil Association: he had spoken at a conference held on Lord Northbourne's estate in the summer of 1939 and been guest of honour at a two-week Kinship in Husbandry conference in 1950. When Faber published his books *Soil Fertility* and *The Earth's Face* in 1947 they featured introductions by Eve Balfour and Sir George Stapledon respectively. Pfeiffer had settled in the USA long before the war, and Balfour visited him there on her tour of America in 1951. LAWRENCE HILLS was another pioneer who made early contact with the biodynamic movement, reading about Steiner in 1942 and visiting MAURICE WOOD's farm the same year. (However, Hills did not accept the movement's assumptions and for many years challenged the biodynamic farmer JOHN SOPER in particular to consider that he might be wrong.) Despite these connections, Dr Carl Mier gave the impression of feeling neglected by the Soil Association after its first decade, complaining that fundamental principles and issues had been lost sight of, and arguing that the biodynamic approach could 'make a vital contribution to the understanding of the theory which lies behind organic practice'.[14]

A study of the first twenty years of the BAA's journal *STAR AND FURROW* shows that it took a considerable interest in the broader organic movement. Replacing a small journal called *Notes and Correspondence* which had run since 1931, *Star and Furrow* came into existence in 1953. Its aim was 'to encourage the free exchange of ideas and experience among those who work with, or are interested in, the agricultural teachings of Rudolf Steiner'. It did not promulgate official utterances of the Association, and its range of contents was not concerned purely with the esoteric: it showed awareness not just of what was happening in the organic movement, but of national policy: a column entitled 'Signs of the Times', which was a digest of items from the press about developments in food and farming, ran for many years. There were plenty of references to Soil Association personalities.

The biodynamic farmer George Corrin mentioned David Yellowlees, farmer-brother of Dr Walter Yellowlees, in an article which showed the BAA's interest in FINDHORN as early as 1956. The journal reviewed Ben Easey's book on organic gardening and Newman Turner's *Fertility Pastures;* Deryck Duffy wrote about the market gardener Roy Wilson and reviewed Jorian Jenks' *The Stuff Man's Made Of;* George Corrin reviewed Friend Sykes' *Modern Humus Farming,* the 25-year report on the Haughley Experiment, and Robert Waller's biography of Sir George Stapledon; he also reviewed Stapledon's *Human Ecology,* which Waller edited. Waller in fact addressed the BAA on the topic 'Reality and the Whole' in December 1964, at the close of his first year as the Soil Association's Editorial Secretary, and he wrote for *Star and Furrow* on 'The Philosophy of Laurence Easterbrook', whom he much admired, following Easterbrook's death in 1965. Another regular reviewer, like Corrin a very experienced farmer, was John Soper, who attended the Soil Association's 1964 Attingham Park conference on the theme of 'Life Threatened' and reviewed Aubrey Westlake's book of that title four years later. Early in the 1970s he wrote at some length on *The Ecologist's Blueprint for Survival* and reviewed *Just Consequences,* a volume on health and nutrition featuring several leading figures in the Soil Association.[15]

Helen Murray; John Soper; Deryck Duffy; David Clement; Alan Brockman; John Davy; Patrick Holden; Nick and Ana Jones

We can now consider some figures whose lives connect the BAA with the wider organic movement in particularly significant ways.

First among them is Helen Murray (now Mrs Helen Zipperlen), whose memories of the organic movement go back to the earliest days of the Soil Association. Her parents Charles and Elizabeth Murray had found themselves responsible for an estate in Wester Ross on the death of Charles' mother in 1933. This estate included wide expanses of rock and heather, crofting villages, declining fisheries and many ageing local people trying to cope with the economic depression. Frustrated by his inability to sell the estate, and being turned down by the Royal Navy on health grounds, Charles Murray sought useful work in England, organized forestry camps in Devon and, as a result of illness, came into contact with the doctors George Scott Williamson and INNES

PEARSE, who had established the Pioneer Health Centre in South London before the war. This meeting led him to read Eve Balfour's *The Living Soil,* and to correspond with her. Scott Williamson and Pearse visited the Murrays in Scotland and there was much discussion of how they might co-operate after the war; to some of this discussion Helen, in her teens, was privileged to listen. Although the plans came to nothing for her father, who died suddenly in 1945, Elizabeth and Helen became actively involved in the organic movement. When a mailing-list error resulted in a postcard being sent to the late Charles Murray, inviting him to attend the post-war re-opening of the Pioneer Health Centre, Helen decided on impulse to accept, and visited Scott Williamson's flat in Hyde Park Mansions (which was soon to become the Soil Association's London headquarters). Hearing that Helen was in London, Eve Balfour commanded her to visit Haughley 'at once', which she did. Balfour was about to start her student-farmworker programme, and Helen Murray joined it in October 1946. She spent a year on this scheme, a further year as a paid farmworker at Haughley and then a year or so on the Wiltshire farm of E.R. Cochrane, who had recently published his study of cattle breeding, *The Milch Cow in England.* In the early 1950s she took a two-year Diploma in Agriculture at Edinburgh and, about this time, was appointed the Soil Association's Scottish Regional Officer, visiting Scottish members to report on their work. She recalls that she knew little herself, but learned a good deal from those whom she visited and passed on ideas. She was a member of ROBERT STUART's Hopes Compost Club and wrote for *Mother Earth* on his work at Gifford in East Lothian.

Around 1954, Helen Murray visited the biodynamic community at Clent in Worcestershire and, a year or so later, on her way back to Scotland from a conference in Oxford, gave Dr Mier a lift to Botton Village at Danby in North Yorkshire, where he had just started a new venture: a collaboration between the Camphill Community Schools organization and biodynamic farmers, to pioneer a suitable life-style for handicapped adults. Murray stayed overnight and was asked by the community's leader Peter Roth to stay at Botton. What was taking place at Botton seemed to her a logical development of the organic idea. 'For some time I had been asking — what if, instead of being a muck-mystic-minority, the organic movement was "the way things are done" — what would it look like? [It would require] a major change of thinking/doing in three areas: financial/economic, social, spiritual.' At Botton Village, Helen Murray felt that she saw the seed of such a change. In October 1955 she returned to Botton and stayed a year, then went back to Haughley to work for Eve Balfour. But she commuted almost every weekend to Botton and started building a house there which by late 1960 was habitable. In the end, the pull of Botton Village proved more powerful than life at Haughley, and Helen Murray settled there and helped plant trees on 'those denuded moors'. She met her future husband there and in 1963 settled with him in the USA, continuing to work for Camphill Communities: first in New York State and subsequently in Pennsylvania, where, at the time of writing, she still lives. Helen Murray was succeeded as Eve Balfour's assistant by Mary Barran (later Heron), who, by coincidence, bought a farm with her husband Giles Heron in the 1970s close to Botton Village. The Herons, although not Steinerians themselves, had a good deal to do with Botton, and have praised the community there in Giles' book *Farming with Mary*. The biodynamic farmer Vivian Griffiths, who spent time at Botton in the 1970s, recalls the Herons as a 'beacon' for local farmers during a period when things were particularly hard in that district.[16]

Helen Zipperlen's letters contain many valuable and vivid memories of Haughley and of some of the Soil Association's founding members, including Scott Williamson, C. DONALD WILSON and Professor Lindsay Robb: also of the Attingham Park conferences held in the 1950s, occasions which she recalls as full of warmth and intellectual stimulation. She found the Soil Association environment far from dogmatic; on the contrary, it was one in which it was possible to ask philosophical questions about, for instance, the nature of matter. The problem was, that there were too many philosophical theories on offer, and Murray felt that only Anthroposophical and biodynamic ideas provided a sufficiently comprehensive framework for dealing with all the ramifications of the physical, social and moral questions which concerned her. The organic philosophy would not fit into existing social, economic or religious structures, so one had to seek out its implications: 'what *would* the world be like if it "went along" with "organics"?' Her experience of life at Botton Village provided the beginnings of an answer to this question. The entire social organism required making healthy, and the organic movement was not simply about producing organically grown food: it implied a transformation of society. Evidently Helen Murray felt that the Soil Association was rather narrowly focused in comparison with the Steiner movement.[17]

Murray formed what she terms 'a quiet alliance' with George Corrin, in an attempt to bring biodynamics and the Soil Association together. Corrin farmed in Montgomeryshire when Helen Murray knew him: he was a consultant for the BAA, a position he held until 1986, and he spoke on the course on organic husbandry which DR ANTHONY DEAVIN of the Soil Association's Epsom Group organized in April 1973. But their efforts met with little success: it was 'one of those riddles of destiny that it was not easy, but for no obvious reason'. The riddle is probably explicable by the very fact of the complex esoteric system which underpins biodynamic methods of cultivation, gives the Anthroposophical movement its coherence and discipline and will only ever appeal — despite its practical achievements and potential scientific interest — to a tiny percentage of the population. Sir Albert Howard felt that there was simply no need to subscribe to Steiner's ideas as the basis for organic cultivation, whose validity could be established empirically.[18]

We can now look at other figures who provided bridges between the BAA and the Soil Association between the 1950s and the 1980s. John Soper had worked as an agriculturalist for the Colonial Service, being awarded the CBE for his achievements: during the Second World War he was captured by the Japanese; following the war he worked in Tanganyika from 1949 to 1958, as Deputy Director, then Director, of Agriculture. Once back in Britain, he lived in Hampshire then moved to the BAA's headquarters at Clent. He did not devote himself to the biodynamic movement until his return from Africa, though he and his wife Marjorie (who was for a time the editor of *Star and Furrow*) had been impressed by what they read about Steiner in Rom Landau's best-seller *God is My Adventure* twenty years earlier. John Soper became Honorary Secretary and Treasurer to the BAA, wrote about its work in the Soil Association journal and was a member of the HDRA Council. He had known Ehrenfried Pfeiffer and wrote an obituary of Jorian Jenks for *Star and Furrow*. Vivian Griffiths considers one of Soper's main achievements to have been his study of Steiner's agriculture course, which he developed and adapted for a new generation without compromising its principles.[19]

Deryck Duffy, a South African, was based in Scotland and managed Lord Glentanar's West Hill Estate in Aberdeenshire, where he ran a farming school. Glentanar backed him financially. Later he went to Edinburgh and was involved in marketing organic wholefood. His links with the Soil Association were very close: he was a member of its original Panel of Experts and became, with Friend Sykes, a co-director of the Organic section of the Haughley Experiment. In this capacity, one might say that he was at the heart of the Soil Association, and it is interesting that an exponent of biodynamic methods should have been given the post, particularly as his attitude towards composting caused some dissension in Soil Association ranks. CDR ROBERT STUART, who for most of the 1950s ran a farm composting service in East Lothian, complained in 1955 to Donald Wilson, at that time the Association's General Secretary, that Duffy was a thorn in the flesh, and objected to his presence on the Advisory Panel. Stuart had initiated a composting scheme at the sewage works in Gifford, but Duffy, as a biodynamic farmer, opposed the use of sewage in composting. Stuart publicly connected his own composting with Soil Association's work, yet Duffy, a representative of the Association, refused Stuart's offer to put him in touch with organic growers, apparently telling him that 'he would rather eat vegetables grown with artificials than those grown on [Stuart's] compost'. When Stuart offered Duffy compost entirely free of sewage, Duffy rejected this too, saying: '[W]e do not consider *anything you* make to be compost' [emphasis in Stuart's letter]. According to Wilson, most Soil Association members considered the biodynamic attitude to municipal compost unreasonable. Stuart pointed out the contradiction involved in his own compost being linked with the Soil Association's name, when the Association's local representative repudiated what he was making. This brief correspondence, which happens to have survived, indicates the existence of tensions between the biodynamic movement and the Soil Association, even in the case of a biodynamic farmer who was centrally involved in the Association's activities. Stuart was on what we might consider the more secular wing of the organic movement, and was outspoken in his dislike of those whom he termed 'cranks', whose presence in it he believed stood in the way of convincing practical farmers that organic methods were economically viable.[20]

David Clement had been committed to Anthroposophy since 1930, and farmed at Broome, near Clent, from 1939 until 1986. In the postwar years, Broome Farm became a leading example of a biodynamic family farm and was the headquarters of the BAA during the 1950s and '60s; there was also a herb nursery there. According to Vivian Griffiths, who knew him well, Clement played an important role through helping maintain a clear picture of biodynamic practices. Maye Bruce had confused matters by developing and publicizing her own QR (Quick Return) composting methods, a vegetarian variant of Steiner's methods; while Rolf Gardiner was reluctant to recognize Steiner's contribution to the work of the BAA, preferring to give all the credit to his good friend Ehrenfried Pfeiffer. Clement was a good communicator, able to maintain relationships with a wide variety of people. Jorian Jenks visited Broome Farm during the 1950s, and at the end of that decade Clement invited several Soil Association members to join a June 1959 tour of Danish biodynamic farms and market-gardens. In later years, Clement worked with DR KEN GRAY of Birmingham University on composting. Providing evidence for the continuity of the organic movement, he was on the steering committee of BRITISH ORGANIC FARMERS, and put the biodynamic case on the British Organic Standards Committee in the 1980s. Clement did not resign as Chairman of the BAA until 1986, but the sale of Broome Farm quickly followed. For Vivian Griffiths, its closure marks the end of the first chapter of biodynamic work in Britain. Clement said that he had served on the local committee of the National Farmers Union during the period when fertilizer use was so rapidly expanding, and that he understood why farmers felt they had to apply artificials. Nevertheless, he refused to go with them; but he found that nobody wanted to know why. 'Not one farmer ever asked me a question. ... Only when I retired, one of them remarked, at my farm sale, "You were before your time".'[21]

ALAN BROCKMAN might be considered to have come to the biodynamic movement by chance, were it not that he believes that everything happens for a reason. From a farming background, he studied electrical engineering at Faraday House and after qualifying worked for Siemens on high-voltage cable-testing and development. When studying in London he used to pass a Rudolf Steiner library. A friend of his was interested in Goethe, and one day a book on Goethe appeared in the window, so the friend bought it and Brockman himself developed an interest in Steiner's ideas. While doing his national service in the army he attended a study group in Derbyshire, and in 1949 visited Clent and met Dr Carl Mier, with whom he had a long conversation. Brockman found the gardens at Clent particularly impressive. It was not until he began to suffer ill-health in his mid-twenties, though (attributing it to the poor quality of food provided by landladies in lodging-houses), that Brockman dedicated himself to biodynamic cultivation. He returned to his father's farm in Kent to recuperate and began to think differently about agriculture. In 1953 the family had the chance to buy a run-down farm, Perry Court, with forty acres of fruit, forty of arable and 110 of woodland. Alan Brockman took over the orchard and planted three acres of apple trees, which served as an experiment with biodynamic methods; he also put biodynamic preparations in the manure heaps on his father's section of the farm. Brockman says that his father was sceptical but tolerant. Whether the preparations were efficacious, we do not know, but Brockman became convinced that chemical methods were positively harmful. He sprayed pear trees with tar oil and within a short space of time found the ground littered with hundreds of dead earthworms.

In 1954 he visited biodynamic farms in Germany and Holland, and throughout his years as a farmer made biodynamic compost, developing vegetable growing as well as continuing with fruit production. His friend Henry Goulden worked at the Peredur School at Forest Row in East Sussex (run by Siegfried and Joan Rudel, whose ideas on nutrition deeply influenced Brockman), and Brockman supplied the school with fruit and vegetables. Much of his produce went to various Steiner schools, and he sent apples to the Chevallier Guilds of Aspalls for their firm's cider. In the 1960s he also supplied the Soil Association's WHOLEFOOD shop in London, taking produce up by van. He joined the Association about 1964. MARY LANGMAN, who ran the shop, farmed at Bromley in Kent (on the farm which had once supplied the Pioneer Health Centre), and through her Brockman came to know other organic farmers in the county: particularly Cdr Noel Findlay at Hastingleigh, whose farm Michael and Deidre Rust took over in the 1970s, and Simon Harris (who, as a pupil at St Christopher's School, Letchworth, where his father was headmaster, had heard Eve Balfour speak not long after the founding of the Soil Association). A Kent Group was formed, and Findlay soon invited Brockman to stand for the Soil Association Council; HUGH COATES then invited him to join the Standards Committee. Marketing matters were important to Brockman, who was a commercial farmer: he considered marketing and pests the two main problems he had to cope with. The biodynamic movement had established the Demeter symbol of quality in 1928, and Brockman brought its standards to Soil Association discussions to serve as an example; they were partly used to form the Association's policy.[22]

Brockman was actively involved in the Soil Association until around 1976, and so was not caught up in the generational clashes which were provoked by the emergence of the Association's West Wales group in the late 1970s, but he did share the view of the Seventies Generation that the Association seemed rather 'fuddy-duddy', with an over-representation of retired service personnel and wealthy landed figures, who could afford to be active in it. Working farmers and growers found it more difficult to spare time to support it. The Seventies Generation were based predominantly on the western side of Britain, where land was cheaper than in the east, and, as we shall see later on in this chapter, this was a crucial factor in the shift of the Soil Association's headquarters from East Suffolk to Bristol in the mid-1980s. But there was a more fundamental problem for the Association, Brockman felt: it was too diffuse in its concerns and in its membership, containing all sorts of different streams, such as vegetarianism and land reform. The BAA in contrast was and remains much more unified. For Brockman, also, there is no doubt that biodynamic cultivation is clearly superior to organic: biodynamic carrots, for instance, keep fresh longer than organically grown; birds and animals respond positively to the sprays, and there is the evidence provided by experiments with crystallization techniques, of which more shortly.[23] For scientific rationalists, any investigation of the claims which defenders of biodynamic methods make would be a waste of time and money; but another of our representative figures linking the biodynamic movement with the Soil Association was an outstanding scientific journalist, JOHN DAVY.

John Davy was the son of Charles Davy, a journalist with the *Yorkshire Post* and later *The Observer*, who, under the pseudonym Charles Waterman, had written a book on Rudolf Steiner's ideas, *The Three Spheres of Society* (1946). In it, he demonstrated familiarity with some classic texts of the early organic movement and the work of the Pioneer Health Centre, and considered the agri-business approach to nature, which treated the earth as a 'passive reservoir of raw materials', to be self-destructive. John Davy boarded at Abbotsholme School; transferred to the Rudolf Steiner school Michael Hall in 1944; was called up for military service with the Intelligence Corps in Vienna for two years and then studied Natural Sciences at Cambridge. In 1951 he attended an international conference of the Steinerian Christian Community and met his future wife Gudrun there. He followed her back to Germany, later being granted a British Council scholarship at the University of Freiburg, where she too was studying. While at Freiburg, he was invited by DAVID ASTOR of *The Observer* to join the paper as its first full-time science correspondent. No doubt his father's position as Assistant Editor on the paper played a part in this, but it is equally true that John Davy repaid Astor's faith in him. Davy worked for *The Observer* from 1954 to 1970: at first a reporter, he became a respected feature-writer, interviewing figures such as Professor B.F. Skinner and R.D. Laing, meeting Watson and Crick and describing their discovery of DNA. His job alerted him to the importance of the ideas of E.F. Schumacher, who also wrote for *The Observer*.[24]

Throughout this period, Davy lived at Forest Row in East Sussex, a centre of Anthroposophical activity, and occasionally lectured at Emerson College, an institution for adult education based on Steiner's work. In 1970 he left *The Observer* to teach at the college, becoming its Acting Principal. Between then and his early death from a brain tumour in 1984 he took on many responsibilities: college administration, international lecture tours and the role of General Secretary of the Anthroposophical Society among them. He also continued to write periodically for *The Observer,* and he was a member of the Soil Association's Editorial Board from 1969 to 1971.

In the same way that many farmers and growers have adopted organic methods because they disliked what they saw of chemical agriculture, so Davy's investigations into scientific developments led him to conclude that much large-scale technology was environmentally destructive and irrelevant to human needs. The root of this problem, he believed, was philosophical and cultural. It lay in the divorce of 'the imagination from the analytical apprehension of nature' and of the separation of the arts from the sciences. (Davy was an enthusiast for art, drama and music; while in Vienna he had taken lessons on the cembalo from a Viennese concert performer.) In other words, it resulted from the divorce between 'the two cultures', to use C.P.

Snow's famous phrase: an issue close to the hearts of Davy's editorial colleagues at the Soil Association, the poet Robert Waller and the former actor Michael Allaby. It was also close to the heart of Charles Davy, who in 1961 had written *Towards a Third Culture,* suggesting ways in which the breach might be healed. Steinerians attach a great deal of significance to Goethe's ideas (Steiner himself had as a young man been given responsibility for editing Goethe's scientific writings) and both Charles and John Davy saw these ideas as pointing the way to a future in which scientific techniques would aim less at gaining control over nature, and more at gaining a deeper understanding of it. In John Davy's view, the organic movement was 'part of an urgently needed ecological revolution', one of whose purposes was to challenge the purely instrumentalist view of nature. In so far as the philosophy of the Soil Association helped promote that revolution, he was happy to work for it.[25]

Perhaps the best example of the marriage of science and art lies in the biodynamic experiments with copper chloride. If, to a sample of copper chloride, you add juice from, say, spinach grown with nitrogen fertilizer and then photograph it, you will find a particular pattern; if you add juice from spinach grown with compost and photograph it, you will find a more, vivid, harmonious pattern. There is experimental evidence to suggest that juice distilled from biodynamically grown plants will result in patterns still more vivid and highly organized than those from organically grown plants. Such results are certainly interesting: they might even help provide an objective means of determining nutritional quality. In Switzerland, the Research Institute of Organic Agriculture (FiBL — Forschungsinstitut für biologischen Landbau) started in 1978 its DOK trial (D for biodynamic, O for organic and K for conventional farming) of the three different growing systems. The Institute's director was DR HARTMUT VOGTMANN, who in 1981 was appointed the world's first professor in organic agriculture and who has been closely involved with the British organic movement for thirty years through his work for the Elm Farm Research Centre. Dr Anthony Deavin, whom we shall consider in Chapter 7, a prominent figure in the Soil Association during the 1970s, was another scientist with an interest in biodynamic techniques and experiments.[26]

The biodynamic movement made an impact on certain members of the Seventies generation of organic activists. Carolyn Wacher has recalled the influence of Katherine Castelliz, who farmed in West Wales, and even Peter Segger — for many, the epitome of aggressive commercialism and class warfare directed at his elders — was impressed by Steiner's spiritual philosophy, writing about his own experiments with biodynamic methods for *Star and Furrow*. Segger has said that the West Wales growers of the 1970s 'revered the Biodynamic Association and all that Steiner left us.' Dr Nic Lampkin's mother was a Steinerian and he was sent to a Steiner school, though he did not, as an adult, adopt his mother's spiritual philosophy. Perhaps the most significant impact that biodynamic ideas made, though, was on PATRICK HOLDEN, arguably the most important member of the Seventies generation — for good or ill: the debate goes on — in terms of his influence on the organic movement's later direction.[27]

In 1972, Holden enrolled for the one-year agricultural training course at Emerson College: not because of any appeal exerted by Anthroposophy, but because it was the only formal course in any sort of organic farming that he could find. Nevertheless, he was glad of the accident which took him to Forest Row. He studied under the soil scientist Dr Herbert Koepf and had practical farming experience under the tutelage of Jimmy and Pauline Anderson, who ran Busses Farm and also had a shop which sold organic produce. There were many social activities as well, and altogether the year had a profound influence on Holden, in much the same way that Botton Village affected Helen Murray. In Holden's words: 'I had a taste of organic farming being more than just a physically different way of producing food. It had a cultural dimension, a spirit of community and a different attitude towards food quality and vitality. It wasn't taught — it was just experienced.' He took part in singing, plays and religious festivals. Here, one might say, Holden discovered an instance of agri-*culture* rather than agri-industry, to use the distinction with which we began the Prologue.[28]

For a final example of the connection between biodynamics and the wider organic movement, we can now look at the story of Nick and Ana Jones, who, after becoming involved in the organic movement during the mid-1970s were, by the mid-1990s, increasingly committed to a specifically biodynamic approach to their milling business.

Nick Jones' father worked for the British Council, eventually returning to England to settle in Wiltshire. Jones gained early experience of agriculture through working on a local farm during the school holidays, then, at the end of the 1960s, went to Cambridge to study history and history of art. He was interested in the 'alternative' ideas in the air at the time — Buddhism, vegetarianism — and in *The Ecologist's Blueprint for Survival,* which appeared not long before he graduated in 1972. But what seems to have made the greatest impression on him was the bread, home-made with Allinson's Cerea flour, which his landlady served for breakfast. Through a Cambridge friend, Jones met his future wife Ana, who studied art and design at St Martin's College in London, visited the Hebridean island of Harris to learn how to make tweed, and became practised in weaving and dyeing. She also had a link with agriculture, her mother's father having been a farmer.

Nick Jones was interested in a career in arts administration, and after graduating took a job with a theatre in West Cumbria. The memory of his landlady's bread remained strong, so, when he and Ana received a coffee-grinder as a wedding present, they used it for grinding wheat in order to make their own bread. They wanted wheat which was organically grown, and obtained it from Robert Milnes Coates. Here we find one of the many instances of the younger generation of organic activists making contact with members of the older generation. On the face of it, Milnes Coates was typical of the establishment figures who still dominated the Soil Association's upper echelons in the 1960s and '70s: Sandhurst-trained, twenty years of army service (reaching the rank of Lieutenant-Colonel), a Cambridge degree in agriculture in his forties, and a landowner in Yorkshire, running two farms, each of over one hundred acres: one conventionally and the other organically. He is remembered with respect and warmth not just by Nick Jones, but by David Stickland, who stayed with him on a number of occasions, and by Michael Allaby. His influence was one of the reasons for the Joneses' increasing commitment to the organic cause during the 1970s; John Seymour's writings were another. Other factors which played a part included Ana's familiarity with the Cranks restaurant in London and the Wholefood shop in York, where her parents lived.

The Joneses began to look for a smallholding to run; instead, in 1974, they chanced upon a dilapidated mill at Little Salkeld near Penrith. Nick Jones was at that time promoting an orchestral concert with Yehudi Menuhin, who was one of the directors of the Soil Association's Wholefood shop in London and encouraged him to go into the business of bread-making. With the help of a tourist-board grant and interest-free loans from friends (a particularly generous gesture, given the rate of inflation at that time), the Joneses restored the mill and opened their business in June 1975. Through Bill Starling, who worked for the grain company Gleadells, they were put in touch with David Stickland, who was developing his new business ORGANIC FARMERS AND GROWERS (OFG); they bought grain from OFG for two or three years. Stickland visited them, and they visited his office at Stowmarket. Jones does not share the antipathy to Stickland that some of the Seventies Generation still nurse: he saw what he was doing as a sensibly pragmatic approach to a situation in which demand for organic grain greatly outstripped supply, encouraging farmers in the direction of long-term commitment to organic methods. Jones was well aware, from his own financial struggles (both he and Ana brought in additional income from their work in the arts) how difficult it was for followers of the organic way to survive. Nevertheless, Jones felt, when the Soil Association standards were established, that in comparison the OFG Conservation Grade was less stringent and therefore more open to abuse.

The Joneses were committed to the Soil Association and attended conferences, but their location and shortage of money made it hard to participate as fully as they would have liked. It was through Bill Starling, again, that they were put in touch with Alan Brockman and became aware of biodynamically grown wheat. Starling wanted them to try Brockman's wheat, as other millers were unwilling to do so, given its low protein content. But the Joneses found that it produced an excellent loaf despite this apparent disadvantage. They came to know Brockman in the early 1990s and to work closely with him. Other facts that drew them towards biodynamics were references in *RESURGENCE* and visits to the Helios Fountain café-restaurant and bookshop in Edinburgh, run by Jimmy and Pauline Anderson. They took a holiday at Emerson College, and by the mid-1990s were moving towards a full commitment to biodynamic methods and produce, both on the seven-acre smallholding which they now ran, and in their bakery.

While Helen Murray was attracted to the Steiner movement by its community spirit and Alan Brockman by its ideas, Nick and Ana Jones were first drawn by the quality of the bread which biodynamic flour produced, and by the integrity of the methods which produced the wheat. For Nick Jones, biodynamic methods are superior to the merely organic, and adoption of them requires a grasp of the underlying principles. This means that there is much less danger for biodynamics of opportunists adopting a purely business-minded approach, as he now believes has happened to the organic movement. At the end of our period, then, Nick and Ana Jones were starting to move in quite a different direction from the organic mainstream.

****

In Robert Burton's *The Anatomy of Melancholy,* the figure of Democritus regrets: 'I could have here willingly ranged, but those straits wherein I am included will not permit'. [29] Although biodynamic cultivation is just one element of the 'organic network', its eighty-year history in Britain would, if thoroughly researched, provide material for a full-length volume in its own right. One hopes that the project will, before much longer, be taken up: ideally, by someone who is neither prejudiced against Steiner's system nor an adherent of it. For the present, though, we must be content with drawing a general conclusion from the facts presented in this section. The adequacy of this conclusion can then be tested by future historians. It seems clear that the biodynamic movement has played a significant role in the development of the British organic movement: not because its ideas are essential to the organic philosophy, but through the activities of figures such as those discussed above; through the quiet influence of Steinerians such as Katherine Castelliz, David Clement and Siegfried Rudel on certain younger members of the organic movement who became prominent in it; and through the part which the biodynamic movement played in helping to establish organic standards. To determine exactly how significant the biodynamic presence has been requires research and debate, but it was there throughout the period covered by this book. In Alan Brockman's view, there have been — and remain — many people in the organic movement sympathetic to Steiner's ideas but who are reluctant to admit to being so, because there is a lack of understanding of esoteric ideas — by definition — and a certain stigma attached to being open to them. Perhaps, as the organic movement became more commercially minded from the late 1970s onwards, and involved with government policy during the 1980s, it was felt best to let one's esoteric leanings remain little-known.

One need not be surprised that various members of the organic movement should be drawn to its biodynamic variant, since there was a very substantial overlap of ideas and concerns: the belief in the virtues of composting; opposition to the techniques of industrial agriculture; a commitment to small farms, to ensure a personal relationship with the land; a concern for wholeness; a sense of the earth's 'elastic limit' (to use Alan Brockman's phrase); a rejection of reductionist science; a responsiveness to aesthetic values; an emphasis on the social dimension of farming, and an underpinning spiritual/religious philosophy. (Some of these issues will be more fully discussed in the final chapter.) In biodynamic thought, the farm is conceived as an organism, able, ideally, to support itself from its own products: this ideas goes beyond what is generally deemed necessary for a non-biodynamic organic farm, though there are those in the organic movement sympathetic to it, and it is at least arguable that it is a logical implication of the organic approach, with its emphasis on the cyclical as opposed to the linear. Similarly, the Steinerian emphasis on community, which appealed so much to Helen Murray and Patrick Holden, has been an aspect of organicist practice too, if one recalls the initiatives of Rolf Gardiner, or the Pioneer Health Centre, or the 'counter-cultural' communes of the 1970s. In the case of the Steiner movement, though, the schools and the Camphill Community projects have proved themselves capable of long-standing success. This may be because those who run them have a disciplined commitment to a set of ideas which, however curious they may seem to those who do not subscribe to them, ensure a clarity of purpose in those who do. It is these ideas which are likely to prove a stumbling-block to members of the organic movement who are otherwise in sympathy with much of what the biodynamic movement represents. If, however, it could be firmly established through the crystallization experiments that biodynamic methods always ensure the production of vivid and harmonious patterns; and if it were possible to conclude from these patterns that the food produced by biodynamic methods was superior in quality and nutritional value to food grown in other ways, then this would suggest that Steiner's cosmology should not be simply written off as the system of a brilliant but deluded mind. [30]

Post-war initiatives

Apart from the BAA, the longest-lasting organization promoting organic farming is the Soil Association, founded in 1946 and, since the late

1990s, much the most high-profile of the various pro-organic bodies. In the immediate post-war years, it was one of several bodies hoping to influence the direction of British agriculture. Before we examine its history in more detail, we shall consider the other representatives of the organic cause which flourished for a while before fading away in the 1950s. I have written elsewhere about Dr Siegfried Marian's *Soil Magazine* and about the journal *Rural Economy*, and shall not go into further detail in this volume; similarly, there is no need to look again at the importance of the *New English Weekly*, which ceased publication in 1949. The Council for the Church and Countryside will be referred to in Chapter 9, and the Duke of Bedford's fringe organization the British People's Party in Chapter 8.

We shall look at Sir Albert Howard's short-lived journal *Soil and Health* and the slightly longer-lived Albert Howard Foundation of Organic Husbandry, then at the journal's adoption by the SCOTTISH SOIL AND HEALTH GROUP. Also associated with the Albert Howard Foundation was Newman Turner's Somerset farm Goosegreen, near Bridgwater, where he ran the Institute of Organic Husbandry and from which he published his journal *The Farmer*.

Soil and Health and the Albert Howard Foundation of Organic Husbandry

The Soil Association would not have existed without Sir Albert Howard's work and his influence on so many of the Association's founder members, but he refused to throw in his lot with it. The reasons appear to be that he disliked the idea of its scientific research being subject to the control of laymen; that he believed the Haughley Experiment would be too expensive to run in a scientifically convincing manner; that he considered demonstration farms a more worthwhile project; and, it is rumoured, that he was a rather abrasive character, not very clubbable, who preferred to run his own show.[31] He took over his friend Lionel Picton's *News-Letter on Compost* while Picton was busy writing *Thoughts on Feeding*, and re-named it *Soil and Health*; he was quick off the mark, the first issue appearing in February 1946, three months before the Soil Association's inaugural meeting. He edited it until his death in October 1947 and the last issue (March 1948) was a Memorial Number, containing many tributes and much interesting information about his career.

It seems that the organic movement has always believed that the tide is about to turn in its favour. Howard was typically sanguine in his claim that the moment had arrived 'when the general public is prepared to set in motion among themselves and to choose what amounts to a new outlook on all matters relating to the soil, nutrition, and health, to formulate their own decisive policy, and to see that it is adopted.' (Howard even believed that the Agricultural Research Council was freeing itself from 'the NPK mentality'.) *Soil and Health* would encourage this sense of purpose by publishing 'all information pertinent to the spread of ideas on the restoration of our soils to adequate fertility and on the connection between a fertile soil and improved standards of health in crops, animals, and man'.[32] It is hard to see any significant difference between this statement of aims and the third of the Soil Association's Objects (see p. 101 below), while the material which Howard printed in *Soil and Health* would have been quite at home in *Mother Earth*. He wrote much of it himself, but Soil Association stalwarts like Newman Turner, Aubrey Westlake and E. Brodie Carpenter also contributed.

The topics covered in *Soil and Health* included restoring life to a farm's dead soil; the value of earthworms and of earth closets; the case for composting; the use of hyacinths for water purification; the nature of health and disease in plants and, more philosophically, the role played by disease in the natural order; the value of sewage sludge for agriculture and horticulture; the de-naturing of bread; the necessity for animals as part of a mixed farming system; bad farming methods as a cause of flooding and of the drying up of springs; and criticism of the government's ill-fated Groundnuts Scheme in East Africa. Within less than two years, Howard and his wife LOUISE E. HOWARD produced more than 250 pages of detailed, serious material making an intellectual and practical case for the rule of return and for the mixed farming which Astor and Rowntree were at the same time dismissing as 'muddled thinking'. *Soil and Health* remains a treasure-trove of early organicist thought.

After his death, a group of Howard's friends established the Albert Howard Foundation of Organic Husbandry. At the Inaugural Meeting in March 1948, Howard's widow was elected President; his former colleagues in India, Hugh Martin-Leake and E.F. Watson, were appointed joint Honorary Treasurers, and the Trustees included

2. THE ORGANIC ALTERNATIVE: FARMING

Newman Turner and C. Langley Owen, a surgeon who put his Sussex estate, Sharnden Manor, at the Foundation's disposal as headquarters for five years. The list of invited members makes interesting reading; it includes the judge and Buddhist scholar Christmas Humphreys; the nutritionist Sir Ernest Graham-Little; the physician J.E.R. McDonagh, and two Scottish doctors, A.G. BADENOCH and Angus Campbell, whom we shall meet again shortly. There was much overseas support: from America, the farmer-writer Louis Bromfield and the organic publicist Jerome Rodale; and from Australia the nutritionist SIR CEDRIC STANTON HICKS, with others from Central America, Southern Africa and India. Lord Douglas of Barloch, the Governor of Malta (former Labour MP F.C.R. Douglas), who had helped draw up the Soil Association constitution, was another invited member. The Foundation adopted Howard's policy of establishing demonstration centres where visitors could observe the practical benefits of organic methods. At Sharnden Manor, Owen constructed a small compost sewage plant and developed a system of continuous composting through the agency of pigs, while another Sussex farmer, a Mr Ridley, gained good dividends from applying organic methods to his dairy herd. Two of Howard's most prominent disciples offered their estates as demonstration centres: Newman Turner, and F.C. King, whose gardens at Levens Hall in Westmorland received more than two thousand visitors in the summer of 1948. Owen, Ridley and Turner also took on students. In Welwyn Garden City, a Mr V. Poliakoff offered his private garden as a centre, and his home as 'an intellectual focus for the discussion of first principles'. Centres were also established overseas. But Louise Howard was aware of the risks attendant upon undue ambition, making it clear that 'The aim of this Foundation is to do good work rather than a vast quantity of work'.[33]

The Foundation issued a number of publications primarily offering advice on, and examples of, compost-based cultivation. Martin-Leake and Watson wrote booklets on composting and the use of sewage and other refuse, while Louise Howard and her late husband's former secretary, Ellinor Kirkham, edited booklets under the general title *Our Answer on the Land,* a reference to Howard's determination to prove the chemical school wrong by showing the practical success which the Indore Process could achieve. These booklets provided instances of such success in various parts of the world, including from behind the represented: earthworms, biological pest control, the value of seaweed as a fertilizer, the dangers of chemical dusts and sprays, the quality of wheat flour, the threat of soil erosion, and the value of humus in improving soil moisture. The journal also offered healthy recipes and some detailed book reviews.

Producing this worthwhile and substantial journal — 64 pages per issue — was an onerous task, and in 1951 *Health and the Soil* ceased publication, though it made a brief comeback with a Special Number for the 1955 Highland Show, edited by Cdr Stuart. This issue alone would serve as a valuable introduction to the organic case as it was made in the 1940s and '50s, examining as it does the relationship between agriculture and medicine, the ecological importance of trees, the quality of bread, the role of earthworms and the health of plants. Lady Howard discussed the balance of Nature, Jorian Jenks raised the question of national self-sufficiency, and Stuart's son Mick asked a question which would take on increasing importance for organic farmers and growers in years to come: 'But Does It Pay?' (Without, it must be admitted, providing any answer which would satisfy an accountant.)

Stuart also established the Hopes Compost Club and ran it for many years; its several hundred members included some of the organic movement's most prominent figures.

Newman Turner and *The Farmer*

We have seen, above, that Newman Turner was active on behalf of the Albert Howard Foundation. At Goosegreen Farm, he established the Institute of Organic Husbandry, 'the World's first centre of Instruction and Advice on Organic Husbandry' and its accompanying journal *The Farmer*, which he ran and edited for ten years. The son of a Yorkshire tenant farmer, Turner studied agriculture at Leeds, where he was taught by Dr I.H. Moore, later the Principal of Seale-Hayne agricultural college, who believed that chemicals had the power to make fertility last forever. Turner worked for the Potato Marketing Board and developed a parallel career as a journalist, writing frequently for *Farmers Weekly* during the war. His pacifism — he was a member of the Society of Friends and much influenced by the Rev Dick Sheppard — led to him being put in charge of a farm for conscientious objectors. He bought the farm after the war and, having read Howard's *An*

Agricultural Testament and corresponded with him, set about applying Howard's methods at Goosegreen. Like most of the organic pioneers, his approach to farming was based on a strong religious faith. 'It is patent,' he wrote, 'that the basis of nutrition and health is biological, and those who seek to make it chemical have yet to satisfy us that they know better than God'.[36]

Turner died at the age of 50 in 1964, of a hereditary heart disease; given the amount of work that he put in to *The Farmer* and the Institute of Organic Husbandry, it is remarkable that he lasted as long as he did. The journal stood for the promotion of a return to the natural basic principles of farming and living, its ideas being drawn from the editor's practical experience.[37] Like all the organicist journals of this period, it was packed with substance; each issue contained more than forty pages of small print. To take the Summer 1949 edition as a representative example, we find Turner's reflections on the work of the recently deceased Dr Lionel Picton; articles on silage, hedge-laying, pasteurized milk, surface tillage, the effects of copper on soils, bee-keeping, the value of herbs, animal breeding, the Hunzas, heart-disease, and the value of trees for fighting drought. Three major environmentalist books were reviewed, and there was a reference to the Institute of Husbandry stand at the Shrewsbury Royal Show, at which Turner, the horticulturalist F.C. King — whose journal *The Gardener* was incorporated into *The Farmer* — and the herbalist JULIETTE DE BAIRACLI LEVY were available for consultation.

In the autumn of that year, Turner arranged a public meeting at Friends' House in London on 'The Importance of Whole Food': the speakers on that occasion were, in addition to Turner himself, his associate Derek Randal, the doctor and childbirth expert Cyril Pink (a regular contributor to Edgar Saxon's journal *Health and Life)*, the novelist and farmer Col. Robert Henriques, and the Welsh farmer Stanley Williams. Williams, with his wife DINAH WILLIAMS, was one of Turner's visitors at Goosegreen; some other interesting names in the visitors' book for the late 1940s/early '50s are the radio producer Robert Waller, later the editor of *Mother Earth;* his colleague Victor Bonham-Carter, whose book *The Survival of the English Countryside* (1971) would influence the organic grower Peter Segger; Elspeth Huxley, who would later write a condemnation of factory farming, *Brave New Victuals;* Fyfe Robertson of *Picture Post;* Simon Harris, later a prominent Soil Association member, and Richard de la Mare of Faber and Faber. It was de la Mare who persuaded Turner to write books; *Herdsmanship* and *Fertility Farming* are still referred to by organic farmers as valuable sources of organic principle and practice.

Turner moved to Ferne, near Shaftesbury in Dorset, in 1953, to Lady Hamilton's estate, where the intention was that he should use his skill in curing cattle disease to establish an animal hospital. This did not work out, and in 1958 he moved to Letchworth, qualifying as medical herbalist. We shall come across him again later in this capacity, and also as a pioneer of organic marketing.

The Soil Association

Outlasting all other organic organizations except the Biodynamic Agricultural Association, the Soil Association was founded as a response to the interest aroused by Eve Balfour's book *The Living Soil* (1943). The number of enquiries it generated seemed to demand the establishment of a clearing-house for the exchange of information. Along with Balfour, George Scott Williamson of the Pioneer Health Centre and the farmer Friend Sykes were the leading spirits in creating the Soil Association to serve this purpose. They organized a Founders' Meeting held on 12 June 1945, inviting figures who might prove generous with time and money; the body held its inaugural meeting on 30 May 1946. Interestingly, the Association's objects did not contain the term 'organic' or any of its equivalents. They were: (1) to bring together all those working for a fuller understanding of the vital relationships between soil, plant, animal and man; (2) to initiate, co-ordinate and assist research in this field; and (3) to collect and distribute the knowledge gained so as to create a body of informed public opinion. The Association was founded as an educational and research body with charitable status, and a limited company. Yet one cannot help feeling that this apparent neutrality is somewhat misleading, since its 109 founder members — among them not just farmers but doctors, dentists, horticulturalists, engineers and journalists — were overwhelmingly sympathetic towards the organic interpretation of the relationship between soil, plant, animal and man. Perhaps the clue is in the word 'vital', which makes a particular assumption.

To give a full history of the Soil Association's first half-century would require a book in itself: one which would tell a story of constant financial struggle, of internecine feuds caused by personality clashes and differences over policy, of attempts to position the Association in the van of the environmental movement, of debates over its very purpose, of splits, offshoots and reconciliations, of ventures into commerce, of clashes between principle and pragmatism ... and, through it all, a determination to survive and persist in presenting the case for organic cultivation and the national importance of agriculture, and a refusal to bow to conventional agriculture, with its government backing and extensive financial resources. This book will concentrate more on how the Association made its case for an integrated philosophy of soil and health, and less on the internal political wrangling which depleted its energies, though some of the major divisions and upheavals require examination. Much of the material will be covered in later chapters: in particular, in those on health and nutrition (Chapter 4), science (Chapter 7), horticulture (Chapter 3), environmentalism (Chapter 6) and the Seventies Generation (Chapters 3 and 5). The present section will first offer a general survey of the Soil Association's history.

From 1946 to the early 1970s

The first period of the Soil Association's history covers roughly a quarter of a century, from 1946 to the early 1970s. It comes to an end with the relinquishing of the Haughley Experiment in 1970, the retirement of LORD BRADFORD after twenty years as President, to be replaced by E.F. Schumacher, and the retirement of Robert Waller from the editorship of the journal. The Association's work during this period can be considered according to three topics: the Haughley Experiment, which will be covered in Chapter 7; the journal; and various forms of promotional and educational activity, including that of Regional Groups.

The name of the Association's journal, *Mother Earth,* was proposed by Dr George Scott Williamson, who believed that the phrase accurately captured the ineluctable fact of humanity's dependence on the soil for its sustenance and therefore its very existence. It is not impossible that this title reduced the publication's credibility in the eyes of scientists and the uncommitted, so from April 1968 it was dropped altogether, the journal remaining simply *The Journal of the Soil Association*. Its format remained the same until the October 1972 issue, the last which Waller edited. Older Soil Association members tend to be nostalgic for *Mother Earth,* and anyone who does not prefer garish design values to intellectual substance can understand why. Not that *Mother Earth* was unattractive: its cover and typeface were of an appealingly restrained elegance, and it was printed on high-quality paper. But it is the richness of the contents that is so striking in retrospect: not merely the range covered, but the depth in which topics could be explored, and the space granted to contributors, enabling them to develop an argument or to describe their farming experiences in detail.

To list the topics it dealt with would simply mean listing every area of organicist concern, so instead we shall summarize the contents of a representative issue from half-way through this period: that of July 1959. What strikes one first is the journal's sheer generosity and its respect for the reader's intelligence: few photographs and 88 pages of close print, seven of them devoted to Editorial Notes which discuss both the organic movement's aims, principles and research, and public responses to developments in agriculture and food production. There is a World Survey of news items on chemicals in farming and food, followed by some longer articles about insecticides. Eve Balfour writes a second letter (twelve pages) about her Australian tour, enabling readers to learn both of her work on the Association's behalf and of organic initiatives on the other side of the world; while back home, Ethelyn Hazell describes the work of the Association's Devonshire Group in exhibiting at the County Show. Re-printed from *Health for All* is an article by Dr R.F. MILTON on a baby's need for whole foods. The journal grants space to cover work being undertaken by the recently established HENRY DOUBLEDAY RESEARCH ASSOCIATION (HDRA), and even to summarize — from the heart of enemy territory — the views of Rothamsted chemist Dr G.W. Cooke on soil fertility. Book reviews cover the topics of humus farming, ecology, water, Max Gerson's cancer therapy, gardening, and conservation of natural resources. As in all issues from 1951 onwards, there is a photograph and brief biography of a leading figure in the Association: in this instance Richard Whittaker, founder of the very active Lancashire Group. Such was the journal which Soil Association members received four times a year. During Waller's editorship it was reduced in size for financial reasons, but still offered more than sixty pages of wide-ranging and organic and environmental material, with an infusion of more philosophical and cultural items reflecting Waller's individual perspective.

From March 1967, the Association also produced a monthly news-sheet called *SPAN,* an acronym derived from the Association's concern for the relationship between Soil, Plant, Animal and Man, whose inaccuracy presumably stemmed from the desire that it should not be confused with a particular brand of tinned meat. The title also suggested a bridge, since *Span* represented an attempt to remedy 'a feeling that communication between members and between members and Head Office is inadequate', its aim being 'to present news items and short, practical articles of general interest'.[38] Edited by Michael Allaby, it replaced the *Members' Notes* pamphlets and a short-lived bulletin which had been sent to Group Secretaries. The first eight issues were rather amateurish, typed booklets; thereafter, it had something of the appearance of an undergraduate or 'underground' publication, but containing a lot of information about organic and environmentalist matters. Appearing more frequently than the journal, it was better able to cover topical news items.

Again, let us examine a typical issue, from around the mid-point of its run: February 1970. The front page provided a report from the Oxford Farming Conference, giving readers an idea of what the food industry would require of farmers if it had its way; Magnus Pyke was there, still busy providing 'miracle stories' of the triumphs awaiting consumers. Harcourt Roy wrote with prophetic fervour about waste and pollution, James Thorburn attacked battery farming, and there was a major article on 'Science and Survival' by John Davy. Jim Worthington gave advice on poulty-keeping, and Brian Furner on greenhouse food crops and what the seed companies were offering gardeners for the forthcoming summer. There was news of a Royal Commission and of a research grant for composting. Book reviews, readers' letters, news from the Groups, and recipes all found a place. In this monthly publication there was far more material than can be found today in the Association's thrice-yearly *Living Earth*.

By the mid-1950s, the Association's membership was around 3,500 from more than fifty countries. Eve Balfour herself was a major reason for the growth of overseas membership, undertaking three tours of

2. THE ORGANIC ALTERNATIVE: FARMING

North America during the 1950s and also visiting Australia, Italy and Scandinavia. She regularly toured the British Isles, encouraging the work of local branches, and there seems little doubt that the Soil Association's survival and expansion during its first couple of decades owed a great deal to her energy in establishing warm personal relationships with members.

The Association exhibited regularly at the Chelsea Flower Show and agricultural shows; displays at the Royal Agricultural Show included exhibition plots. There was always a sufficient interest in the organic way to ensure that the Association's mail order department was busy dispatching re-prints of journal articles, and books on organic growing and nutrition.

Once in a while, the Association produced educational films to communicate its ideas. *The Cycle of Life* (1950) was a study of the rule of return, presenting the Association's ideas on the relationship between soil and health. Of course, it looks hopelessly old-fashioned today, its black-and-white footage — and some of the people featured in it — reflecting the age of Ealing Comedies from which it emerged; but it presents a well-structured and at times dramatic summary of the organic movement's ecological philosophy, effectively linking environmental issues to the concerns of ordinary citizens. One feature of especial interest is the appearance in it of Sir Robert McCarrison. In 1969 another film, *The Secret Highway,* rather over-ambitiously attempted to demonstrate the damaging environmental effects of modern farming, with particular reference to hedgerows. Although written by the noted ruralist George Ewart Evans, the script fails to achieve the lucidity and economy of the earlier film, and some Soil Association members regarded it as an embarrassment. Given that at one point it shows a group of conservation volunteers enjoying themselves by singing 'One Man Went to Mow', this is hardly surprising. Shortly before its end, the film bizarrely offers an inspirational shot of a USAF troop-carrier taxiing down the runway at an East Anglian airbase. Not long after this debacle, Soil Association Council member MARGARET BRADY made the film *Our Daily Bread,* a charming production on the subject of wholemeal flour and home baking, which featured her with her grand-daughter and was first shown at Haughley's 1970 Open Day.

Regional Groups were very active during this period, and seen as an important part of the Association's work, acting as 'ambassadors' throughout the country and enabling members to support each other. They exhibited at agricultural and horticultural shows, organized conferences, film shows, meetings and wholefood suppers, and co-operated with other bodies concerned about environment and nutrition. Riccardo Ling has recalled the sense of excitement he felt as a member of the thriving Middlesex Group, whose members in the 1950s included Dr J.W. Scharff, Dr Norman Burman and Mrs Winifred Savage, wife and then widow of Edgar J. Saxon, editor of *Health and Life*. At Leicester, the Group was founded by two dentists, Kenneth Rose and Everard Turner: they persuaded the local council to introduce a Dano composting system and sell the compost it produced. Chairman of the Leicester Group was J.L. Beckett, Leicester's City Engineer and Planning Officer, who had studied under Professor Patrick Abercrombie, Professor of Town Planning at University College, London and a founder of the Council for the Preservation of Rural England. The Leicester Group instituted production of an organic 'Rearsby loaf' at a small local mill. Michael Rust, who farmed in Leicester during the 1960s, and later took over Cdr Noel Findlay's farm in Kent, succeeded Beckett as Chairman. The Epsom Group, which gained a reputation for its successful fund-raising activities, was founded by Elizabeth Moore in 1963 and organized meetings whose speakers included Rolf Gardiner, Ruth Harrison and E.F. Schumacher. The farmer Ralph Coward and surgeon Laurence Knights were prominent in the Wessex Group. There were other notably energetic Groups in Lancashire, Dumfries and Galloway, Northumberland, Oxford, Devon, and Worcester, and both *Mother Earth* and *Span* reported their activities in detail. The Groups also provided an excuse for the Soil Association's Membership Secretary, Joy Griffith-Jones, to take off on jaunts around Britain which she wrote up in a *Span* column called 'On the Road'. Part travelogue, part social diary, her contributions could not easily be mistaken for the adventures of Jack Kerouac in his novel of the same name, and were regarded by some as disappointingly frivolous. In Joy Griffith-Jones' defence, one might say that her travels nevertheless helped to keep the Association's headquarters in touch with the regions, and to convey some sense of community.[39]

The Association's message was also communicated through the writings of individual members. Ruth Harrison's *Animal Machines* (1964) was a major contribution to the case against factory-farming, while KENNETH MELLANBY of Monks Wood Experimental Station wrote on *Pesticides and Pollution* for the Collins 'New Naturalist' series. John Davy, as mentioned earlier, was Science correspondent for *The Observer*. The case of another *Observer* contributor, the economist E.F. Schumacher, provides a curiosity. It appears that the Association was unaware of the fact that this prominent environmentalist was one of its members (he had joined in 1951) until Michael Allaby and Robert Waller came across his name among the files at Haughley and decided that he must become more directly involved in the Soil Association's work: a decision that resulted in Schumacher becoming President in 1971. The Association's involvement in the retail trade through its Wholefood shop will be looked at in Chapter 5.

From the early 1970s to the mid-1980s

The second period of the Soil Association's history, from the early 1970s to the mid-1980s, was very unsettled. Schumacher, Eve Balfour and the new General Secretary Brigadier A.W. Vickers formed an Action Group to guide the Association through its serious financial difficulties, and in a speech given in June 1971 Schumacher outlined his ideas on how it should develop. He urged the Association to demonstrate that organic farming was viable, and to gather and communicate all available evidence on the subject of wholeness. It was time to incarnate the Association's values in the world of commercial interests and to challenge the ideas on which that world was based. It is an irony of Schumacher's influence that he should have discouraged the broader environmental perspective of Waller and Allaby, who both transferred their allegiance to *The Ecologist*. With the ending of its responsibility for the Haughley Experiment, the Association's role as a research organization pretty much disappeared, and the journal dispensed with the format which it had maintained since its inception. *Span* did not long survive Allaby's departure. The change in tone from the Waller/Allaby era was heralded by the comments of the journal's new editor David Stickland in the December 1972 edition, which brusquely dismissed the 'obscure, meaningless phrases' supposedly characteristic of Waller's editorship in favour of an approach easily comprehensible to 'the ordinary reader'; one which was far more practical and commercial but would not neglect the organic philosophy.[40]

After 25 years of stability, the journal underwent a variety of transformations in the next decade or so. It started ambitiously as a monthly, simply entitled *The Soil Association*. Following Stickland's controversial departure in late 1974 to run Organic Farmers and Growers, it was issued every two months and then, from 1975, became the *SOIL ASSOCIATION QUARTERLY REVIEW (SAQR)*. These were somewhat austere publications and not until 1982–83, following what was in effect a coup organized by a younger generation of organic activists did the journal, re-constituted as a magazine, offer once more a wealth of material to its readers. Let us examine two contrasting examples of Soil Association publications, one from the late 1970s and the other from the mid-1980s.

The *SAQR* for September 1978, edited by Joy Griffith-Jones, devoted its first five pages to a US survey of the motivations and practices of organic farmers in the Mid-West, re-printed from the journal *Compost Science*. This survey concluded that organic farmers were not fundamentally different from conventional farmers, but there was no editorial comment on the significance of this finding. There were a number of items on practical farming issues: sheep dipping, cabbage fly, foggage, grain prices and the dangers of dieldrin. Ruth Stout, 'the mother of mulch' was profiled, and a re-printed article on plants which are supplementary staple crops in the developing world provided an international dimension. A wide range of book reviews occupied five pages and readers' letters were lengthy. There were accounts of Soil Association activities: the BRYNGWYN PROJECT in South Wales and the Chirnside conference week run by Cdr Stuart on the Scottish Borders. One item catches the eye of the contemporary reader: 'When the Oil Runs Out' — a response to a paper given by Sir Kenneth Blaxter at the Oxford Farming Conference, which demanded to know what the government would do to ease the transition to a less fuel-hungry agriculture. (We are still waiting for an answer.) There was much in the journal that twenty-first-century readers would be only too pleased to find in *Living Earth,* but the over-all feeling is somewhat utilitarian; the philosophical values, social and cultural dimension, and to a considerable extent the issue of health, had faded right into the background.

Six years later, the *SAQR* for September 1984 was an altogether more attractive and inspiring publication with a wider range of subject-matter, edited by a group which Sam Mayall's grand-daughter GINNY MAYALL chaired and whose other members were Francis Blake, David Greig, Dr David Hodges and Peter Segger. The international dimension was now represented by a paper on 'A Food Policy for the Third World' which Claude Aubert had presented at a Nature et Progrès/IFOAM conference. Three articles attacked the methods and effects of agribusiness. Robin Jenkins of the Greater London Council challenged one of its main objectives by pointing out its sheer inefficiency of energy use and analysed the network of seeds interests controlled by Royal Dutch Shell; David Hodges argued that the chemical approach to farming was eroding British soils, and Dr Jean Monro examined the physical effects of pesticides. The *Review* reported how the Association had promoted the alternative at the Royal Show, and looked at those who were practising it: trainees at WORKING WEEKENDS ON ORGANIC FARMS (WWOOF), Nick and Ana Jones in Cumbria, and the Biodynamic Agricultural Association (BAA). David Hodges presented technical abstracts, Roy Lacey wrote on gardening, DR PETER MANSFIELD on nutrition and health, and Gail Duff on recipes. There were several topical, contextual items culled from the national press, and updates on the Association's own news. Peter Segger wrote a long review of Tory MP Richard Body's *Farming in the Clouds,* and Eve Balfour of *Alias Papa,* Barbara Wood's biography of her father E.F. Schumacher. Balfour also contributed an obituary of Richard Whittaker, prime mover in the Lancashire Group. The philosophical dimension had not been reinstated, but the new editorial group were providing a more holistic range of topics than had been evident a few years earlier.

The fifteen years or so of this second phase saw far-reaching changes in the Association, resulting from a generational clash which became evident during the 1970s and will be discussed more fully in the following chapter. Although the Association's re-structuring in the early 1970s was attended by optimism — the Association was 'coming alive again', according to the Christmas 1972 message in *Span*[41] — membership remained stubbornly around 4,500; even the fact that its President was the author of *Small is Beautiful,* or that the guru of self-sufficiency, John Seymour, was a supporter seemed to make little difference, though their books helped promote organic philosophy and practice. Another great literary success of this period was Lawrence

Hills, but his reputation benefited the HDRA, which he founded and ran (see Chapter 3) rather than the Soil Association, in which he was also active. In the 1980s, Richard Body's attacks on agri-business and government subsidies attracted considerable attention; but since he was an ideological free-trader and Thatcherite his impact on the organic cause may have been equivocal.

Two of the Regional Groups were particularly important during this period: the Epsom Group, which through its member Dr Anthony Deavin instituted courses in Biological Husbandry at Ewell Technical College and attracted a younger generation to the Soil Association; and the West Wales Group, established in the mid-1970s, whose effects on the Association as a whole were to prove far-reaching.

We shall look at the Seventies Generation in more detail in Chapters 3 and 5. The point to note here is its impact on the Soil Association's involvement in the world of marketing. Many members of this generation were determined to make a living as farmers or growers using organic methods, and decided that the best way to survive financially was to obtain premium prices from consumers who sympathized with their 'green' approach. Development of organic marketing necessitated the creation of standards to protect consumers, and the first Soil Association standards document appeared in 1973.

Perhaps the turning-point of this second phase of the Soil Association's history was the AGM held in Edinburgh in 1982, which led to the resignation of Eve Balfour (who had taken on the role of President, following Schumacher's sudden death in 1977) and of her successor, the then-President Lord O'Hagan. The two Vice-Presidents, Angela Bates and Eric Clarke, also stood down. The Association's direction was now largely determined by people who wanted to promote the organic case more vigorously and publicly, and to launch campaigns. They attacked government food policy, being particularly scathing in 1983 about Food for Britain's slogan 'Naturally British', and in May the following year made a media breakthrough when BBC Radio 4's programme 'On Your Farm' visited Patrick Holden and Peter Segger in West Wales.[42] Not only did this lead to over three hundred enquiries and to the character Tony Archer converting to organic in the radio soap opera *The Archers*, but senior officials from the Ministry of Agriculture were sufficiently interested to make the journey westwards themselves. In 1985, the Association's offices moved westwards as well, to Bristol, largely through the influence of the Wales and West Country groups. Eve Balfour had started farming at Haughley 66 years earlier, so this was a significant break with the past.

From the mid-1980s to the mid-1990s

Between the mid-1980s and the mid-1990s, the Soil Association's magazine changed its name and, at one point, its format; it also began to display, gradually at first, the tendency to prioritize design over content that would become such a prominent feature of its production in the twenty-first century. The *Soil Association Quarterly Review* was re-named *Living Earth* from the beginning of 1988, the editorship passing from Graham Harvey to Geoffrey Cannon. Its aim under the new title was 'to preserve the zeal of the *Quarterly Review* and to revive the thought in *Mother Earth*.' It would not turn its back on the past, but would remind readers of 'the words of wisdom already published in the Journal'. After an absence of sixteen years Robert Waller was brought back to write a column and serve on the editorial board. To the disgust of some readers, the cover of the first *Living Earth* featured a photograph of John Gummer, Tory Minister of State at MAFF, with his two children and a green Soil Association symbol superimposed on his coat.[43]

For two years, from Spring 1993, *Living Earth* was published in conjunction with Tim Lobstein's *Food Magazine* in the belief 'that common causes are best promoted by combining forces', and for financial reasons; it contained a greater emphasis on food technology, nutrition and 'matters beyond the normal realm of organic issues'.[44] In the summer of 1995, under Robin Maynard's editorship, *Living Earth* went solo again, citing as its reasons for the divorce editorial difficulties and the feeling that both the Soil Association and the Food Commission could promote their own agendas and serve their members better with distinct publications.

Fifty years after the first appearance of *Mother Earth,* the Autumn 1996 issue of *Living Earth* presented a considerable visual contrast with its predecessor, though it addressed — albeit more superficially — several of the perennial organic concerns. The early *Mother Earth* had been a product of the Age of Austerity, and refused to carry advertisements; the mid-'90s *Living Earth* was a product of the Age of

Consumerism and welcomed 'relevant' advertising because the revenue would enable greater expenditure on campaigning. Colour photos and advertisements had not yet found their way into the magazine, but there were plenty of (small) photographs. A full transcript of the Prince of Wales' Lady Eve Balfour Memorial Lecture was enclosed with the magazine, though fifty years previously it would have been printed in full in the journal itself. The magazine concentrated on issues of sustainable forestry, a topic which would remind older readers of the close links which once existed between the Soil Association and Richard St. Barbe Baker's Men of the Trees organization. Other articles looked at seed saving, 'mad cow' disease, the hidden costs of industrial agriculture and, on a more consumerist note, the commercial success of the Yeo Valley 'brand' of yogurt, whose producer Tim Mead urged the organic movement to get more involved with the supermarkets. Some encouraging news items revealed that the Worldwide Fund for Nature was backing organic farming and that some urban agriculture projects were proving successful. Links with the past were provided by an article on Faber and Faber's agricultural editor Richard de la Mare and Mary Langman's obituary of Eve Balfour's sister Lady Kathleen Oldfield.

The Association's promotional and educational activities during this period shifted primarily to involvement in government policy, the development of a marketplace and instituting various campaigns. By 1986, annual UK turnover of organic produce exceeded £5m. — tiny by today's standards but a sign that the organic message was starting to attract a wider audience — and the major supermarkets were in discussion with the Soil Association about supplies. Patrick Holden, Peter Segger and Lawrence Woodward had established contacts with officials at MAFF, and when the UK REGISTER OF ORGANIC FOOD STANDARDS (UKROFS), created in 1987, published its standards in 1989, they closely followed those of the Soil Association. The Association campaigned — over-optimistically — in 1990 for twenty per cent of British farmland to be organic by the year 2000, supported Common Ground's campaign to save orchards, and urged the production of safe meat. Campaigns for sustainable forestry and community-supported agriculture, and against genetically modified organisms, were to follow. The Association also sought reform of the Common Agricultural Policy, which favoured intensive farming methods.

At the end of our period, the Local Groups were still active.⁴⁵ By 1996 there were 44 of them, of which nearly one-third had been founded in the 1990s. Some contained only a handful of members, others anything up to two hundred; the overwhelming majority of them had links with environmental groups, and 75 percent were affiliated to the HDRA. But the most remarkable statistic in a survey of all the groups was that only a quarter of Local Group members were also members of the Soil Association itself. It seemed that the Association was still failing to attract as members many people who were sympathetic to its aims and outlook, but various concerns and food scares which arose in the late 1990s would help change this situation.

British Organic Farmers

The younger generation who gained control of the Soil Association in the early 1980s were also involved in some other initiatives. Lawrence Woodward became Director of the Elm Farm Research Centre, which we shall look at in Chapter 7, and several Soil Association members were prominent in establishing the ORGANIC GROWERS ASSOCIATION (OGA) in 1981, which we shall look at in Chapter 3, and British Organic Farmers (BOF), which was founded the following year and some aspects of which we shall consider here. BOF's establishment relates to Erin Gill's insight that the Soil Association was not primarily a farmers' organization. BOF resulted from a long-term recognition in the organic movement that there existed a need for 'a serious and professional farmers' group' to act as a forum for exchanging ideas and for instigating projects which would help develop organic agriculture in the UK.⁴⁶ In March 1982, a BOF steering committee was set up, its members being Hugh Coates, the Aylesbury farmer and miller; Patrick Holden; Lawrence Woodward; David Clement of the Biodynamic Agricultural Association; the miller and baker Michael Marriage, and Richard Mayall, son of Sam Mayall and father of Ginny. Stuart Donaldson, who provided office facilities, was Secretary and Treasurer.

In the summer of 1983, OGA and BOF began publishing the journal *NEW FARMER AND GROWER (NFG)*, launched with the assistance of Marriage, Holden, Woodward, Peter Segger, the

West Wales grower Charles Wacher, tomato-grower Douglas Blair, and the editor of *The Archers,* journalist Graham Harvey. It declared its purpose was to serve the practical and technical needs of organic producers in Britain, and it fulfilled this with articles on — for instance — farmyard manure, treatment of mastitis, the implications of EC Regulation 2092/91 for organic producers, the value of co-operatives, the importance of earthworms, sheep dipping, green manures, the opportunities for organic meat production ... and so on. But *NFG* was also concerned to present the case for organic farming as the alternative to agri-business and food industry technology, and in issue No.6 presented a six-point 'Charter for Agriculture':

1. Ensure all production and management of farm resources are in harmony rather than in conflict with the natural systems.
2. Use the developed technology appropriate to an understanding of biological systems.
3. Rely primarily on renewable energy and diversification to achieve and maintain soil fertility for optimum production.
4. Aim for optimum nutritional value of all staple foods.
5. Encourage decentralised systems for processing, distribution and marketing of farm products.
6. Strive for an equitable relationship between those who work and live on the land, and by maintaining wildlife and its habitats, create a countryside which is aesthetically pleasing to all.[47]

Ralph Coward — in a piece which would have fitted comfortably into the *New English Weekly* forty years earlier — urged a change in land policy which would enable those who valued good husbandry to rent small farms, thereby helping regenerate rural society; and Chris Mair, in an article about the Mayall family, reminded readers of those 'who helped form the backbone of the movement as we know it today'.[48] RICHARD YOUNG pointed out the contribution that organic farming could make to environmentalism, and Barry Wookey reported on the interest shown in organic methods by third-year agriculture students at Reading University.

But how could those who were interested in farming organically overcome the financial and technical problems they faced? To try to answer this question, BOF issued a substantial booklet sponsored by

Barclays Bank, *Organic Farming: An Option for the Nineties,* with an introduction by Patrick Holden, who argued that organic farming offered an integrated response to a range of environmental issues, but identified three factors which had deterred potential converts: 'lack of premium incentive from the market place [;] an unfavourable economic environment during conversion, due to lack of government incentives [and] the absence of an advisory and research framework of support'.[49] This situation was changing, and to demonstrate that organic farming was now beginning to be respectable, the booklet was endorsed by Sir Simon Gourlay, President of the National Farmers' Union.

One intriguing item in the Autumn 1994 issue of *NFG*[50] reveals that both Holden and Gourlay belonged to something called the Gay Hussars Dining Club, a clique which was changing its name to the Agricultural Reform Group and whose other members included environmentalist JONATHON PORRITT; the Cambridgeshire 'barley baron' Oliver Walston; Hugh Raven of the SAFE (Sustainable Agriculture, Food and Environment) Alliance, and Fiona Reynolds, Director of the Campaign for the Protection of Rural England. The mind boggles at trying to imagine what agricultural reforms might have been unanimously accepted around that particular dining-table.

Mary Langman, a Soil Association founder-member and friend of Eve Balfour who nevertheless sympathized with the aims of the Seventies Generation and acted as mentor to Patrick Holden and Lawrence Woodward in particular, considered that by the late 1980s the producer side of the organic movement was the most 'alive' and *NFG* a more interesting paper than the Soil Association's *Quarterly Review*. RICHARD YOUNG has recalled the stress of putting it together, when its writers and editors were already fully committed to their farms and market gardens; but it is precisely this sense of personal commitment and involvement which gave the journal its liveliness: the feeling that it came 'from the ground up' and that anyone involved in the BOF or OGA could contribute to it.[51]

In the mid-1990s, BOF and OGA, who had long been sharing the offices at Colston Street, Bristol with the Soil Association, merged with the larger organization to become its Producer Wing. The Association had instituted what it called 'Project Unity' early in 1992, intending to merge with BOF the following year, though it was not until the BOF/OGA AGM at Ryton in November 1994 that BOF voted to proceed towards integration with the Soil Association. The Shropshire dairy farmer Ed Goff condemned the merger in a scathing satire, 'Plain Tales from a Slurry Tower', arguing that BOF should remain independent and act as a strong professional union representing all organic farmers.[52] But the boat had sailed, BOF disappeared, and *NFG* became the Soil Association's quarterly magazine *Organic Farming*. We shall return to *New Farmer and Grower* in the following chapter.

Permaculture

During the autumn of 1982, two courses on PERMACULTURE were run in the UK: one was a five-day introductory course in Sussex, run by John Quinny; the other, a two-day Permaculture Design course held in Cumbria. In the wake of these events, the Permaculture Association of Great Britain was established the following February at an Inaugural Annual General Meeting, held in Nottingham and organized by Penny Strange. The first issue of the Association's newsletter appeared the same month. It defined Permaculture as 'a system of organising the landscape in a self-sustaining way. [It] integrates ideas from the fields of organic farming, renewable energy technology, forest farming and the fund of human experience. The underlying ideology is: use your common sense.' Like organic husbandry, Permaculture rejected conventional agriculture as environmentally destructive and grossly inefficient in its energy use. Its name implied *perma*nent agri*culture,* and its approach was based on four criteria for a sustainable form of cultivation: that the system should produce more energy than it consumed; that soil should be created rather than lost; that nutrients should not be lost, but recycled; and that regional food needs should be met locally.[53]

Permaculture's principles had been developed in Australia during the 1970s by Bill Mollison, then in his forties, and his younger associate David Holmgren. Mollison's was a varied and colourful background: after working as a baker and running away to sea, he became a technical officer with the CSIRO (Commonwealth Scientific and Industrial Research Organisation) Wildlife Survey division in the late 1950s and spent ten years or so in remote outposts, also during that time gaining an academic education. As early as 1959, the thought occurred to him that it would be possible for humans to build systems displaying a wide variety of complex interactions, but it was not until he studied the ecology of the Tasmanian rain-forests that he began to think more seriously about this matter. In the late 1960s he began teaching at the University of Tasmania, and in the early 1970s experienced a fit of revulsion against conventional society, becoming what might be termed a 'drop-out', withdrawing to the bush and building a house on a mountain outside Hobart. He was no hermit, though, and among those who sought him out was David Holmgren, who became his sounding-board and helper as the principles of Permaculture took shape. This process was largely complete by 1975, and in that year Mollison gave a radio talk on his ideas, as a result of which he received around nine thousand letters. After a sabbatical in Britain he returned to Australia, reduced his university commitments and gradually gained complete independence, living precariously from designing Permaculture systems.

Permaculture One: A Perennial Agriculture for Human Settlements, written with Holmgren, appeared in 1978 and was reviewed by Lawrence Hills in *New Ecologist* that autumn. Hills saw the book as a 'thesis' or 'concept': well expressed and of wide interest but as yet incomplete because not tested in practice. Perhaps rather loftily, he criticized the authors for their use of the academic jargon of environmental studies, which was 'no substitute for first hand knowledge'. He had doubts about the application of Mollison's ideas to British conditions, and regretted that Permaculture was catching on among students and other beginners in the fields of agriculture, horticulture and forestry. Practical experience was now essential: the 'concept' needed to be put to the test in a variety of countries and appropriately modified in a variety of conditions. Then, Hills prophesied, in ten years' time one of these 'two young Tasmanians' could 'produce a really worthwhile book on the subject'. Sadly, Hills was not able to comment on Mollison's *magnum opus Permaculture: A Practical Guide for a Sustainable Future*, which was published in 1990, the year Hills died.[54]

During the 1980s, Permaculture was indeed put to the test, not just in Australia but in North America and Europe. Mollison claimed to have trained around one thousand people between 1981 and 1986, despite his own personal lack of money, because he felt that the capacity to teach Permaculture skills must be spread as rapidly as possible. His aim was simple but far-reaching. '[A]ll I want to do,' he avowed, 'is to re-green the Earth'. And elsewhere he wrote: 'Perhaps we seek the Garden of Eden, and why not?' He implied in this latter suggestion that his purpose was to steer a course between the destructive, energy-expensive practices of agri-business on the one hand and the drudgery of peasant farming on the other. The idea was later taken up by Anthony Wigens of the Country College at Alford in Lincolnshire, who appealed to 'The Eden Factor' as 'a powerful image of how we should live' — that is, supported by fruit trees, vines, shrubs, vegetables and grass. But, as anyone who read Mollison's books would discover, this was hardly a call to primitivism: successful Permaculture required much conscious study of ecosystems in order to create a complex, integrated and responsive pattern of mutually beneficial planting. A Permaculture system made use of many disciplines: ecology, energy conservation, landscape design, urban renewal, architecture and geographical location theories. In Patrick Whitefield's view, there developed another strand to Permaculture. Whereas permanent agriculture aims to create 'edible ecosystems', the principles can be extended to all areas of human activity: hence the idea of *perma*nent *culture*'. Taken as an approach to sustainable design systems, Permaculture can be applied to building, town planning, water supply and commercial transactions. 'It has been described,' says Whitefield, 'as "designing sustainable human habitats"'.[55]

We shall concentrate here on the features of Permaculture which bear more specifically on food production, and on Permaculture's relationship to the organic movement. It is interesting to note that Mollison and Holmgren referred (as F.H. King had done in *Farmers of Forty Centuries*) to Chinese agriculture, with its observance of the rule of return, as superior to Western techniques, and quoted Kropotkin on the need for regional self-sufficiency. They favoured a people-intensive (but not labour-intensive) agriculture, offering a varied diet, making maximum use of renewable resources, and remaining independent, as far as possible, from commercial systems of distribution. They argued that this was not the same as 'self-sufficiency', which in their view had about it something of a fortress mentality. Permaculture implied co-operative communities, or 'polycultures', of people with different skills. Its basic characteristics were small-scale, intensive patterns of land-use; diversity of species and habitats; a commitment to long-term processes spanning generations; wild or little-selected species of plant and animal as integral elements of the system; integration with agriculture, animal husbandry and forestry; and use or transformation of various forms of marginal land. (Bruce Marshall, who had reclaimed land in southern Scotland through the use of earthworms, clover and shelter belts was particularly admired).[56]

At the end of their 'thesis', Mollison and Holmgren presented 'The Permaculture Tree', a diagram of their 'holistic system, a synthesis of disciplines translated into real effects', showing how 'Knowledge flows to Productivity'. One can see why an experienced practical horticulturalist such as Lawrence Hills would have been sceptical about the imposition of a pattern of academic disciplines on the complex variety of eco-systems, but the Permaculture Tree is impeccably organic in its representation of elements interlinked with each other in an aesthetic pattern, albeit one of geometrical formality. Mollison's 1990 guide to Permaculture was far more detailed and its many diagrams reflected the practical experience which he and his followers had gained in the intervening dozen years.[57]

Some of the activities of the British Permaculture Association during that period are recorded in *Permaculture News,* which, as we have seen, was first published early in 1983. Like its contemporary publication *New Farmer and Grower,* the journal conveys the enthusiasm of a group of people promoting something fresh and important; it was a journal produced and written 'from the ground up' by Association members. Penny Strange was its first editor; she was succeeded after four years by Jolyon Fillingham, and he in turn was succeeded by Graham Bell, who established the first '"up front" Permaculture learning site in Scotland' at Coldstream. Among the other leading figures during the 1980s were Helen Woodley, Mike Roth, Andy Langford, Bernard Honey, Rod Everett and Sylvia Miller. *Permaculture News* addressed both theoretical and practical matters. Its first issue contained a summary by Mike Roth of Permaculture's 'root and branch philosophy', with its 'trust in natural systems' and its pragmatic preparedness to use any suitable technique in the creation of a 'robust self-regulating environment'. 'Nature doesn't need to work', declared the journal's Spring 1989 issue: all needs could be supplied from within a properly designed Permaculture system, surpluses could be shared and the destructive impact of conventional food-production methods reduced to a minimum. Permaculture appears thus as a mixture of science and, according to Pooran Desai, Taoism: a working with the flow of nature; a recognition of the intrinsic connectedness of the world, which awareness counterbalances our egoism and acts as an antidote to greed. Desai saw the Japanese farmer Masanobu Fukuoka as a model of Permaculturists, since he combined observation of natural processes with minimum intervention in them. Nevertheless, in order to gain maximum benefit from natural processes, work was required to establish the systems, and *Permaculture News* devoted space to various means of doing so. It looked at the possible benefits of gorse and broom, and at clearing bracken, mulching, aquaculture, conservatories, soil processes, purification of water, the multiple benefits of poultry, foggage, seed saving, fibre plants, compost toilets, the role of birds, and the importance of agro-forestry. Digby Dodd considered that a quantum leap in agro-forestry production was essential and proposed the establishment of a Temperate Products Institute, to study 'the total yield of *all* natural products per hectare', research the productivity of individual plant species, and develop cultivation techniques and appropriate technology. MAFF should become MORE (the Ministry of Renewable Resources). The wider social dimension to Permaculture principles was not ignored: the newsletter looked at LETS (Local Exchange Trading Systems), ethical investment issues and the generation of trade in local communities. Permaculturists were, as one would expect, at odds with the money economy, and Graham Bell hoped to find 'ways of returning to a society where value is measured in something stronger than money'. And there was a particular interest in city farms, as we shall see below.[58]

What, then, was Permaculture's relationship to the organic movement? Mike Roth addressed this question in the first issue of *Permaculture News,* concluding that organic farming and Permaculture were not in opposition and that Permaculture, taking a flexible approach to possible techniques, was happy to consider all the principles and practices of organic cultivation. A couple of issues later, though, Digby Dodd was less prepared to compromise, stating bluntly that 'Organic husbandry as currently practised will be totally inadequate'. Marthe Kiley-Worthington, herself a practitioner of what she termed 'ecological agriculture', expressed her fear that behind all the late-1980s 'euphoria' about organic agriculture, lay the reality that it was just another form of high-input cultivation. Writing in *New Farmer and Grower* in the summer of 1992, Peter Harper from the Centre for Alternative Technology in mid-Wales saw Permaculture as 'a universal philosophy of sustainable living', much broader in scope than the 'narrow and pragmatic' organic movement. Permaculture had a 'very ambitious programme, of which food production is only a part'. To Harper, it seemed that the organic movement emphasized high productivity through more ecologically natural practices; Permaculture emphasized 'sustainable ecological systems while trying to improve their productivity'. But he foresaw 'a time of coming together' and, four years later, suggested that Permaculture was complementary to organics and that so far the yields from Permaculture had been disappointing.[59]

More recently, Patrick Whitefield has suggested that Permaculture differs from organic growing in that it is more about design, whereas organic growing is more about method. Permaculture, says Whitefield, is a broader concept, being concerned with a good deal more than simply growing food. It takes a wider view, considering all inputs into a system and their relationship to the whole. Like Peter Harper, Whitefield regards Permaculture and organic cultivation as complementary, but believes it is not essential to use organic methods in order to practise Permaculture. (One feels that Whitefield must have in mind a rather narrow definition of 'organic', since non-organic methods would hardly be compatible with the principles of Permaculture.) While admitting that many organic practitioners take a wider view, Whitefield argues that this wider view is intrinsic to Permaculture but only incidental to organic practice. The organic movement had paid little attention to no-till methods of food production, and food from overseas could be deemed 'organic' despite having been airfreighted thousands of miles.[60]

Since this study of the organic movement is more concerned to show links and similarities between people and organizations than it is to emphasize their differences, we can usefully apply knowledge of the organic movement's history to an understanding of the relationship between that movement and Permaculture. The criticisms which Permaculturists make of the organic movement seem to be based on the assumption that it is overwhelmingly concerned with food production, marketing and, in the first decade of the twenty-first century, the creation of a brand. This is hardly the whole truth, but the assumption is not without foundation, given the fact that the organic movement's supposed success over the past two decades seems generally to be measured by the number of 'organic' products sold, chiefly through conventional retail outlets. This misunderstanding is possible because, one might argue, the organic movement has narrowed its sights and become less aware of its own history and philosophical basis. When Patrick Whitefield compares the wild woodland with the wheat field, we might recall that Sir Albert Howard began *An Agricultural Testament* with a study of the ecology of a forest and of its soil's rich fertility. Similarly, the principles of Permaculture which Whitefield enumerates include recognizable organicist principles: the importance of diversity, the value of the small-scale, responsiveness to local conditions, and the importance of the ecological dimension: the study and planning of how things work together as wholes. As I suggest at the end of this book, the principle of holism could be regarded as one of the 'bedrock' concepts of the organic philosophy, and the idea of health as wholeness can be found in organicist writings from the movement's earliest days. Permaculture is based on an awareness that things have many causes and many consequences, and that there is no getting rid of anything from a system: all will come back. This is precisely why Howard so disliked the narrowly focused analytical approach of the 'laboratory hermit', why organicist thinkers have always opposed reductionism, and why the Soil Association, particularly when Robert Waller and Michael Allaby ran its editorial department and again in the 1980s when the Seventies Generation took control of the *Quarterly Review,* was so keen to point out the polluting effects of industrial agriculture. And of course the fundamental principle of organic cultivation is the Rule of Return of wastes to the soil. Permaculture has made a point of addressing wider social issues — it offers, for instance, a way of re-populating the countryside — and is an alternative to centralized, bureaucratic systems. But the organic movement was interested in such questions right from the word go, as the *New English Weekly, Rural Economy* or the writings of Massingham in the 1940s testify, or the writings of Schumacher and GEORGE McROBIE thirty years or so later.[61]

It is also worth considering some of the personal connections between Permaculture and the organic movement. As we shall see later in this chapter, P. A. YEOMANS, who developed the KEYLINE system and influenced Bill Mollison, was much admired by Eve Balfour, who visited him during her Australian tour of the late 1950s.

He was also an influence on 'forest gardener' ROBERT HART, who, in a television interview with Jonathon Porritt, confessed himself 'haunted' by the title of Yeomans' book *The City Forest*. Hart, in turn, inspired British Permaculturists. Another British inspiration was the organic dairy farmer ARTHUR HOLLINS, whose techniques of low-energy cultivation Adrian Myers described in *Permaculture News*. Hollins considered it wrong to expose the soil, preferring to cover undisturbed soil with decaying organic matter. Simon Pratt rated Hart and Hollins, along with Bruce Marshall, as the people who had most to teach. In Bristol there was a project which impressed the American Permaculturist Sego Jackson when he visited it in 1985: this was the Windmill Hill City farm, an experiment in 'social anarchism' run by the former Dartmoor farmer David Gordon. Hollins and Gordon, along with Ron Lee, were close friends of Robert Waller, who gave them considerable support and encouragement. Bill Mollison had in fact taught a course at Windmill Hill in 1982, and both he and Jackson gave courses to members of the National Federation of City Farms and Community Gardens, an organization which actively promoted Permaculture.[62]

We can note some other points of contact from the pages of Permaculture journals: Margaret Watchom, in an article on 'Health and Soils', referred to Eve Balfour's emphasis on the *living* soil, and Steven Read, writing on 'Health and Illness', supported the organicist view that disease takes hold more easily of organisms which are not positively healthy. Among the titles offered for sale by Eco-Logic Books were the 'organic classics' *The Rape of the Earth* by Jacks and Whyte, *Soil and Civilisation* by Edward Hyams, and books by Louis Bromfield; while reviewer Tim Bastable was highly positive about Nic Lampkin's textbook *Organic Farming*, describing it as a must for anyone venturing into farming or horticulture.[63]

Although it falls outside the period which this book covers, it seems to me important to refer to the work of Rob Hopkins, Permaculturist and leading spirit in the Transition Towns movement. Here we have a clear instance of Permaculture influencing the organic movement, given the enthusiastic way in which Patrick Holden in particular has responded to Hopkins' ideas. Might there be a case for the view that Permaculture represents a reassertion of, or insistence on, aspects of the organic movement which have become overshadowed by the emphasis, during the past twenty years, on marketing and brands? Permaculture is certainly a part of the organic network, but one which has opted for an attitude of uncompromising indifference towards the world of conventional publicity and commercial systems, concentrating instead on what might be termed 'communal self-sufficiency'. We shall see in Chapter 5 that many in the organic movement were unhappy about getting close to conventional marketing systems: this was a dilemma that Permaculturists never needed to face. Permaculture was happy to remain on the margins while the Soil Association headed for the mainstream, and to anticipate a time when — as Schumacher in the 1960s and as the winter of 1973–74 had in different ways provided warning — an oil-dependent agriculture and food distribution system would no longer be viable. Permaculture, in keeping uncompromisingly to organic principles, may yet prove pointedly relevant to an economy in which fossil fuels are likely to become scarce and expensive. In 1993, Robin Jenkins reported on the fifth international Permaculture conference in Copenhagen, for the Soil Association magazine *Living Earth*. Under the rather patronising title 'Permaculture Comes of Age', Jenkins offered a brief summary of Permaculture's principles and drew the conclusion that these principles embodied 'what self-sufficient peasants have done for many generations'.[64] Just as F.H. King, for example, Albert Howard and H.J. Massingham had pointed out many years before.

Education and training

Back in the early 1970s, there was a particular problem for those who wished to become organic cultivators: where could they learn the necessary skills? When Patrick Holden decided in the early 1970s that he wanted to be an organic farmer, the one place he could find which offered a course was the Rudolf Steiner foundation Emerson College at Forest Row in Sussex, where the training was in biodynamic methods only.

The organic husbandry courses at Ewell

In July 1972, the Soil Association ran a course in Biological Husbandry at Ewell Technical College in Surrey, where Dr Anthony Deavin was director of scientific research; the first of several such courses to be held there during the 1970s. The programme for the second course, in April 1973, gives a good idea of the range of these events. Its aim was 'to give the necessary minimum knowledge of the science of soil biology as a base on which the practical experience of organic farming members could be discussed.' The impressive list of speakers consisted of Deavin himself on plant growth and soil fertility, and Dr Helen Fullerton of Glasgow University on soil structure. Eve Balfour spoke on natural ecosystems in biological husbandry and, reflecting her love of the American naturalist Aldo Leopold's writings, on the need for a 'land ethic': a form of farming based not on the profit motive but on 'an understanding of the living soil'. Other experienced farmers also spoke — Sam Mayall, George Corrin of the BAA, and Mrs Dinah Williams — and one day was spent visiting farms: either Loseley Park in Surrey, or Cdr Noel Findlay's farm near Ashford in Kent, which had been completely organic ever since 1949. Dr Lambert Mount of the McCARRISON SOCIETY related biological husbandry to health, and the Soil Association's President, E.F. Schumacher, placed organic farming in a worldwide economic context. A few months before the oil crisis of that autumn, he pointed out that farming and food production 'would have to turn to methods that did not rely so completely on fuel oil and the alternative was a return to mixed traditional husbandry'. The practical examples provided by farmers like Corrin, Findlay, Mayall and Williams indicated that such an approach could be economically viable.[65]

Working Weekends on Organic Farms (WWOOF)

Another initiative dating from the early 1970s was Working Weekends on Organic Farms (WWOOF). Founded by SUE COPPARD, its aim was to enable would-be organic farmers and growers to gain practical experience and learn from those more knowledgeable. Coppard was from an urban background, but visits to her aunt and uncle's farm in Sussex had given her a love of the countryside. While working in London in the early 1970s she found herself wishing she could undertake some part-time farm work, as a way of getting into the countryside, preferably in the company of like-minded people. She approached Michael Allaby, who suggested she contact John Davy at Emerson College. She visited Davy, who liked her idea and arranged for the farm managers to give her, and other possible volunteers, a trial weekend. She raised some interest through an advertisement in *Time Out* and, as the trial at Emerson proved successful and enjoyable, the idea began to grow. Allaby wrote about Coppard's initiative in *Span*, as result of which she was interviewed by *SEED* magazine (see Chapter 5). Things rapidly gathered momentum and Coppard found herself dealing with farmers who wanted help from volunteers, or who, ideally, were willing to offer training, as well as with potential volunteers who wanted to learn how to farm organically or simply to have a break from urban life. WWOOF's newsletter expanded and a 'fix-it-yourself' list came into being so that more experienced members could make their own arrangements directly with farmers, often for longer periods than just weekends. By 1990 about 120 farms and smallholdings were participating, and there were around a thousand members. Like the Organic Growers' Association, WWOOF saw organic farming as having much wider implications, aiming 'to help bring about a happier, more ecological society'.[66] In 1984, WWOOF published the first-ever *Directory of Organic Organisations,* compiled by Coppard and Chris Mager, containing around 250 entries and providing a valuable service to the movement.

Dr Nic Lampkin and Aberystwyth

The Emerson College courses were esoteric; the Ewell courses were valuable but short and did not survive the 1970s, and farmers who took on WWOOFers did not necessarily give their volunteers any systematic education in the ways of husbandry. There remained a need for properly run courses offered by official institutions, the problem being that the demand for such courses was limited. Nevertheless, by the end of the 1980s, as demand began to grow, some had been established. Robert Brighton, Principal of Worcester College of Agriculture, started running courses there in 1986, offering five-week block-release courses on the Principles and Practice of Organic Agriculture, while Derbyshire College of Agriculture and Horticulture provided an organic training option as part of the BTEC National Diploma in Agriculture. In 1987, the Agricultural Training Board worked with BOF/OGA on a training needs assessment exercise, as a result of which it developed, in conjunction with the ORGANIC ADVISORY SERVICE (OAS), a range of courses for farmers wishing to convert to organic, and on composting, soil fertility, pest control and other topics for those who were already farming organically. At university level, Reading offered an organic option on its B.Sc. in Agriculture, but the first full degree course in organic agriculture was provided by Aberystwyth. Dr Nic Lampkin was the leading spirit in this advance.

Lampkin came from a strongly agricultural background: his parents were agricultural researchers and he had uncles who were (conventional) farmers. His mother was an Anthroposophist and he was sent to a Steiner school, though he did not as an adult maintain a commitment to Anthroposophy and abandoned its spiritual perspective. Organic farming appealed to him for ecological reasons, as an example of the way in which humans can live with and enhance ecosystems. He took a degree in Agricultural Economics at Aberystwyth and went on to do his doctorate on the economics of converting to organic, with the help of funding from the Milk Marketing Board. His supervisor was the soil scientist DR V.I. STEWART (see Chapter 7). From 1985 he was a lecturer in Farm Business Management, and in 1986 began writing his book *Organic Farming* (1990), which Francis Blake, reviewing it in *Living Earth,* described as 'a serious, comprehensive, logical, scientifically correct exposition of organic agriculture', not to be easily dismissed by the conventional agricultural establishment. Lampkin was appointed Director of the Centre for Organic Husbandry and Agroecology, a project in which Aberystwyth collaborated with the university at Bangor, Carmarthenshire College, the Welsh Agricultural College and the Agricultural Development and Advisory Service. Its progress was interrupted by Lampkin's absence in New Zealand during 1991–92, during which he was team leader for a research programme into organic farming and agroecology. The Centre also had difficulty in attracting direct funding, and by the mid-1990s was largely defunct. Having eventually completed his doctorate in 1992, Lampkin set about increasing the amount of teaching on, and research into, organic farming at Aberystwyth. By the time of its Tenth Anniversary Issue in 1993, *NFG* was able to announce that the University of Wales at Aberystwyth was offering the UK's first opportunity for students to specialize in organic farming at degree level, as part of a degree in Rural Resources Management. A new three-year B.Sc. course would include modules on sustainable agriculture, organic crop and

Iron Curtain, where Dr Werner Grussendorf had farmed organically in the Russian zone of Germany before going into exile in 1951.

After five years, the Foundation merged with the Soil Association; Louise Howard's book *Sir Albert Howard in India* was also published in 1953, and the Soil Association received money from the Foundation towards the establishment of the Albert Howard Memorial Lectures. This was not entirely the end of a discrete Howard initiative, as Louise Howard immediately launched *The Albert Howard News Sheet,* the first of which appeared in May 1953 and the last of which (No. 100) in December 1964, by which time she was nearly 84. These foolscap, typewritten pamphlets were an example of 'that form of propaganda which relies on a bringing forward of facts': facts about soil science, municipal and other forms of composting, the effects of agricultural chemicals, and various items on nutrition.[34] In the end, Lady Howard graciously directed those of her subscribers not already Soil Association members to become so.

The Scottish Soil and Health Group

The story of *Soil and Health* magazine did not end with its Memorial to Sir Albert Howard. Instead, it passed into the hands of the Soil and Health Group in Edinburgh, whose leading lights were the doctors previously mentioned, A.G. Badenoch and Angus Campbell, and the farmer Lt-Cdr R.L. Stuart. The title was changed to *HEALTH AND THE SOIL,* but the format remained the same, and the journal continued to work closely with the Albert Howard Foundation, Lady Howard writing a long introductory essay for the first issue (Summer 1948) about her late husband's character and achievements, and stressing the need for awareness of 'The wholeness of Nature'. In describing itself as 'devoted to the interests of the health of soil, plant, animal, and man', the journal placed itself extremely close to the Soil Association, and continued to publish material which would have been at home in *Mother Earth,* and which in many cases was written by prominent Association members: the horticulturalist W. E. SHEWELL-COOPER, for instance; the market-gardener Roy Wilson and the scientist DR NORMAN BURMAN. In the Summer/Autumn 1950 issue, Sir Evelyn Howell described 'A Visit to Haughley' he and his wife had made the previous year.[35] Many familiar topics were livestock production, organic farming as a business — Lampkin was particularly determined to make the case for its economic viability — and organic farming's relationship to society at large.[67] From 1995 onwards Lampkin worked on establishing a European curriculum, and in 1999 Aberystwyth launched its full degree course in organic agriculture.

By 1995, the provision of courses in colleges had expanded somewhat, with organic courses of various kinds being offered at Dundee College of Further Education; Lackham College in Wiltshire; Otley College at Ipswich; Pershore College of Horticulture, and Shuttleworth College at Biggleswade in Bedfordshire. Broomfield College, near Ilkeston in Derbyshire, was feted by Laura Davis as 'Derbyshire's Organic Academy', because it offered a Higher National Diploma in organic farming. Various advances had been made, but Robert Brighton was justified in sounding a warning note about the possible consequences of FE colleges having been removed from local authority control in 1993: henceforth, the FE system would be a business, and agricultural colleges have indeed suffered as a result of this policy.[68]

Some overseas connections

Although this book deals only with the British organic movement, some overseas figures require mention — necessarily brief — on account of their connections with it and influence upon it. We saw above that Eve Balfour much admired Aldo Leopold; her talk at Ewell in April 1973 also referred to US soil scientist W.A. Albrecht, whom she had heard address the 1958 convention of the Natural Food Associates in Memphis. The farmer and novelist Louis Bromfield, who transformed Malabar Farm in Ohio from a state of dereliction to one of productive fertility, gave the Albert Howard Memorial Lecture in 1955. In 1958, the Lecture was given by the Australian nutritionist and soldier Sir Cedric Stanton Hicks, who had close links with the British organic movement.

Another influential figure from Australia was the farmer P.A. Yeomans, inventor of the 'Keyline' system, whose work Hicks supported and who contributed to the success of Eve Balfour's Australian tour of 1958-59. Yeomans applied his experience of mining engineering to the land in developing a system of ecological management which preserved and utilized the maximum possible amount of moisture in the soil. His achievements were also admired by the soil scientist Dr. V.I. Stewart, who learnt of it from Eve Balfour and applied its principles in the Bryngwyn Experiment in South Wales (see Chapter 7); Robert Hart, advocate of 'forest farming'; biodynamic farmer George Corrin, and Soil Association agricultural advisor R.W. Widdowson. The Soil Association's *Quarterly Review* featured a major article on Keyline principles in its December 1979 issue. As mentioned above, Yeomans' system is also one of the strands in Permaculture.[69]

Closer to home, the Italian agronomist Prof Tommaso del Pelo Pardi was a Soil Association member who, like Yeomans, combined an engineering background with a gift for land reclamation. He developed the system his father had begun and established a training farm near Rome in the post-war years. In 1960 he spoke at the Soil Association's conference at Attingham Park, Shropshire, and, during the same visit, gave a demonstration of his methods at the Welsh farm of Mrs Peggy Goodman, organized by the Country Landowners' Association. More than twenty years before Cuba was forced to do without agricultural inputs from the Soviet Union, del Pelo Pardi was appointed the first-ever honorary member of that country's Academy of Science of Soils.[70] (Here is an interesting example of the organic movement's political eclecticism: del Pelo Pardi, who had fought in the Italian army during the Second World War, accepting an honour from communist Cuba and being praised by Fidel Castro.)

Dr Hartmut Vogtmann, who in 1981 was appointed the world's first professor of organic agriculture, at the University of Kassel, worked closely with Lawrence Woodward and the Elm Farm Research Centre. From France, the dairy farmer and grassland scientist André Voisin joined the Soil Association and impressed its members with his books *Soil, Grass and Cancer* and *Rational Grazing*. Robert Waller saw Voisin as an agricultural philosopher in the mould of his hero Sir George Stapledon. Waller also forged links with the farmer Christian de Monbrison, although there were some who felt that Waller's motives were less concerned with Monbrison's organic farming than with the opportunity the friendship provided for him to disappear to his beloved France. Nevertheless, Monbrison's work was also admired by Angela Bates and Mary Langman. Langman was fluent in French and played an important part in establishing a connection with the Soil Association's French equivalent Nature et Progrès, whose meetings Angela Bates and Elizabeth Murray also sometimes attended. Langman was active in the establishment of IFOAM.

How the case for organic cultivation was made

These, then, were the organizations and journals which took on the might of agri-business and government policy; there were many books, too. It all amounts to thousands of pages by farmers, soil scientists, biologists and agricultural philosophers.

They presented their case using two different tactics, one negative and the other positive: by attacking the reductionist philosophy and damaging effects of the agri-business approach, and by presenting the holistic philosophy and beneficial results of organic husbandry. Sir Colin Spedding, who chaired the committee which established the standards for UKROFS but who, despite his interest in agricultural systems, does not favour organic farming, considers the organic movement to have damaged its own cause by its assaults on the practices of fellow farmers. In contrast, Cotswolds farmer Richard Young believes that the movement fails to draw sufficient attention to the weaknesses of the agri-business approach and the harm it causes. During the short-lived yoking together of the Soil Association's *Living Earth* with *Food Magazine,* some Association members felt that the latter's relentless, well-informed attack on food technology was too depressing, and wanted a more positive outlook reflecting the organic movement's positive message.

Throughout the five decades from the mid-1940s to the mid-1990s, organicist writings provided ample material both for those wanted an attack on agri-business and for those who preferred an encouraging vision of the alternative. As we now live in a period when the case for organics seems at times to amount to little more than pointing to an increase in retail sales of organic products, it is worth recalling that there exists a wealth of material from the twentieth century, much of it written by competent farmers and scientists, which puts together a coherent case for organic husbandry: one worthy of consideration, whether or not it is ultimately accepted. We shall be looking at the scientists in Chapter 7; let us conclude the present chapter with a survey of the case — negative and positive — for organic methods of farming, as made in the fifty years following the end of the war.

Negative, first: industrial-chemical techniques of farming exhaust the soil. One of the key texts for the organic movement was *The Rape of the Earth* (1939) by G.V. Jacks and R.O. Whyte, a study of soil erosion throughout the world. Agri-business had yet to hit its stride, though the United States 'Dust Bowl' showed the damage which a purely extractive approach to fertility could bring in its wake. The Soil Association's journals monitored erosion in different parts of the world, and drew to their readers' attention the evidence that both nitrogenous fertilizers and agricultural machinery could damage soil structure.[71] During the 1980s Dr David Hodges produced a series of articles on soil erosion for the Association's *Quarterly Review,* and the Association launched a Soilwatch campaign in 1987, arguing that the UK's high rates of erosion resulted from conventional agricultural practice: monoculture, heavy machinery compacting the soil, intensive use of artificials, and the removal of hedges and woodland.

Two boosts to the case against industrial farming were given by books published in the first half of the 1960s. First came Rachel Carson's *Silent Spring* (1963), with much grist to the organic mill: powerful evidence for the destructive effects of chemicals on wildlife. Concern about pesticides pre-dated Carson's book, though: Laurence Easterbrook had been raising anxieties about 'trigger-happy spraying' in his *News Chronicle* column back in 1959, as had John Davy in *The Observer.* The sinister events in 1963 at Smarden in Kent, where fluoroacetamide poisoning caused by wastes from an insecticide factory led to the death of every living creature on a two-acre stretch of land, provided another piece of evidence to be used against chemicals. Books by John Coleman-Cooke and Kenneth Mellanby kept up the momentum.[72]

The organic movement has never eased off in its fight against agricultural chemicals. The Pesticide Action Network was founded in 1984; Nigel Dudley, Richard Body and David Hodges maintained the pressure in the late 1980s to such effect that the industry launched a counter-attack. Much of the organic movement's objection to agricultural chemicals has been based on their putatively damaging effects on health, but from a purely farming and gardening point of view the dangers are the damage to wildlife and bio-diversity, and, as Kenneth Mellanby pointed out, that pests and weeds will develop resistance to chemical attempts to do away with them. The treatment of animals under conventional farming systems is another perennial theme in the organic case against those systems. The first quarterly issue of *Mother Earth* devoted four pages to the state of animal health, and during the 1950s the journal drew attention to the detrimental effect of nitrogenous fertilizers on dairy cows, to the dangers of using hormone supplements for fattening stock, and to the stresses imposed on pigs and hens by intensive rearing and production systems.[73]

In 1964, what might be termed 'the *Silent Spring* of factory farming' appeared — appropriately, with a foreword by Rachel Carson. This was *Animal Machines,* by Soil Association member Ruth Harrison, a gruelling examination of the conditions which animals and poultry endured, and one which aroused a passionate response from consumers hitherto ignorant of the methods used to produce their food. The government appointed Professor Brambell as chairman of a committee to study the question. Its report appeared at the end of 1965, but its recommendations were not implemented. The continued mistreatment of animals, with its consequence that they required ever-increasing dosages of medicine, enabled the organic movement in the 1980s to criticize the emergence of an 'Animal Pharm'. The aesthetic case against agri-business was powerfully expressed by Marion Shoard: it reduced bio-diversity, destroyed landscape features and imposed unsuitable buildings. Ethics and aesthetics are philosophical areas in which it is difficult to find firm ground, but there were more hard-headed considerations of resources and economics which could be brought to bear. Schumacher identified the vulnerability of industrial agriculture back in the 1960s: how could a system so heavily dependent on finite resources hope to survive? Sheer practicality demanded an examination of alternatives. Then there was the economic cost of climbing on 'the chemical treadmill': farmers' indebtedness to banks, caused by the need to invest in costly equipment and to pay for the necessary inputs of fertilizers, chemical controls and antibiotics; and the cost to the taxpayer of government subsidies, which Sir Richard Body so pugnaciously itemized in the 1980s and '90s.[74]

It may be unjust to suggest that the organic movement has taken a certain perverse pleasure in discovering and communicating the latest

2. THE ORGANIC ALTERNATIVE: FARMING

crimes and misdemeanours of industrial farming, or, during, the mid-1970s, in pointing out that an economy dependent on oil would be forced to change its ways and live more frugally. Nevertheless, there seems little doubt that the movement's higher profile since the late 1990s owes a lot to the various food scares of the late-twentieth century, and that agri-business has done some of the movement's work by providing for it a negative example of agricultural practice. Only some, though; the movement was able to capitalize on the delinquency of agri-business because it already had a body of thought and practice on which to draw. Good farming, the organic movement has always argued, is the care, not the exploitation, of the soil. The rule of return must be observed; if it is, soil grows rich in the humus which lends it life. The challenge, then, is how to ensure this maintenance of, and increase in, humus, enriching the precious few inches of topsoil on which human survival depends. The American agricultural scientist F.H. King, in his classic *Farmers of Forty Centuries,* first published in 1911, recorded the approach used in the Far East: the return of all forms of biological wastes to the soil. The Hunza tribesmen of India's north-west frontier employed very similar methods. Applying a Western scientific technique to enhance traditional methods, Sir Albert Howard developed his Indore Process of composting, whose success in various countries provided evidence in favour of humus farming. In Britain its chief early exponents included Friend Sykes, Newman Turner and the market-gardener Roy Wilson. Compost and farmyard manure were clearly valuable substances; the problem was how to obtain sufficient quantities of them. Mixed farming was essential (though there have been one or two interesting experiments with stockless farming systems), as was the encouragement of the earthworm population, given their capacity to improve soil structure and augment its fertility through their casts. In a largely urban society, use could be made of town wastes, and the organic movement undertook a good deal of work with municipal composting projects, which will be examined more fully in Chapter 7. Taking advantage of the national coastline was also a possibility; the benefits of seaweed were frequently promoted.[75]

The organic case was strengthened by the achievements of those whose experiments and experience suggested that organic methods were viable both agriculturally and economically. Notable instances included Shropshire dairy farmer Sam Mayall, who converted to organic in 1949 as a result of comments made by his student son Richard; and, in the 1970s and '80s, the Wiltshire arable farmer Barry Wookey, whose example proved large-scale farmers might adopt the organic approach as effectively as smaller farmers. Then there were the grassland improvers, who showed that lush pasture could be created without heavy doses of bagged nitrogen: Arthur Hollins at Market Drayton in Shropshire, and the West Country farmers Ron Lee and David Gordon.

The work of Stanley and Dinah Williams in North Wales — and later of their daughter Rachel Rowlands — and of brother and sister Richard and ROSAMUND YOUNG in the Cotswolds, showed that the organic approach could benefit animal health. As for plant health, the argument ran that if plants were grown in a humus-rich soil they would develop greater resistance to pests and diseases than those grown in a soil depleted of its minerals and infused with chemicals. Weeds have always been a problem for organic farmers, though the market-gardener F.A. Secrett had discovered that increased soil fertility could drive out certain perennial and deep-rooted weeds, while another successful organic farmer, Robert Milnes-Coates of Yorkshire, believed that the way to deal with weeds was to smother them with heavy cropping.[76] As for weeds in grassland, observation of freely grazing animals indicated that they valued plants which farmers might wrongly have considered surplus to requirements. Instead of aiming to control nature, an organic farmer seeks to study its pattern; 'weeds' may be beneficial to animal health, or act as guides to the improvement of soil conditions.

Back in the 1940s, Donald P. Hopkins, a tireless challenger of the organic school's claims, proffered the view that organic farming appeals to those of an aesthetic disposition. It was a shrewd comment, and not merely because many of the leading organic pioneers were indeed artistically inclined. In this connection, Lawrence Woodward's apparently paradoxical suggestion that Sir Albert Howard's approach to farming was not truly organic takes on a particular significance. Woodward considers Howard's agricultural ideas to be based on an input-output model in which compost and farmyard manure substitute for chemical fertilizers. Whether or not this is fair to Howard (who attached much importance to the symbol of the Wheel of Life), Woodward is surely right to emphasize that organic farming aims, ideally, to achieve a self-sustaining system. As we shall see in Chapter 9, the essentially aesthetic concepts of balance, harmony, and unity in variety have been important in organicist thought. Organic farming, with its awareness of ecological systems, works to create a virtuous circle, where all parts contribute to the whole. This can be seen in its purest form in the biodynamic concept of the farm as an organism and in the ingenious synergies created by Permaculture. Instead of the economic efficiency of the orthodox school, the organic movement aimed for what Jorian Jenks termed 'biological efficiency', which he defined as 'the production of more and better nourishment by the more skilful management of natural resources'.[77]

****

Completely against the trend of post-war agriculture, with minimal financial resources, only a modicum of government support, and in the face of much hostility and mockery, many exponents of the organic approach demonstrated during the second half of the twentieth century that it could be made to work, and that even in this hostile agricultural climate it was a viable alternative to conventional agriculture. But we should ask, at last, why 'conventional' or 'orthodox'? Writing in *New Farmer and Grower,* the land-use consultant Joy Greenall pointed out that so-called 'conventional' agriculture, dependent on fossil fuels, was a mere blip in the history of farming, and wanted to know why farming based on biological systems and working with the balance of nature, seemed to frighten so many in the world of agriculture.[78] It is not clear whether the question was intended rhetorically, but one answer must certainly be that it threatened the vested interests of those who had done very well under the post-1945 dispensation and whose products would no longer be necessary if biological methods came to be widely adopted.

3. The Organic Alternative: Gardeners and Growers

Given that most people's gardens are very small areas of land, the case for domestic organic gardening may not seem particularly compelling. What harm can a few slug pellets, chemical sprays or patent fertilizers do when applied on such a microcosmic scale?

Quite a lot, was the answer given by ALAN GEAR of the Henry Doubleday Research Association (HDRA) in the early 1990s:

> When we burn our garden rubbish we are contributing to atmospheric pollution and global warming, in exactly the same way as a farmer who burns the straw on his fields. When we liberally scatter fertilizer granules around the garden, there is every possibility that excess nitrate will make its way underground to join the excess fertilizer run-off from farm land. Wildlife is poisoned by pesticides sprayed on to crops by the gardener who uses fungicide on his roses or who sprays greenfly on his broad beans, just as wildlife is destroyed by modern farm practices. True, the scale is incomparably smaller, but consider that there are 18 million gardeners in the UK covering an area of a million acres. Collectively, what we do in the garden has enormous impact, for good or otherwise.

This is the negative case for organic gardening. The positive case is of course the same as it is for organic farming: 'health for soil, plants and people', as Gear's mentor Lawrence Hills expressed it in his autobiography. With increasing public concern about 'the potential long-term risks to health from consuming food containing pesticide residues', growing one's own fruit and vegetables organically became an appealing alternative, the idea being boosted by the mid-1980s television series *All Muck and Magic?*, which featured Alan and JACKIE GEAR and attracted more than three million viewers.[1]

3. THE ORGANIC ALTERNATIVE: GARDENERS AND GROWERS

Organic gardening in the 1940s and '50s

The HDRA is the supreme success story of British organic gardening — in fact, if measured purely in terms of numbers, of the British organic movement as a whole — and we shall consider it more fully below. As it was not formally established until 1958, let us look first at how organic gardening was promoted during the 1940s and '50s. Already in existence for nineteen years when the Soil Association was founded, the National Gardens Guild (NGG) was chaired for a quarter of a century by Lady Frances Seton, one of the Association's founder members. It stated its aims as: 'To grow flowers and vegetables in urban areas and villages. To plant waste spots and waysides with flowers, and to unite all classes in restoring living beauty through gardening'. (Since the NGG was founded the year after the General Strike, one wonders if its implicit purpose was to avert any further threat of revolution through the creation of flowerbeds. One might also discern the prospect of 'guerrilla gardening' in the second of its aims, though Lady Seton and her fellow guild-members would surely always have asked permission first.) The NGG's publication *The Guild Gardener* was a highly regarded horticultural journal with an emphasis on organic husbandry, exhibiting impeccable organicist 'care for the problem of increasing soil fertility to produce the maximum of pure health-giving foods for plants, animals and men'. Contributors included Richard St. Barbe Baker, the crofter Roy Bridger and H.E. Witham Fogg, who would later write regularly for the Soil Association journal.[2] Sir Albert Howard's work inspired gardeners as it did farmers. F.C. King, the head gardener at Levens Hall in Westmorland was one of his most fervent disciples: he worked for the Albert Howard Foundation of Organic Husbandry and as editor of *The Gardener*, which was printed as part of Newman Turner's journal *The Farmer*. Ben Easey paid tribute to the Indore Process of composting in his 1955 book *Practical Organic Gardening*, a very detailed and comprehensive study of the topic which included more than fifty pages on the problems posed by diseases, pests and weeds, and was favourably reviewed in *The Observer* (22.1.1956) by Vita Sackville-West. The word 'practical' in the title carries somewhat barbed overtones: at the end of the book, Easey aims a shaft or two at 'the "Whitehall warriors of the compost army", well out of harm's way, [who] have expected the practical organic gardeners ...to advance regardless, spurred on only by war-cries and a belief in "that which is right"'. Easey was essentially a scientist, interested in experiments and in conclusions reached on the basis of experience, and he regretted the lack of data on organic methods: 'No one has yet recorded in detail ... what changes of the obvious sort the gardener converted to organics may expect to encounter, or has yet methodically contrasted a five-year-old organic garden with an orthodox one.' There were few statistics on pests and diseases in relation to organic gardens, or on the effectiveness of organic methods of their control. The HDRA, which Lawrence Hills established as a registered charity three years later, undertook work of this nature. Hills collaborated with Easey on the production of his book for Faber. Howard also influenced Dr W.E. Shewell-Cooper, a prolific writer of books on gardening. Before the Second World War, Shewell-Cooper was Superintendent of Swanley Horticultural College in Kent, met Howard while in that post and started reading his work. In 1950 he bought a twelve-acre estate at Prior's Hall, at Thaxted in Essex, where, the Soil Association journal *Mother Earth* tells us cautiously, he ran 'what he claims was the first organic horticultural college': the Horticultural Advisory Bureau and Training Centre, with four specialist instructors for a dozen students and an advisory service run by post, visit and lecture.[3] Shewell-Cooper used no inorganic fertilizers, but generated around one hundred tons of compost annually from vegetable wastes, straw, goat manure, applied mulches, sedge peat and fish manure. He was particularly interested in 'no-dig' techniques of cultivation. His students went on to hold horticultural posts in various parts of the world.

A Soil Association Council member from 1946–47 to 1962, Shewell-Cooper founded his own organization, the GOOD GARDENERS' ASSOCIATION (GGA) in 1963 — it still runs today — aiming to encourage organic methods. He published a monthly newsletter, and at Arkley Manor in Hertfordshire, to which he moved in 1960, members were able to see the practical application of his methods. By 1950 he had been awarded the MBE. George VI requested him to write a book on the royal gardens: titled *The Royal Gardeners,* it was published in 1952 just after the King's death. He could boast, among various honours, of being a Chevalier de l'Ordre National du Mérite and a Doctor and Fellow of Vienna University's Horticultural College. Later in his career, Shewell-Cooper's relations with some of his organic colleagues and with the Soil Association became problematic. Alan Gear considered Shewell-Cooper pompous and obsessive, but always very interesting. Shewell-Cooper clearly had a talent for what we would today term 'networking', and for gaining publicity; John Wheals, Soil Association Treasurer during the 1970s, believed that his skill at marketing himself in fact ensured that the organic message reached a wide audience.[4] Shewell-Cooper's son Ramsay was a Soil Association Council member in the 1980s. He had trained with the noted organic market-gardener F.A. Secrett, the French horticulturalist Professor Coutenceau and at Cambridge University's Royal Botanic Gardens.

Ten years after the end of the war, Ben Easey was able to offer the details of more than thirty organic gardens and nurseries — some private, some commercial — and of seventeen suppliers of organically grown seeds, plants, manures and equipment of particular interest to organic gardeners. Two County Councils — Middlesex and Dumfries — were among the names, as was the Viscount Newport (later Lord Bradford), the Soil Association's President. Two firms stand out as instances of continuity in the organic movement: the seed company Hunter's of Chester, and Chase Protected Cultivation Ltd, at Chertsey in Surrey. Both still exist today (the latter as Chase Organics). Hunter's has a link with one of the earliest organic pioneers, as its founder James Hunter worked with Robert H. Elliot of Clifton Park near Roxburgh on improving grassland during the 1880s. J.L.H. CHASE was a specialist in gardening with cloches, which his father had invented in 1912, and wrote books on his methods for Faber. Chase began experimenting with organic methods in 1942, was one of the Soil Association's founder members and later served on its Council. He was, inevitably, influenced by Howard, and to his gardens at St. Ann's Hill came Laurence Easterbrook, Eve Balfour, the biodynamic pioneer Ehrenfried Pfeiffer and the nutritionist H.M. SINCLAIR, among other organic notables. At his 1954 Open Day, he was able to point to the success of organic methods, as his own results were still steadily improving after ten years without chemical fertilizers. What was disparaged as 'muck mysticism', he believed, would become the textbook knowledge of the next generation. He was a friend of the herbalist Juliette de Bairacli Levy (see Chapter 4) and his nurseries helped her by undertaking various experiments. The recently established HDRA exhibited at Chase's Open Day in June 1960. Chase continued to be actively involved in the organic movement through to the 1980s, contributing articles on South American agriculture and environmental problems to the Soil Association journal, and speculating on the prospects for 'Agriculture in the Eighties'. He spoke at an IFOAM conference in the summer of 1984, advocated the benefits of seaweed and left a legacy to the Soil Association after his death in the early 1990s. Chase is one of those who, though now largely forgotten, provided an important thread of continuity from the early years right up to the dawning age of organic consumerism.[5]

Another important figure for organic gardeners was Arthur Bower, the so-called 'compost king', a commercial grower at Wisbech in the Fens from 1912 to 1964 and subsequently a consultant horticulturalist. His interest in organic growing stemmed from a chance meeting with Sir Albert Howard at a farm machinery show in 1941, the story of which Bower amusingly recorded. Howard visited Bower's holding several times to advise him on farmyard manure and composting, and Bower was so convinced by the results 'written on the land' that his farm went all out for composting. For Bower, the practice came first and the philosophy later, deduced from empirical evidence, and he believed that scientific investigation of such results would put to flight all accusations of 'moonlore, muck and magic'. It should be emphasized that Bower was chairman of his county's NFU horticultural committee and a highly respected market gardener whose holding, during the 1950s, was inspected by a host of growers and advisory and research officers. Bower would later influence another fenland grower, John Butler, who ran a smallholding near Boston in Lincolnshire and was commissioned by Robert Waller to interview him. Bower's various remarks on the nature of composting opened up a new dimension in Butler's own organic thinking, of which more later in this chapter.[6]

One other figure is worth noting here: F.A. Secrett of Milford, Surrey, one of the country's leading commercial horticulturists, who had, at one time, been Adviser on Vegetable Production to the Minister of Agriculture. In 1950, Secrett gave a lecture to the Royal Society of Arts in which he outlined the indispensable functions of humus. He was not then, as he later became, a Soil Association member; he was recounting the experiences of more than four decades working the land, experiences which led him to sympathize with the organic philosophy.

Lawrence Hills, the HDRA and the Gears

We can now turn to the extraordinary career of one of the organic movement's most influential characters, Lawrence D. Hills.

I have summarized the story of Hills' entry into the organic movement, through contact with older gardeners and Sir Albert Howard, in *The Origins of the Organic Movement*. The full story can be read in Hills' autobiography *Fighting Like the Flowers* (1989), published the year before his death, which demonstrates the enormous amount of experience, as both horticulturalist and writer, which enabled him to establish the HDRA as Britain's largest organic membership body. During the war, Hills had become friends with Richard de la Mare, as a result of which de la Mare employed him for 25 years as one of his readers at Faber and Faber. One of the many manuscripts he handled was Newman Turner's *Fertility Farming*, and Turner later became President of the HDRA.

The HDRA was set up in 1954 as a research station to investigate the value of Russian comfrey, although it was not legally constituted as charity until 1958, when it produced its first newsletter and began recruiting members. Henry Doubleday (1813–1902) had introduced comfrey into England in the 1870s, and Hills described it as the crop which changed his life, a plant unique in its yield and medicinal qualities. Newman Turner used it as part of his alternative veterinary treatments, and in 1955 he participated in the 'comfrey race' which Hills organized, with a record British yield of over 67 tons 16 cwt. per acre. Hills began to expand the HDRA's horizons when it occurred to him that gardeners could carry out research into chemical-free methods on their own plots. Skilled at using the Press to spread his ideas, he sent to the *Daily Telegraph* a letter which was printed as a news item and drew a hundred responses. Hills soon had fifty experimenters lined up, but needed an organization to which they could belong. In October 1958 the HDRA was constituted as a charity whose chief aim was to foster the study of Russian comfrey and of improved methods of organic farming and gardening. 1958 was a turning-point for Hills in another respect: David Astor appointed him gardening correspondent of *The Observer*, and the money he earned as a journalist helped fund the HDRA's work.

Looking through the HDRA newsletter, which first appeared

in the autumn of 1958, one is again reminded that this was an age when substance took priority over design. That Hills' interests went well beyond comfrey was evident from the second issue: the trial ground at Bocking in Essex, fifty yards from which he lived with his parents, was 'the cheapest and most unusual Research Station in the World, with its trial plots "replicated" in farms and gardens in many countries.' Hills had recruited 105 members, 22 from overseas; ten years later, membership was nearly 1,500. The newsletters record the research undertaken by members on a wide variety of topics: among them green manuring, composting techniques, the use of perfumes as pest repellents, no-dig potatoes, hedgehogs, sowing by lunar phases, Japanese pumpkins, liquid comfrey manure as a tomato feed, rhubarb as an antidote to club root, the use of sawdust, traps for codling moths, dealing with slugs, spraying with seaweed, and the potential of soap as a pesticide. As Hills pointed out in the second newsletter, Fisons was spending £750,000 per annum on pest control research, to discover chemical answers, but 'the organic ones are waiting to be found'.[7] There were an increasing number of growers eager to point the way, with the result that the newsletter became ever more bulky — and more informative, containing a wealth of practical advice.

Early in 1959, Hills wrote that the HDRA was 'not a Social body; we are concerned with research'; but some of his campaigns carried strong social implications. From the outset, he was concerned about the possible effects of nuclear fall-out on animals and plants, and opposed the large amounts of government money spent on the space programme. He keenly supported the case for municipal composting, producing a guide to 'good sludge' which proved remarkably popular. A resolution at the 1964 AGM, that pesticides should be more strictly labelled, was taken up by the Labour MP Mrs Joyce Butler as a Private Member's Bill, passing into law three years later as the Farm and Garden Chemicals Act. And, like many in the organic movement, he opposed the fluoridation of water, arguing that preventive measures against tooth decay were preferable.

Hills wrote many letters to the national newspapers, about the dangers of agricultural chemicals, or the value of toads and ladybirds as pest controllers, or — one of the HDRA's major campaigns — the dangers of bonfire smoke. He found this strategy an effective means of recruiting new members. The oil crisis of 1973–74 provided an ideal opportunity to point out the need for greater national self-sufficiency in food production, and his 'Dig for Survival' campaign attracted a huge response, with the HDRA office receiving at one point nine hundred letters a day. During the 1960s, through the influence of HILDA CHERRY BROOKE, whom Hills married in 1964, the newsletter began to devote more space to nutritional matters, with Cherry, as she was known, contributing a regular feature, 'Housewives' Help'. (Cherry had cured Hills of coeliac disease through changing his diet.) The HDRA worked with Michael Allaby of the Soil Association to produce the *Wholefood Finder,* which appeared, after various frustrating delays, in 1968.

Hills' health had greatly improved by the early 1970s, which was just as well, since his writing and the HDRA were making ever-greater demands on his energy, and he had now added a place on the editorial board of *The Ecologist* to his other commitments. HDRA membership doubled between 1973 and 1976, passing 6,600, and members continued to support the various experiments. The newsletters of this period reveal interest in garlic, paper composts, lucerne, beetles, couch grass, moles, the use of water hyacinths as purifiers, the need for more allotments, and companion planting. There is a strong sense of collaborative enterprise about the newsletter's contents, and a commercially disinterested desire to create a body of knowledge of successful organic methods. But to some, the very variety of members' plots posed a problem. As early as the fifth AGM, in 1963, there had been objections to the lack of scientific rigour in the experiments, and these were reinforced two years later. Disagreements over whether the Bocking trial ground should concentrate on research or demonstration remained an issue and led to serious disagreements on the HDRA Council in 1976.

By the early 1970s, Hills was requiring more help with the work at Bocking, and was concerned that knowledge of organic methods should be passed down to another generation. From 1971 he began to have students living in a caravan on the site, and at the beginning of 1974 took on Alan and Jackie Gear as full-time deputies, whose work was supplemented by a shifting population of students and WWOOFers. The Gears, who were both scientifically trained, brought greater order to the experimental data and established a laboratory (though this was subsequently sold and converted into a library). Dr Anthony Deavin and, later, Dr Bill Blyth acted as honorary scientific advisors. Only another list can do justice to the HDRA's activities in the late 1970s. Hills himself travelled the world speaking about its work, local groups were established, membership experiments on lead contamination in vegetables were carried out, the Vegetable Seed Library was set up in 1978, investigations were begun into drought-resisting plants for Third-World growers. Publications included Lawrence Hills' Penguin book on *Organic Gardening* and Alan Gear's slide set on pest control. And, although Hills is adamant in his autobiography that the HDRA was not a rival to the Soil Association, he may well have taken some satisfaction when in 1977 it became the largest of all the organic membership groups. (According to a source who worked at Haughley that year, the Association's General Secretary Brigadier Vickers dismissed Hills as 'just a bloody journalist', a remark perhaps fuelled by envy of Hills' success at a time when the Soil Association was struggling to make headway.)[8] In 1978, Pauline Pears joined the HDRA as a research assistant, which in Jackie Gear's view was one of the best things that happened to it.

Despite the HDRA's success, the Gears were beginning to feel frustrated. The trial ground at Bocking was really too small by now, and Jackie was keen to establish a national demonstration centre, to prove that organic gardening could be done well, and did not need to be a mess. She felt that the HDRA still spent too much time preaching to the converted, and needed to go out to the public more vigorously, persuading by example. At first, the Gears considered keeping the headquarters at Bocking and taking on a demonstration centre somewhere else, in a tourist area, but after discussions with the National Trust about the walled garden at Kingston Lacy in Dorset fell through, they decided on the more radical option of moving the entire operation to a more geographically central site. Ryton, on Coventry's eastern outskirts, fitted this requirement and, at 22 acres, was the right size; in other respects it was, in Alan Gear's description, 'horrible': bleak and windswept, with no earthworms in the poor soil, and riddled with the stumps of dead elm trees, which had to be blasted out with explosives. By the standards of contemporary risk-assessment it was 'a dreadful gamble' and one which could have sunk the HDRA completely, given the amount of debt which the organization accumulated. Some Council members resigned over the issue, but Lawrence and CHERRY

HILLS, although respectively in their seventies and eighties, supported the move. The site's unpromising nature proved advantageous in one respect: the local authorities approved the HDRA's plans, possibly on the basis, the Gears believed, that nothing could be worse than what was already there.

The HDRA moved in July 1985 and had to be open for business within a year. With additional help from some Council members and unemployed youngsters on a Manpower Services Commission scheme, this was achieved. During the year, in January 1986, Lawrence Hills retired and Alan Gear became the HDRA's Executive Director. The gardens, in their most rudimentary form, were opened on 5 July, with extensive media coverage. Further coverage was to follow when, in the autumn of 1986, Channel Four television came back to the Gears with the news that an earlier proposal for a series was now slated to be broadcast the following May. The Gears pointed out that certain biological imperatives rendered this schedule problematic, and gained a few weeks' grace, the series in fact starting in July 1987. It featured the Gears, Pauline Pears, Sue Stickland and Bob Sherman, and attracted between three and four million viewers; the accompanying booklet sold more than seventy thousand copies, putting it in Channel Four's top five most popular programmes. The series earned the HDRA some much-needed money and led to a surge in membership: from 7,500 at the end of 1985 to 18,000 five years later, following two further series. A financial deficit of more than £15,000 in 1984 was turned into a surplus of nearly £7,500 by the end of 1990. The gamble had been justified and the Hillses lived to see the success. Cherry died in 1989 and Lawrence the following year; regrettably, he never wrote the second volume of autobiography which should have followed *Fighting Like the Flowers*.

During its first ten years at Ryton, the HDRA expanded its activities enormously. It forged links with Coventry Polytechnic (which became Coventry University in 1992) in order to collaborate on research; erected a research glasshouse; created a Bee Garden and ornamental kitchen garden; re-launched the Heritage Seed Programme in 1992; with the help of a Department of the Environment (DoE) grant established a consultancy to advise local authorities on composting; constructed a reed-bed sewage treatment works; set up a Local Authority membership in 1993, which gained over one hundred members by the year's end; introduced rare breeds of livestock (pigs, sheep, hens) at Ryton; set up in 1996 a long-term project to monitor the conversion of horticultural units to organic, and the same year, again found DoE support, this time to fund the project 'Go Organic in School Grounds'. By the mid-1990s more than eighty HDRA Groups were active , and in 1995 the organization opened another site, at Yalding in Kent. It became increasingly involved with overseas projects: in collaboration with Durham University it set up the Drought Defeaters programme, which helped plant nearly one million trees in the Ethiopian highlands. In Ghana an Organic Agriculture Network was set up, and in 1995 a Third World Organic Support Group was launched to help fund overseas work. The gardens at Ryton also became known for their catering: the cafe was soon granted a place in *The Good Food Guide.* To deal with all this work, staff numbers rose from thirty employees and another thirty or so community programme workers in 1987, to more than ninety full-time equivalents by 1996.

Lawrence Hills, a professional writer, had always made publications a priority, and this aspect of the HDRA's work was maintained and increased, with a series of *Step-by-Step* guides to practical organic gardening being added to each year, and Alan Gear's *New Organic Food Guide* and Sue Stickland's *The Organic Garden* appearing in 1987. *Thorsons Organic Consumer Guide* (1990) by David Mabey and Alan and Jackie Gear sold more than 30,000 copies, and in 1995 Pauline Pears and Sue Stickland had their book *Organic Gardening* published in the Royal Horticultural Society gardening series.

The HDRA exhibited at many shows, including the Royal Show and Health Show Olympia, and aroused royal interest at the 1988 Chelsea Flower Show when the Queen and Prince Philip visited its stand. The following year the Prince of Wales made his first of many visits to Ryton, an occasion which generated remarkable publicity. Perhaps much more remarkable than the Prince's visit, though, was the visit in 1990 of a group of the Ministry of Agriculture's Chief Scientists, and in 1992 of Nicholas Soames, the Minister of Food. Also in 1990, Ryton hosted an open day for commercial organic growers, under the auspices of the National Institute of Agricultural Botany. Media visitors over the years included Patricia Gallimore of *The Archers,* Radio 4's *The Food Programme,* and the cooks Marguerite Patten and Sophie Grigson.

According to Alan Gear, Lawrence Hills, although a Soil Association

Council member for many years, had considered the Council something of a waste of time, and preferred to 'paddle his own canoe'. Inevitably, though, some prominent Soil Association members were HDRA supporters and served on its Council. Newman Turner was the HDRA's first chairman and then President until his death in 1964 (though he didn't continue on the Council after his death, as the HDRA's own chronology rather spookily indicates). The Faber agricultural and horticultural editor Richard de la Mare, Soil Association gardening correspondent H.E. Witham Fogg and nutritionist DORIS GRANT were Council members in the early years. Later, the scientists Dr Anthony Deavin and Dr David Hodges were on the Council, as was the West Wales grower Charles Wacher. Earl Kitchener, a member of the Wholefood Trust (see Chapter 5) was HDRA President for more than thirty years.

The fiftieth anniversary celebrations of the HDRA (by that time re-named Garden Organic) in July 2008 featured a touching exhibition of some of Lawrence Hills' books, a typewriter of the kind on which he produced his innumerable articles and letters, and photographs of the Bocking trial ground. In a nearby hall, a short film of his life played on a loop; it contained a wonderful shot of him on a wintry day, resplendent in a tall Russian fur hat, pedalling in stately fashion on a 'sit-up-and-beg' bicycle. It is a world away from the organic millionaires, Feasts of Albion and 'yummy mummies' so prominent in the contemporary organic movement. Anthony Deavin described Hills as utterly devoted and 'a well of optimism', qualities which inspired the Gears and enabled them to endure years of financial privation. It was on such a basis that one of the organic movement's greatest successes was built.

Other notable organic gardeners

Hills' dominance should not be allowed to obscure the contributions of other organic gardeners. Despite the HDRA's success, Soil Association publications continued to feature gardening columns, written at various times by Brian Furner, Roy Lacey, H.E. Witham Fogg and Arthur Barritt. Brian Furner, of Erith in Kent, gardening correspondent for the Soil Association in the 1960s and '70s and a Fellow of the Linnean Society of London, had rejected chemical methods in 1950, his outlook transformed by his reading of Balfour, F.H. King and other organic pioneers, particularly Sir Albert Howard, whose widow Lady Louise he met and conversed with. Furner depended for his livelihood entirely on his lectures and writings about the food crops he produced, which therefore had to be of high quality. He edited a revised edition of Ben Easey's book. Around the mid-1970s, Furner was replaced by H.E. Witham Fogg, a founder member of the HDRA who served on its Council for the first ten years.

Arthur Barritt, the *SAQR*'s gardening correspondent in the early 1980s, was an advocate of the manuring virtues of seaweed who had worked as an experimental officer for the Agricultural Development and Advisory Service (ADAS), after which he returned to his previous occupation of fruit-grower. He had been convinced of the superiority of organic produce on grounds of taste. Barritt was succeeded by Roy Lacey, who favoured organic gardening primarily on moral grounds. The organic pioneers had objected to exploitation of the soil, a stand still valid in the 1980s, said Lacey; but there were now other factors lending further force to the moral argument. By gardening inorganically and buying fertilizers and chemical forms of pest control, you were stepping on to the agricultural treadmill of production, where the drive to increase output at all costs destroyed the health of the soil. Organic methods aimed at optimum production and, as Schumacher had put it, demonstrated non-violent humility towards the complex harmony of the natural world. In 1988, the publishers David and Charles brought out Lacey's *Organic Gardening,* the first book to be given the Soil Association's official stamp of approval and one which 'firmly point[ed] organic growing in the direction of the latest scientific understanding, ecological awareness and technological advance'.[9] The year before Lacey's book appeared, Geoff Hamilton's *Successful Organic Gardening* had been published. With Hamilton, we enter the era of the organic gardener as media celebrity; Bob Flowerdew and Monty Don would follow in his wake.

While figures like Geoff Hamilton helped take organic gardening into the main stream, Jack Temple, a market-gardener in Surrey, headed in the other direction. Temple had suffered ill health as a young man, until early in the 1940s he became a professional grower using organic methods. During the 1970s he became gardening correspondent of *Here's Health* magazine, and he spoke at the Soil

Association's Ewell courses on biological husbandry; in the mid-1980s he produced *The Here's Health Guide to Gardening Without Chemicals*. He was a friend of Alan and Jackie Gear until he — in their phrase — 'went nutty', going down the route of a particularly bizarre form of alternative healing. In 1984 he helped sponsor the Festival of Mind, Body, Spirit and, towards the end of his long life — he died in 2004 — became a guru to Cherie Blair: a tragic anti-climax for someone who had for more than thirty years demonstrated the viability of organic horticulture.[10]

To conclude this section, there is one other gardener, of a specialist nature, who should be mentioned: Robert A. de J. Hart, advocate and practitioner of forest farming and gardening. Originally a journalist, Hart had been much affected by his grandmother's Christianity and his own interest in Gandhi and Kagawa. From the latter, a Japanese Christian reformer known as 'the Japanese Gandhi', he learned the importance of fodder-bearing trees as a means of combining conservation with food production. Hart also became familiar with the writings of the organic pioneers: Howard, McCarrison, Stapledon and, particularly, H.J. Massingham. As a journalist, he decided to concentrate on the issue of world hunger and, in order to gain practical experience of food production, bought a smallholding in Somerset before moving to a farm near Church Stretton in Shropshire. In the mid-1980s he started a self-sufficiency project on the western side of Wenlock Edge, which became an example of agroforestry, influenced by Yeomans' Keyline system. Hart had floated this idea in the *Soil Association Quarterly Review* in 1979, originally calling it 'ecological perennial horticulture', a development of the forest farming system he had worked on with J. Sholto Douglas and one which would concentrate on the use of perennial plants and on controlled systems of plant association, or 'ecological, symbiotic systems of organically intensive cultivation'. Hart defined a forest garden as 'a system which imitates the multi-storey structure and diversity of the natural forest, in which every single plant has some useful function to perform'. All very well, one might think, for those fortunate enough to own land in a spacious county; but by the mid-1990s Hart's methods were being applied in inner-city areas too, 'spearheading a rural invasion of the townscape' in North London and Birmingham. As we saw in the previous chapter, Hart was a major influence on Permaculture in Britain.[11]

Organic growers

'No matter how large the organic sector gets, farmers and growers will always be at its heart,' Graeme Matravers assured readers of the Soil Association journal *Organic Farming* in 2008.[12] The magazine's title would not have reassured growers, though its strapline had for many years described it as the Association's 'journal for organic agriculture and horticulture'. An interesting change of strapline took place, however, between the Autumn 2007 and Winter 2007/08 issues: on the cover of the latter, the journal was described as being 'for the new farmer and grower'. The Organic Growers' Association (OGA) and British Organic Farmers (BOF) had been set up in the early 1980s as a result of dissatisfaction with the Soil Association's perceived lack of interest in the problems of farmers and growers, and had produced the journal *New Farmer and Grower (NFG)*, which, after BOF/OGA merged with the Association in the mid-1990s, was re-titled *Organic Farming*. By 2006, organic growers were again feeling that the Association was failing to support them and, following a conference at the Royal Agricultural College, Cirencester in December that year, established the Organic Growers' Alliance, which soon began publishing its own journal *The Organic Grower*. The Soil Association was clearly rattled by this development, and the change of strapline smacked of an attempt to evoke *NFG*'s heyday as a vibrant, grass-roots production and imply that *Organic Farming* was still its spiritual heir. There was a certain feeling of déjà vu about the founding of the new Alliance, as a number of the people involved had been instrumental in giving impetus to the original OGA a quarter of a century earlier. Before we look at the OGA's history, which is inextricably linked to the Seventies Generation of activists, we shall consider someone who took up growing in the 1960s and overlapped with them: John Butler.

John and Shirley Butler

Butler in fact provides a link with Robert Hart, whose book *Forest Farming* (written with J. Sholto Douglas) he reviewed for the *SAQR*, praising its recognition that the secret of successful farming is to set the stage, keep the balance and let nature do the work. This mystical tendency of organicist thought was strongly marked in Butler, to

3. THE ORGANIC ALTERNATIVE: GARDENERS AND GROWERS

the extent that he regarded organic cultivation as a means by which God could be experienced, through closeness to nature and being sensitive to life's wholeness. The task of composting was in fact 'the process of raising matter to a higher level of consciousness ... Whatever encouraged life was good ... ' It would be entirely wrong to conclude from these ideas that Butler was some sort of impractical dreamer. As a young man, he had served in the army as a cavalry officer, been a cowboy in Australia and begun a degree course at Wye College in Kent, which he abandoned because of its emphasis on reductionist science and economic values. A 'socialist' period in Peru followed, when he worked in the Andes as an agricultural volunteer, helping re-forest a denuded valley. His family owned a farm at Bakewell in Derbyshire and he farmed there until it had to be sold, then, in the mid-1960s, moved to a great-aunt's three-acre smallholding in Lincolnshire. He set about living off this land, without capital or machinery, using the tools already there; he bought some sheep, and experimented with growing vegetables. Butler attended a school of meditation in London once a week, found that other members were interested in buying his organically grown produce, and began to specialize in this area. His father had been a Soil Association member and, when young, Butler had read *Mother Earth,* particularly enjoying any contributions by Professor Lindsay Robb, who combined scientific expertise with a religious perspective on environmental problems. In 1968 Butler himself joined the Association and, with Spartan dedication to the 'velvet' soil at Bicker Fen, created what Brigadier A.W. Vickers, the Association's General Secretary, described as 'a little showpiece' of organic growing. The market for organic vegetables in London at that time was wide open, and Butler found himself able to command high prices for his attractive goods, produced entirely by hand. His work even impressed his local branch of the NFU, who visited in 1976 for a farm walk.[13]

Through the meditation school, Butler met Shirley, his first wife, who shared the hardships and whose commitment was an essential part of the success which, for several years, the smallholding enjoyed. On 21 June 1972, John and Shirley Butler experienced the distinction — if that is the right word — of being featured in both the *Daily Mirror* and *The Sun* as the eccentric smallholders who, as the latter publication expressed it, ate 'Chickweed Omelette and Juicy Nettles'.

Weeds were a problem in the rich soil, so the Butlers turned this to their advantage by growing chickweed and selling it to Craig Sams in London, for his wholefood business. The nutritionist John Yudkin gave their meals of sorrel salad and boiled charlock his approval, but only Shirley was a vegetarian: John enjoyed the fat bacon his pigs provided. More serious media coverage of the Butlers' methods of cultivation came later in the 1970s, when Don Haworth produced 'Three Acres at Bicker Fen' for BBC2, broadcast on 8 August 1975 in a series called 'Living on the Land'. The Butlers never gained — or sought — the celebrity attained by John and Sally Seymour, but they were contemporary exponents of 'self-sufficiency', their year's accounts for 1974 showing that they 'had lived almost free, and … earned £1,600, out of which the only big expense [was] the cost of running their van'.[14]

Butler wrote about his smallholding as well as cultivating it; he contributed progress reports to the *SAQR* in the 1970s and regular longer articles, under the title 'The Changing Seasons', for *Seed: The Journal of Organic Living* (see Chapter 5), run by Craig and GREGORY SAMS, throughout the first half of that decade. In the winter of 1974–75, the editorial staff published their own letter to the Butlers, praising them as 'the pioneers whom we and others shall one day follow. If two who love God and work as hard as you … shall be defeated by the wretched rules of the modern market-place, then all of us who believe in your way of life shall share your failure'. John Butler served on the Soil Association Council in the late 1970s and early 1980s and might have been expected to ally himself with the new influx of farmers and growers elected to the Council during that period, but this was not the case. He disliked the often confrontational atmosphere which developed, and did not share the younger generation's intense interest in marketing. Contrary to their view, he was by 1981 sceptical about the existence of 'a vast, unsatisfied market for organic produce', preferring to sell to private agents, avoiding shops, getting a better price and maintaining a personal connection with the buyer. He considered the whole topic of marketing dreary, and recommended 'anyone who has a field of vegetables to sell, to cultivate a patient indifference as to his reward'.[15] This was not advice likely to be heeded down in West Wales.

West Wales growers

Peter Segger was the first of the Seventies Generation to be elected to the Soil Association Council, in 1977; he was regional representative for the West. Formerly a businessman in the frozen fish trade, Segger had become concerned about methods of food production and the separation of rural and urban life while living in Oxfordshire in the early 1970s. Victor Bonham-Carter's book *The Survival of the English Countryside* (1971) influenced his thinking on these questions. (This provides an interesting link with an older generation of organic activists. Bonham-Carter dedicated the book to Robert Waller, a long-time friend with whom he had worked for BBC Radio in Bristol in the 1950s. Historian of Dartington Hall, Bonham-Carter had written a study of *The English Village* in 1952, and was chairman of the Soil Association's Editorial Board and a Council member in the early 1970s.) In 1974 Segger sold his business and looked for a small farm, eventually finding one at Cilcennin near Cardigan Bay. With little if any knowledge of farming, he decided that he wanted to produce wool, so he bought a flock of pedigree sheep and also started growing vegetables, relying heavily on books to teach him the practicalities of organic production: he found Lawrence Hills' *Grow Your Own Fruit and Vegetables* (1971) especially valuable. At first he knew of no-one to approach for help and his efforts were 'very amateur'; but there was the Soil Association, and in 1975 he established its West Wales Group in order to draw on the Association's experience and bring together any other farmers or growers in the same predicament. The group became 'a vibrant social and agricultural organization' with a membership of around 140. Environmentalism was making an impact and, according to Segger, 'All of us were influenced by Schumacher — his book [*Small is Beautiful*] was seminal.' John Seymour, the apostle of self-sufficiency (see Chapter 6) belonged to the group.[16]

Charles and Carolyn Wacher became members of the West Wales Group after moving to Aeron Parc near Tregaron in 1977. Carolyn has described herself and her late husband (Charles died in October 1986) as 'very middle-class' Londoners who worked for charities in the early 1970s before spending four months in the United States looking at back-to-the-land communities. They were impressed by

Charles Reich's *The Greening of America* and *Getting Back Together* by Robert Houriet, and on returning to England joined WWOOF; Charles Wacher worked for Lawrence Hills and as a volunteer on Patrick Holden's farm near Tregaron. Wacher's family had money, which enabled him and Carolyn to buy Aeron Parc outright and provide themselves with good equipment for farming it. They were 'NOT hippies' [Carolyn's emphasis], nor did they aim to emulate John Seymour; although idealistic, they wanted to be more commercial than Seymour, and they managed to gain credibility in the eyes of their neighbours by selling their produce for several years at the local market. Francis Blake, in his obituary of Charles, described the Wachers' holding as one of the most beautiful and complete examples of an organic holding in Britain.[17]

Nearer to the coast than the Wachers, David Frost had started farming in 1976 at Llanrhystud. His background was academic and left-wing: he held a research post at Regent Street Polytechnic in the late 1960s, looking at housing and community issues. There was a lot of interest at Regent Street in environmental questions, and Frost cites Frank Fraser Darling's 1969 Reith Lectures, *Wilderness and Plenty,* as a formative influence. A keen gardener, he joined the HDRA and studied its journal closely; in the early 1970s he transformed his suburban garden into an allotment and started considering the possibility of communal living. After an interlude in France, Frost moved with his wife Anne to Wales in the mid-1970s and began growing vegetables; he then bought a local business with a delivery van and a market stall in Aberystwyth. Although to some extent influenced by John Seymour's writings, Frost, like the Wachers, had to be commercially successful in order to survive. A mutual friend put him in touch with Peter Segger, who became a source of produce for Frost's business and drew him into the Soil Association's West Wales Group. He met the Wachers and bought from them too. The business expanded, selling organic and very fresh produce. In their own garden the Frosts grew crops to sell locally, at the Aberystwyth market, in their Salad Shop, and to hotels and restaurants, but also began to send produce further afield through Peter Segger's Organic Farm Foods business based in Lampeter. For fifteen years they grew a wide range of fruit and vegetables: in some years, as many as forty.[18]

The Tolhursts and the Schofields

By no means all the younger generation of growers were based in West Wales. Iain and Lyn Tolhurst were running a holding in a bleak part of Cornwall, near Roche, concentrating on strawberries, when Iain became involved with the burgeoning growers' movement in the late 1970s. His background — Bristol working-class — was very different from that of the West Wales growers. He had left school at fifteen and travelled around the UK finding work as a farm labourer or in the building trade. A gifted craftsman and boat-builder, Tolhurst has always been divided between his love of woodwork and his love of gardening. He was drawn towards agriculture and horticulture by a job he took on a dairy farm near Milton Keynes. It destroyed his romantic view of agriculture, but he stayed for four years and had his own garden while there, selling vegetables and strawberries to a nearby health farm. What he saw of conventional agriculture on the dairy farm convinced him that the industrial approach was destructive of nature: the treatment of animals, which were riddled with disease; the poor soil structure, the lack of earthworms and the consequent erosion. Reading *Silent Spring* in the mid-1970s, when in his early twenties, confirmed what he had instinctively gleaned from his love of the natural world. Around the same time, he joined the HDRA and also turned vegetarian. In 1976 he went independent, buying land in Cornwall with saved and borrowed money and working it with Lyn, whose support was indispensable. It was a bitterly hard period, but, like the Wachers in Wales, they gained the respect and assistance of locals and set up a farm shop. In fact, it was Charles Wacher, in the late 1970s, who contacted him and brought a group of people, including the Somerset growers Francis Blake and his brother John, to see what the Tolhursts were doing. Graham Shepperd had been down to inspect their holding for the Soil Association and communicated his interest. Tolhurst found the Wachers 'a breath of fresh air' in what he considered the stuffy and élitist atmosphere of the Soil Association, but he attended an Association conference at Lackham, in Wiltshire, meeting Dinah Williams and Sedley Sweeny there, and he went on a walk at Barry Wookey's Rushall farm. When the Organic Growers' Association emerged at the beginning of the 1980s, Tolhurst took an active role. Whereas the Soil Association was, he felt, stuck in a time warp, the OGA was more ambitious and dynamic, moving things up a gear or two. Justly or otherwise, Tolhurst thought that the Soil Association was dominated by people with much greater financial security than the growers; the OGA offered the growers a sense of identity, helping to maintain their momentum and motivation, and showing them how to develop a market. In 1988 he took over seventeen acres of land at Julian Rose's Hardwick Estate west of Reading and went on to run a large and successful box scheme and become the first person to attain the Organic Stockfree symbol, having abandoned the use of grazing animals and animal inputs in the early 1990s. The BOF/OGA merger with the Soil Association, at the end of the period covered by this book, led, in Tolhurst's view, to renewed marginalization of growers, and he has been prominent in the Organic Growers' Alliance.[19]

Alan Schofield was another who, like Tolhurst, committed himself to organic methods as a result of his experiences in conventional agriculture, though the seeds of his conversion had been sown by his school biology teacher, who drew the pupils' attention to the interconnectedness of the natural world. As a farmworker in the dairy industry, he applied OPs (organophosphates) to the backs of the cows. Later, he worked in intensive horticulture, with fungicides; the employees suffered reactions which were persistently diagnosed as allergies to the strawberry plants. Schofield turned to working with an organic grower and, having learned the essential skills, bought a green-field site in Lancashire in the mid-1980s. He and his wife Debra have remained among the most prominent organic growers. Shortage of space prevents a detailed account of other leading growers, but they include the brothers Francis and John Blake, and Charles Dowding, in Somerset; and tomato specialists Douglas and Penny Blair at Pilling in Lancashire. Tim and Jan Deane in Devon will be discussed in Chapter 5.

The Organic Growers Association and New Farmer and Grower

Those present at the early OGA conferences still recall the tremendous sense of excitement that these occasions generated. The first National Conference of Organic Growers (pre-dating the OGA's formal establishment) was held at the Royal Agricultural College, Cirencester in January 1980, the result of a meeting of organic growers concerned about their relative isolation and the lack of both a national forum and of any specialist advisory services. Since there was a predicted increase in demand for organic vegetables, it was important to disseminate such knowledge of techniques as already existed and encourage further research and experimentation. The conference was supported by the Soil Association, the International Institute of Biological Husbandry (IIBH) and Organic Farmers and Growers (OFG), and featured a most impressive range of speakers, including Dr Hartmut Vogtmann, then the Secretary General of IFOAM; Claude Aubert, former President of Nature et Progrès; the scientists Dr Ken Gray, Dr Victor Stewart, John Gemmett of the National Institute of Agricultural Botany and entomologist Nigel Scopes; David Stickland of Organic Farmers and Growers, an expert on agricultural marketing; and growers Douglas Blair, Peter Segger, John Stevens (specialist in herbs) and Alastair Rennie (mushroom grower). And all for a mere £28. Among the topics under discussion were weed and pest control, polytunnels and cloches, potting composts and soil fertility. The following year, at the same venue, speakers included Iain Tolhurst; Eliot Coleman, Director of the Coolidge Research Centre in Massachusetts; Dr Glen of the Long Ashton Research Station, and the Soil Association's agricultural advisor R.W. Widdowson. Geoff Mutton emphasized the importance of attractive presentation as a feature of successful marketing, and there was a call for well-defined standards.

The OGA was founded early in 1981, with Peter Segger as Chairman, Charles Wacher as Secretary and Geoff Mutton as Treasurer, while husband-and-wife team Chris and Chris Mair edited its newsletter. In a leading article in the newsletter's second edition, Peter Segger exhibited the OGA's dimerous nature by striking a note of combined idealism and commercial shrewdness in proposing 'A Healthy Revolution', to be based on 'a profound and clear concept of wholeness'. OGA members needed to be more outspoken about their revolutionary activities, which would bring changes in spite of existing patterns; but for their efforts to be successful, they must aim for 'economic and scientifically credible radical horticultural techniques', must record full details of crops grown, be cost-efficient and pay attention to marketing. Segger's subsequent articles maintained this blend of fervour and business acumen: a new spirit was abroad in the organic movement, 'strong, practical, dedicated and unstoppable': but unstoppable only if it could be embodied in 'a hard-nosed marketing strategy'. Like all revolutionaries, Segger found it necessary not only to paint a rosy picture of the future, but to condemn the misrule of those whom he was displacing. In thanking Lawrence Woodward's Elm Farm Research Centre for all its help, he aimed a shaft at the organic movement's older generation by saying that such effort would not have been necessary 'had our movement been more effective and active'. According to several accounts of this period, Segger was far from averse to offending his elders, his behaviour being deemed confrontational and provocative: for Angela Bates, a Vice-Principal of the Soil Association, his attitudes were tantamount to class war. What is not in doubt is that Segger's energy and abrasiveness helped moved things on; that the newsletter dealt with the practical problems which growers had to overcome if they were to survive, and that the OGA provided growers with a strong and encouraging social infrastructure.[20]

Following the formal establishment of British Organic Farmers during the winter of 1982–83, the OGA newsletter was transformed into *New Farmer and Grower (NFG)*, a more sophisticated production and one whose pages tell us a great deal about the spirit which infused the organic movement in the 1980s and early '90s. The main feelings conveyed may best be described as early optimism increasingly tempered by wariness, and of generosity in co-operation. Despite John Butler's scepticism, it did seem that the market for organic produce was expanding in the early 1980s. '1985 will come to be seen as a watershed for the organic movement,' *NFG* opined in the autumn of that year, with 'organic' becoming a household term, synonymous with healthy food. All major supermarkets were showing interest, and the organic movement took pride of place at the Grow Show Conference in September. 'Organic Growing: How I Make It Pay' was the topic addressed by Peter Segger, Iain Tolhurst, Douglas Blair and Donald Cooper; Cooper was instrumental in forging links with the supermarkets. 1985 saw such a surge in demand that, thought Charles Dowding, maintaining supply might prove a problem in 1986. On the other hand, down in Dyfed organic growers were producing large volumes of vegetables which they were having problems distributing. Chapter 5 will look in more detail at the issue of marketing, which was a constant and major concern of the OGA and of BOF. Growers became aware, as the demand for produce increased, of the dangers which success posed for their ideals. MAFF's interest in organic standards was exciting in that it meant the possibility of official support for organics; it was dangerous in that it might mean, in Lawrence Woodward's words, 'being taken over, swamped and diverted'. This dilemma was debated freely and in depth in *NFG*'s democratic pages.[21]

If 1985 had been a 'watershed' for the movement, 1987 was 'a major milestone', as the government became involved with organic standards. By this time, too, improvement grants were available to organic growers, under an ADAS scheme started in October 1985. But — and *NFG* was always realistic about the wider situation, which it monitored meticulously — at the beginning of 1989, a staggering 95 per cent of organic vegetables sold in the UK was being imported, and the Secretary of State for the Environment, the late, unlamented Nicholas Ridley, was letting it be known that he considered organic produce a means of 'ripping off' the public. As the UK moved into recession in the early 1990s, the OGA reported that many growers were barely operating at a profit, and were being paid for vegetables at below production price. In March 1991 a one-day seminar at Pershore College of Horticulture asked the question: 'Marketing: Which Way Forward?', and discussed that perennially elusive panacea, co-operatives, as well as Community Supported Agriculture.[22]

The early optimism had faded considerably by this time, and wariness, or downright dissatisfaction, dominated. The editorial to *NFG*'s Autumn 1991 issue reported that MAFF and the Royal Agricultural Society of England remained committed to a large-scale approach and that the former's commitment to organics was 'derisory', with a mere £3.5m. allocated for research and development: 0.4 per cent of the government's research budget. In the following issue, John Selwyn Gummer, in whom the organic movement had once invested a certain amount of hope, was dubbed 'Seldom Glimmer' in a mocking cartoon, and Tim and Jan Deane reported on the demise of the Somerset Organic Producers co-operative, whose chief problem had been a lack of new members. Despite the disappointments caused by British political institutions, the *NFG* editorial in the Summer 1992 edition greeted the forthcoming EU Regulation (2092/91) as 'one of the most important and exciting developments in the history of the organic movement' — before enumerating the various problems with it. The optimism was exceptionally short-lived, for by the autumn of 1992 the editorial was describing the Regulation as disruptive and reporting that small-scale growers were being forced out of business all over northern Europe. By 1996, the situation had changed again, and a MAFF marketing seminar held at Ryton discussed the huge demand for organic produce and the shortage of UK growers. The *NFG* editorial in that summer's edition blamed MAFF itself for its failure to offer decisive economic incentives, which had led to organic producers being swamped by demand. By the time the OGA merged with the Soil Association, the early optimism had been sorely tried by fluctuating commercial fortunes and the insecurity they brought, a process which *NFG*'s pages chart in close detail.[23]

Also evident in *NFG* was the sense of community among growers which the OGA engendered, and the co-operative sharing of experience. There was much practical advice. Early issues included articles on cabbages; lettuces (Peter Segger); windbreaks, strawberries, peas and rhubarb (Iain Tolhurst); cauliflowers and celery (Charles Wacher); garlic (Laura Davis), and tomato production and pest control (John Dalby). As time went on, growers visited each other's holdings and reported on progress and methods, with the OGA organizing farm walks and barn dances. The now noted writer Joy Larkcom contributed an article on salads and Pam Bowers profiled Malcolm Sensby, the first mushroom-grower to be granted the UK organic symbol. Laura Davis addressed the persistent problem of dealing with pests, emphasizing the need to study the inter-relationships between weeds, insects and plants. John Dalby visited the nursery of Alan and Sandra Payne in the Vale of Evesham. David Frost contributed a piece on plant raising and organic composts.[24]

NFG also flagged up the potential dangers of biotechnology as early as the spring of 1991, when Phil Harris of the HDRA looked at the issues at stake and urged the organic movement to formulate a viewpoint to counter that of the biotech industry. An editorial in 1994 expressed with exquisite understatement its doubt 'that the *modus operandi* of the financial and commercial giants is accidental, impartial or benign when it comes to the acquisition and control of resources on which the future of humanity ultimately depends'; the same issue published a piece on this topic by Vandana Shiva and promoted the HDRA's Heritage Seed Programme. Of interest to genealogists of the organic movement is a profile by Judy Steele of Hampshire growers Steve Forster and Mandy Wright: Mandy's father Alec had been foreman on Mary Langman's farm at Bromley Common in Kent, which had once supplied the Pioneer Health Centre in Peckham.[25]

A particularly striking feature of *NFG* is the sense that it was open to any member of OGA or BOF and that dissident views could be expressed at length. We shall examine in Chapter 5 some of the debates which were conducted in *NFG* on the vexed topics of involvement with mainstream retailers and government bodies. But one also senses that the OGA, although the senior partner in terms of years, became BOF's junior in terms of importance; farming comes to take up considerable space in *NFG* as the years pass. Nevertheless, the OGA's place in organic history is secure and of major significance. Early in the twenty-first century, six of its leading figures — Francis Blake, Ric Bowers, Alan Schofield, Peter Segger, Iain Tolhurst and Carolyn Wacher — reflected on the role it had played and drew the following conclusions.

Firstly, the Soil Association did not at that time (the late 1970s) provide what growers needed; it was chiefly oriented towards farmers, with little to offer growers in the way of technical advice, let alone commercial strategies. In contrast, the OGA was a dedicated association which recognized the special needs of growers and provided specialist information. In Peter Segger's words, it 'brought together a very large body of grower experience with the objectives of self-improvement and to share this on a low-cost and open basis with all organic growers'. Iain Tolhurst saw it as having a wider significance, bringing organic production 'into the real world', starting the development of markets, and removing the old-fashioned image of organic farming. The OGA's main achievements included giving the growing sector a sense of identity, publishing an informative journal, helping establish achievable standards, starting grower-funded trials of organic composts and weeding-machines, developing packaging (with the 'Food You Can Trust' logo), producing the first organic products directory in 1982, organizing farm walks, and persuading *The Grower* magazine to take organic growers seriously.[26]

It is interesting that three of those who responded to the questionnaire identify one of the main items in the OGA's legacy as its impact on the Soil Association: its annual conferences, which grew out of the OGA/BOF conferences at Cirencester; its Producer Services Department, and even, in Tolhurst's view, its continued existence, which the new wave of organic growers and farmers had made possible. He added, though, that by the turn of the century there were some who maintained that the organic movement would be better off without the Association. As we have seen, Iain Tolhurst was one of those who by 2006 would feel that the Association was again neglecting growers' interests.

4. The Organic Alternative: Health and Nutrition

The Soil Association's name expresses that body's true concern by implication only: it is health, to which end the soil is a means. Eve Balfour's book *The Living Soil,* whose success gave rise to the Association, was sub-titled 'Evidence of the importance to human health of soil vitality, with special reference to post-war planning'. Published in 1943, in the wake of the previous year's Beveridge Report on social insurance, *The Living Soil* was very much of its time in its desire to create a better society than that which had condemned so many British families to poverty and malnutrition in the 1920s and '30s. The end of the war would offer the opportunity of a fresh start in health policy as it would in agricultural policy; for the organicists, the two policies would ideally have been united. For if it was true that nutritionally valuable food could best be provided by a humus-rich, 'living' soil, then it followed that organic methods of cultivation would act as a form of preventive medicine. This line of argument had been formulated just before the war in the *Medical Testament* of the Cheshire doctors, led by Lionel Picton, and it helps to explain why the early organicists greeted the advent of the National Health Service (NHS) with anything from reserve to downright hostility. Just as the 1947 Agriculture Act took farming policy in a direction remote from that which the organic movement wanted to see, so the NHS, effective from July the following year, did the same for health policy; the movement has ever since considered it a 'national sickness service' on account of its neglect of nutrition and preventive medicine, its prescription of 'a pill for every ill', and its lack of interest in defining a clear conception of what health actually is. As the chemical and machinery companies have been to agriculture, so the pharmaceutical companies have been to the NHS. In the early years of the twenty-first century, government attempts to influence the nation's diet often meet the accusation of 'health fascism': it is assumed by those who raise such a cry that health is a matter purely of individual choice and no business of the state. Given that the early organicists were themselves generally resistant to centralized state power, why were they so concerned about the nation's health? Various reasons can be identified.

The importance of health: eugenics, economics and the fall from wholeness

In *The Origins of the Organic Movement* I drew attention to the eugenicist strain to be found in the thought of figures such as Dr Alexis Carrel and the Earl of Portsmouth; but if the term 'eugenics' is used in a broader sense than that which indicates selective breeding alone then a good deal of organicist thought can be deemed eugenicist, its aim being improvement of the race or national stock through sound nutrition. This strain can be found in the pamphlet *C3 ... or A1?* issued by the Pioneer Health Centre; in the Cheshire doctors' fear that the British would become 'a C3 nation'; in the title of WESTON PRICE's study of dental health, *Nutrition and Physical Degeneration;* in T. H. Sanderson-Wells' advocacy of a 'sun diet' to ensure a vigorous population; in Lionel Picton's belief, expressed in *Thoughts on Feeding* (1946), that wholemeal bread would increase the fertility of British women; and in the importance which the organic movement has attached to F.M. Pottenger's experiments with cats, which suggested that degeneration of stock might be reversed — in that species, at least — by changes in diet. Edgar J. Saxon's journal *Health and Life* at times exhibited a similar strain, calling for 'a virile, eager and rooted nation'. The Scottish journal *Health and the Soil* featured on its front cover during 1948 and 1949 an idealized picture of a small blond boy, evidently representing the purer stock which organic cultivation would produce: a human equivalent of the abundant yellow corn next to which he sat with his faithful dog.[1]

Scarcely any of the early organicists called for selective breeding, for which they were taken to task by the Nazi sympathizer Anthony Ludovici, a close friend of Portsmouth and of Rolf Gardiner, in his 1945 book *The Four Pillars of Health*. He criticized the organic school — and in particular Dr G.T. Wrench, author of *The Wheel of Health* (1938), a study of the North-West Frontier's Hunza tribesmen

4. THE ORGANIC ALTERNATIVE: HEALTH AND NUTRITION

— for stressing the importance of diet and neglecting the question of breeding. In January 1946, DION BYNGHAM reviewed Ludovici's book in *Health and Life,* agreeing that the organic school was failing to investigate 'the complete requirements for health' and apparently condemning with Ludovici the random mating and cross-breeding which led to 'a proliferation of biological and human misfits'. He returned to this debate in an article the following month. Although Byngham assisted Saxon on *Health and Life,* he tended to the fringes of the organic movement, working with Siegfried Marian on *Soil Magazine* and with the Duke of Bedford on *Peoples Post.*[2]

Related to this eugenicist concern about the nation's future was a more purely economic consideration: the cost of poor health to government finances. Balfour quoted a 1936 report on Britain's health services, which had identified the cost of ill-health as at least £300m. a year. This sort of argument would surface from time to time in later decades, as it did, for instance, in a lecture by Dr Hugh Sinclair, who advocated research into nutrition because of the cost of the NHS, and in a discussion on absenteeism in industry during one of Cdr Robert Stuart's Soil Association Weeks, in 1973.[3] A healthy workforce benefits the economy. Not merely that, but the NHS itself proved a drain on Treasury finances, whereas a policy of preventive medicine would have reduced the amount of taxpayer subsidy. Again, the simplicity of a virtuous organic circle was offered as an alternative to the vicious circle of complexity sent spinning by a combination of the state and big business.

Then there was what might be termed the religious case for health. It is important to recall the centrality to organic philosophy of the concept of a natural order, which Balfour stressed in the first article to appear in *Mother Earth,* 'Why it Happened'. 'Left to herself,' Balfour wrote, 'Nature always produces order. It is man who causes chaos by his persistent attempt to resist or ignore natural laws ... Nature's biological laws ... are manifested through active processes of intense vitality.' The result of ignoring or defying these laws was sterility in the soil and ill-health in the individual. Sir Albert Howard, in his manifesto for a post-war solar age, *Farming and Gardening for Health or Disease,* described health as the 'birthright' of every boy and girl. As I suggested in *The Origins of the Organic Movement,* one can see in organicist thought a lapsarian mythology according to which there has been a decline from a state of harmony with God's laws to the sickness and imbalance which characterize our own era of rapacity. In their visions for post-war reconstruction both Balfour and Howard saw the restraint and removal of any institutions which stood in the way of health 'and that efficiency, well-being, and contentment which depend thereon'.[4] Some examples of such health still existed in remote parts of the world, uncorrupted by western capitalism: the Hunzas of course, or the islanders of Tristan da Cunha. These populations served as reminders of what should be our natural state. For Edgar Saxon, the person of Christ exemplified radiant health: a model of what people might be if they lived in harmony with divine laws. At a less spiritually exalted level, the work of the Pioneer Health Centre implied that poor diet and inadequate social conditions were denying the urban population their chance to reach full potential and express their natural creativity: a loss to the individuals themselves and to the community, which suffered from an accumulated waste of people's talents.

Positive health

But what did the organicists mean by 'health'? Like organic cultivation, it is a positive concept. The mere absence of disease does not, in organicist thought, indicate health, but in Balfour's view our society had come to take 'subnormal physical fitness' for granted. The organic movement has rooted its definition of health in that word's origins in, and associations with, the idea of wholeness, and, indeed with its associated concept of holiness. In his 1959 survey of the organic movement, *The Stuff Man's Made Of*, Jorian Jenks defined health as 'a positive condition of wholeness', dependent 'in large measure on harmonious relationships between soil organisms, plants, animals and men'. In support of this positive approach he referred to the United Nations World Health Authority, whose Constitution's first clause stated that: 'Health is a state of complete physical, mental and social well-being, and no[t] merely the absence of disease or infirmity.' The social dimension to health had been recognized by the doctors George Scott Williamson and Innes Pearse at the Pioneer Health Centre (PHC) (though Scott Williamson objected to the phrase 'positive health' on grounds of the adjective's redundancy: there was no such thing as 'negative health'). Health, wrote Scott Williamson, was 'like a seed' which grew in 'the social soil'.⁵

Given the influence of the Peckham Experiment and its doctors on the founding of the Soil Association, the views of Scott Williamson and Pearse on the meaning of the term 'health' carry significant weight. Speaking at St Andrew's University in May 1971, Pearse admitted health to be 'a vague ill-defined entity', inseparable in public mind and policy from absence of sickness. In contrast, she proposed that health must be regarded not as a *state* but as a *process,* with its own 'functional potency' and 'action-pattern'. Some individuals manifest resistance to pathological agents, a characteristic which Scott Williamson had as a hypothesis attributed to their relationship with their specific environment. The work at Peckham appeared to support this idea, and Scott Williamson and Pearse concluded that health depended upon 'the development and working of a faculty of the organism — *the faculty for mutual synthesis of organism and environment'.* Or, to return to the idea of a natural order: 'In health man observes a different natural law [i.e. different from the laws of pathology], the law of function, for health means living a full functional existence, in which development proceeds according to potentiality'. Health was, in fact, something which could be as infectious as disease, spreading through a community, but a wide-ranging study of its laws had — has — yet to be undertaken.⁶

Philosophers use the phrase 'ostensive definition': demonstrating the meaning of a word or concept by pointing to an example of it. The nutritionist Sir Robert McCarrison provided the Hunza tribesmen as an ostensive definition of health, describing them as a people 'unsurpassed by any Indian race in perfection of physique; they are long lived, vigorous in growth and age, capable of great endurance and enjoy a remarkable freedom from disease in general'. For McCarrison's admirer Dr G.T. Wrench, this description chimed in with his own memory of 'the vigorous and exuberant life of [his] English public school, where everything that really absorbed one's boyish interests was based on a glowing vitality and responsive health'. Wrench, too, identifies health with wholeness, which he defines as 'sound physique of every organ of the body without exceptions and freedom from disease': a somewhat narrower conception than that of the Peckham doctors.⁷

The organic movement's ideas about health have been derived

almost entirely from the work and ideas of McCarrison, Wrench and the Peckham doctors. DR KENNETH BARLOW, writing nearly thirty years after McCarrison's death and nearly forty after the PHC's closure, in his book *Recognising Health* (1988), devoted a chapter to McCarrison and five to the PHC. His definition of health in that book owes more to the latter than to the former. As one would expect, he rejects the orthodox medical assumption 'that the war against disease gives rise to health', just as organic cultivators reject the idea that healthy plants can be produced by waging war on pests. His concept of health is considerably more wide-ranging than Wrench's: a thorough study of the subject would require research into 'the relations between human beings; of the relations between human beings and other forms of life; and finally the relationship between all that is alive, and the planetary habitat': in other words, an ecological approach, in which health consists in human beings growing into their habitats. But those habitats must suit the needs of their biology.[8]

A similar emphasis on the wider social and philosophical context of health could be found in the journal *Health and Life*. Its founder Edgar Saxon died in 1956, but under the editorship of James Gathergood it continued to promote an organic, holistic approach to the quest for vitality. Ill-health was a deficiency state producing symptoms which re-appeared until it was rectified; but health — by which the journal meant wholeness, 'fullness of living, in every part of human nature' — could be restored. It was 'an instinct, an impulse, an energy, that is in us all the time, but is hampered by material deficiencies and derangements, and by mistaken beliefs and wrong ways of living'. This idea of health as a vital force was also adopted by the GP Aubrey Westlake, who, as we shall see later in this chapter, moved into alternative medicine later in his life. He believed he had found the clue to the nature of health in 'the supersensory force — Vis Medicatrix Naturae (the healthy power of Nature)'. In true organic fashion, health was 'a balanced pattern of forces': a balance 'between the forces of matter and the cosmic (supersensory) forces'. Quoting Edward Bach, Westlake attributed health to 'perfect harmony between Soul, mind and body'. Disease consisted in 'an imbalance brought about by an excess, deficiency (due to blockage) or distortion of the forces involved'. Another GP, Peter Mansfield, has defined health most simply and most comprehensively as 'the ability to participate in Creation'.[9]

These concepts of health share certain features despite their differences. Scott Williamson may not have liked the phrase 'positive health', but it importantly conveys the sense that for the organic movement health is something dynamic. It can be conceived in aesthetic terms, as the ideas of balance and harmony suggest; Westlake saw a 'pattern' of health, as, by implication, did the Peckham doctors, with their belief that order will emerge spontaneously from variety in the right environment. These concepts of health also have in common far-reaching social implications. It is not that healthy individuals will create a healthy society; rather, society must be re-arranged in order to enable its members to grow in health through provision for their biological nature. Thirdly, the quality of food is crucial to these models of health, and one of the main purposes of re-arranging society would be to enable the production and distribution of fresh and 'honest' food. Balfour and Howard were uncompromising about this, the former writing that 'the welfare of the people as a whole must in future take precedence over all narrow interests', and the latter that exploitation of land for the purpose of profit must cease. Even Westlake's esoteric approach to health required food containing high-quality proteins in order to maintain the right energy-levels. If the quality dropped, disease followed and its vicious circle could be broken only by 'a restoration of quality protein obtained by a right attitude, intention and action to soil, plant, animal and man'.[10]

The Peckham doctors and their long-term influence

The Pioneer Health Centre and its aftermath

Since the organicist philosophy of health and nutrition is derived very largely from the experience of McCarrison and the Peckham doctors, let us look first at their influence during the half-century which followed the end of the war. Scott Williamson and Pearse were more high-profile during the post-war years than McCarrison, who, despite his academic and international standing had not been given any major responsibility for dietary policy during the war. The re-opening of the PHC in 1946 aroused much interest, including the publication of John Comerford's book *Health the Unknown* (a title which alluded to Alexis

Carrel's popular book of the 1930s, *Man the Unknown)* and in 1948 a visit from H. M. Queen Mary in the company of Prime Minister Attlee and other members of the government. In 1947 the Centre inaugurated a school, and, with Attlee's help, it re-gained the sixty-acre farm at Bromley in Kent which provided organic food for the Centre's catering. But the project was dogged by lack of funding and received no support under the new NHS, its principles being deemed contrary to Ministry of Health policy and its administration unsuited to the established scheme. The noted surgeon Sir Ernest Rock Carling, an authority on radiation and member of the Medical Research Council, was deeply sceptical about the Centre's validity as a scientific experiment, and despite a concerted campaign on the part of the staff and of the families who belonged to the Centre, it was forced to close at the end of 1950. Thus, by a supreme irony, the new NHS proved to have no place for an investigation into the nature of health. Because it 'arose out of an original conception which *extends into a region beyond the ambit of present day Medicine and Hygiene* [it was] not eligible to receive the enabling stamp of Authority'. Scott Williamson and Pearse concluded from the Centre's demise that the Welfare State was essentially a totalitarian system, intolerant of any influence 'that comes from outside its own programming of compelling "care" ... It is not yet ready to consider the possibility that the cultivation of order, ease and virtue in Society, might prove an even greater power for the welfare of the people than the abiding "care" of the administrator'.[11] Scott Williamson died two years after writing this obituary of the PHC, but Innes Pearse lived on until 1978, and Mary Langman, who had worked at the Centre in the 1930s and helped run the farm at Bromley, remained active in the organic movement right into the twenty-first century. Both women influenced a younger generation of organic activists.

The PHC remains one of the great lost opportunities in the study of family and community health provision, an experiment considered significant beyond the organic movement. The anarchist Colin Ward saw Scott Williamson and Pearse as '*the* truly creative figures in 20th-century social medicine', and in 2002 the Wellcome Trust hosted an exhibition on 'Positive Health: The Pioneer Health Centre Peckham and Beyond', seeing something of the Centre's influence in the government's Healthy Living Centres initiative. Within the organic movement, the PHC is of particular interest as one of the

4. THE ORGANIC ALTERNATIVE: HEALTH AND NUTRITION

few initiatives which have attempted to address urban problems, and as evidence for the organicist faith in what is termed an 'emergent order'. The narrow approach to scientific experiment evinced by Rock Carling — a man who believed that atomic radiation could prove valuable as a means of reducing human fertility levels — helped destroy a project which transformed the lives of those who participated. Colin Ward praised Scott Williamson and Pearse in his review of Alison Stallibrass' book about Peckham, *Being Me and Also Us,* and Stallibrass is uncompromising in her view that the experiment was in many ways a success, 'yield[ing] valuable information concerning the nature and needs of humankind' and offering a harmony-from-diversity in which people's health could blossom. Although the building itself, after use as a college annexe, has now been transformed into a block of flats, some of those who worked at, or belonged to, the Centre have kept its ideas alive through the charity the Pioneer Health Centre Ltd, and the support of the Scottish Academic Press.[12]

Pearse was determined that investigation of the hypothesis on which the Peckham Experiment was based should continue, despite the Centre's closure and her husband's death. After many years' work on his notes, papers and draft chapters, and with the help of Douglas Trotter and Allan Pepper, she published *Science, Synthesis and Sanity: An Inquiry into the Nature of Living* (1965), a major work of biological philosophy which required the inclusion of a 27-page 'Dictionary of Quality' to help clarify its many neologisms. The publishers, Collins, printed on the dust-jacket the opinions of four writers, to whom the manuscript had evidently been submitted prior to its acceptance. All four were noted religious philosophers: Charles Raven, Paul Tillich, Prof. H.H. Price and Prof. D.M. MacKinnon. Raven saw the book as an important contribution to the understanding of community, while Price praised its novel approach to biological phenomena. Tillich and MacKinnon, who were both interested in existentialist thought, may have seen in the book an antidote to reductionist or determinist theories of human nature. Kenneth Barlow, who worked at Peckham, explicitly states that the Peckham Experiment was biased to what he terms the 'religious' view of nutrition rather than the 'scientific', by which he appears to mean the holistic rather than the reductionist.[13]

Certainly there was at least one academic who drew theological conclusions from the work at Peckham: this was the clergyman Douglas

Trotter, who, as a young man, had joined the Iona Community founded by the pacifist George MacLeod and been sent to the re-opened PHC in order to study a community in action. There he met Henrietta Hony, a staff member who soon became his wife, and there he had many long discussions with Scott Williamson about the ideas which eventually found their way to publication in *Science, Synthesis and Sanity*. Trotter, who lectured for twenty years at St. Andrew's University, continued his own research into the meaning of Peckham; in the mid-1980s his aim was to make Scott Williamson's thought more accessible, but the final result, eventually published in 2003, was a learned and complex theological work, *Wholeness and Holiness,* sub-titled *A Study in Human Ethology and the Holy Trinity.* Yet, as the main title indicates, even in such a specialist study, a familiar organicist theme can be identified. Through Trotter's post at St. Andrew's, Pearse came to have links with that university and the University of Dundee, while the Scottish Academic Press published her last book, *The Quality of Life,* in 1979, just after her death, and, ten years later, Stallibrass' *Being Me and Also Us. The Quality of Life* is Pearse's final plea for continued research into those problems which the PHC had investigated; it concludes with a distinction which organicist thought often makes: between quantity and quality: between the Space-Time world of the physicists, where entities are non-specific, and the 'world in which we "live": the world of specificity of pattern, unique, ineffaceable', with 'its own content and dynamic; and its own laws and regularities': a creative mode of living where health is to be found.[14]

Pearse's call for continued research was repeated in December 1985, when the Pioneer Health Centre Ltd celebrated the fiftieth anniversary of the opening of the purpose-built centre in St. Mary's Road, Peckham. Professor Peter Townsend's remarks were particularly important. He had visited the PHC as a young married man and, while impressed by its vitality, been sceptical about its experimental validity. More recently, however, he had found the Peckham approach to conceptualizing health 'ever more inspiring', for its recognition of the influence of the extended family on individual health. He also emphasized the wider social context of health and the way in which social institutions — whether market or state — served to reduce individual choice. Like Kenneth Barlow, he believed that 'if we are to achieve better health we must accept the requirement of re-organising society itself'. Another professor, Brian Goodwin of the Open University, considered the Peckham doctors' work of continuing relevance 'to current concepts of health as a dynamic, transformational and integrative process at all levels from the cell to the planet': a truly organic vision.[15]

Ever optimistic, the PHC Ltd looked to 're-invent' Peckham for the 1990s. In the late 1980s, the body held talks with 'health promotion enthusiasts' in Liverpool, looking to apply the PHC's principles to existing or new health and community facilities. There was a growing interest in 'holistic and participatory' approaches to community health, and Alex Scott-Samuel, who chaired the PHC Ltd's Project Team, looked forward to 'seeing "a thousand Peckhams bloom" all around the UK'.[16] We can consider two attempts to apply the principles of the PHC: one from the organic movement's early years, the other a generation later, and both initiated by GPs who were personally and profoundly influenced by the Peckham doctors.

Dr Kenneth Barlow and the Coventry initiative

Kenneth Barlow's life was changed for ever by meeting Scott Williamson and Pearse at the Adler Society in the late 1920s. Through them he met Patrick Geddes, and at their suggestion took up medicine in the early 1930s. Barlow's unpublished autobiography *The Diary of a Doomwatcher* gives a fascinating account of the literary, political and medical worlds in which he mixed during the inter-war period, and of his war-time experiences as a GP in Coventry. It was just outside Coventry, at Binley Common, that he attempted to establish a community which would extend the principles of the Peckham Experiment. The idea was Scott Williamson's: he suggested to Barlow that he should recruit two thousand families in Coventry, build a township or neighbourhood with a 'Peckham' Centre in the middle and have farmland surrounding it to provide organically-grown food. Barlow began raising money for the project before the end of the war and before the passing of the legislation which created the NHS, at a time when he hoped that local initiatives might contribute to the re-shaping of medicine. Barlow's autobiography, and another unpublished typescript which deals with his attempt to establish the COVENTRY FAMILY HEALTH CLUB HOUSING SOCIETY, testify to the enormous amount of energy, diplomacy and fund-raising which the proposed project necessitated, and to his gradual disillusion with local politics in general and the Labour Party, who governed the city council, in particular. Barlow concluded that the Coventry City Fathers did not wish to grant an association of families the right to conduct their own affairs and manage their own circumstances, thereby opening up the way to health and to a coherent, truly democratic community — 'from the ground up', to use a favourite organic phrase.

By the end of the war, the Family Health Club had acquired the necessary land; it owned 350 acres, and had attracted nearly four hundred families. Barlow sold his general practice in order to commit himself completely to the project. At the Club's weekly meetings plans were made for starting the process of family 'overhauls' of health; the member families visited Peckham regularly and were enthused by what they saw there. Barlow, despite his lack of farming experience, set about ploughing the farm land. But, like the PHC itself, Barlow's initiative found itself bogged down in financial difficulties. At one point, the venture was saved by Lord Glentanar, whom Barlow went up to Glasgow to meet, and the Brandon Woods Farm Co. was set up. The families who had joined the Club were still living in various parts of Coventry, but they were keen to have farm produce. Authority, though, required that even to sell itself its own vegetables, the Club required a licence. Eventually, it started its own greengrocery round: 'an attempt by a group of families to employ its own producers' as Barlow described it; a form of co-operative, yet one which had to face powerful antagonism from within the co-operative movement. Barlow drew from his experience of political hostility conclusions which are relevant to Chapter 8, believing that, in having gained political influence and acquired much property, the co-operative movement was working against social and personal health, apparently no longer willing 'to let the inherent powers of the family in its home work upon its immediate environment in any way capable of expanding or enriching the life of the local community'. The PHC's last attempt to secure official support from the authority was made, by Barlow and Douglas Trotter, in 1949. Lord Lindsay of Birker, an early enthusiast for the PHC whose words carried weight in the Ministry of Health, gave them support, and the Vice-Chancellor of Birmingham University, Dr Priestley, was sympathetic. The university's professor of social medicine, however, proved to have no interest and gave no sign 'of believing that the human family in its biological and sociological relationships lay in the very heart of the subject of social medicine'. On the Coventry City Council, a certain Alderman Hodgkinson, Secretary to the city's Labour Party, who thought he had quashed the Family Health Club some years previously, made quite sure of it this time.[17]

Later in her life, Barlow's daughter Joanna felt that Scott Williamson and Pearse had perhaps used him for their own purposes and that, although he had admired them and been genuinely committed to their work, he had suffered professionally on their account. He trained as a radiologist and with some difficulty found himself a place in the NHS, being appointed Director of Radiology at the Ipswich group of hospitals in 1954; but he remained lastingly sceptical of the NHS approach and committed to the organic movement. He was a Soil Association Council member from 1946 to 1951, a founder of the McCarrison Society, which he chaired from 1976 to 1982, and edited the journal *NUTRITION AND HEALTH*. He was Vice-Chairman of the Pioneer Health Centre Ltd, and with Peter Bunyard, one of the younger generation of ecologists, co-edited *Soil, Food and Health in a Changing World* (1981). In 1971, Robert Waller, then a consulting editor for the publisher Charles Knight, arranged the re-issue of Barlow's wartime book *The Discipline of Peace*. As someone strongly interested in philosophy, Waller thought Barlow's book unjustly neglected — as indeed it remains. But it is one of a small number of works — *Science, Synthesis and Sanity* is another — which have tried to provide the organic movement with a philosophical basis.

Dr Peter Mansfield

Twenty years after the abandonment of Kenneth Barlow's project, Innes Pearse profoundly affected another doctor who would go on to be prominent in the organic movement. This was Peter Mansfield, who, having studied medicine at Cambridge, became a Research Fellow in Community Medicine at University College Hospital, looking at the development of a large-scale health centre in London. While researching the doctors' practices which became part of it, he found that the smaller-scale ones were doing a more effective job than the larger, and that one particular single-handed practice was the best of all, having the most contact with patients. This was not what the authorities wanted to hear, so Mansfield abandoned a future as an academic and began his own general practice in Bermondsey, south-east London (where before the Second World War Aubrey Westlake had also been a GP). One of his mentors at University College Hospital had been Dr John Horder, who aroused his interest in the nature of health and put him in touch with Innes Pearse. Mansfield visited Pearse at her home in Sussex every Sunday for about six months, and on one occasion, around 1969, was so overcome by the intellectual vistas revealed by her ideas on health that he was unable to drive or work for several days. Mansfield tracked down everyone who had been involved at the PHC, including Lucy Crocker, who had written *The Peckham Experiment* (1943) with Innes Pearse; and Lord Donaldson, who had funded the PHC so generously in the 1930s. Mansfield now wanted to start a project of his own, based on the principles of Scott Williamson and Pearse, but felt that the idea of a building as a health centre was no longer valid: instead, a whole area should become the site of the experiment, with the GP's surgery as the focal point. In Bermondsey, he set up a gardening project, using the rich soil of the former watercress beds, and involved patients in the work. Mary Langman drew him into the McCarrison Society, and he came to know Gordon Latto and study his methods.

In 1976 he moved to what was considered a fairly needy area, in Lincolnshire, and built up a general practice at Grimoldby near Louth. His practice was unusual: expenses on drugs were very low; he was available at all sorts of times; he ran a smallholding and encouraged patients to be involved and eat the food it produced, using food as medicine in true Hippocratic fashion; he shared Lawrence Hills' faith in comfrey; he was regarded, inevitably, as an eccentric by other Lincolnshire GPs. ('Don't talk to Mansfield — it only encourages him.') He founded the Templegarth Trust, a charity, to encourage the sort of vitality that the PHC had exhibited and be a sort of 'think-tank' for fostering health; membership was open to all, and other professionals — teachers, a dentist, a priest — became involved. One practical outcome was the purchase of a mill for making wholemeal flour. Mansfield was also a child health doctor to many clinics in the Louth area and all the schools, as well as working at the local hospital's paediatric clinic. During this demandingly busy period, Mansfield made contact with another Lincolnshire member of the organic movement, John Butler.

Despite writing some lengthy articles on his project for the Soil Association journal, and serving on the Association's Council, Mansfield did not receive a great deal of assistance from the Association: only from figures such as Lord Donaldson and Mary Langman who had been involved in the PHC. It seems that he did not play a great part in the Pioneer Health Centre Ltd either, raising objections to Allan Pepper's scheme to start a Peckham-style centre in Glenrothes in Scotland. He learned a lot from the McCarrison Society, valued its conferences and came to know Geoffrey Cannon and Michael Crawford, but admired Lionel Picton most of all, seeing his work as 'an achievable life' which serves as an example for all GPs.[18]

The significance of Peckham

It might perhaps be wondered why the organic movement still attaches so much importance to the Pioneer Health Centre. It was an experiment which ran for less than ten years and closed down six decades ago; Kenneth Barlow's attempt to expand it at Coventry never got properly under way and Peter Mansfield's work at Louth was a unique initiative with minimal impact. In fact, one might gain the impression that even parts of the organic movement — despite the Soil Association's Bristol headquarters containing a 'Peckham Room' — prefers now to concentrate on micro issues such as whether a particular organically grown vegetable or fruit contains a higher degree of vitamins or trace elements than its chemically grown equivalent, and to attribute health to the purchase of certain brands of organic food. Nevertheless, it is right to give the PHC prominence. Apart from other considerations, George Scott Williamson was, along with Eve Balfour and Friend Sykes, one of the Soil Association's three founders. He and Innes Pearse inspired a number of people committed to the organic movement during long lives: Richard de la Mare, Kenneth Barlow, Jack Donaldson and Mary Langman, all active from the 1930s to the 1980s or beyond; the Devonshire farmer Ethelyn Hazell; Rosemary Fost, and Douglas and Henrietta Trotter. Pearse, who outlived her husband by 25 years, transformed Peter Mansfield's life and career, and became a mentor to a younger generation, among them Heda Borton, daughter of Cdr Robert Stuart. Barlow, and Scott Williamson and Pearse, produced philosophical works which attempted to base an understanding of human nature in biology rather than mechanics. The PHC's importance also lies in the fact that, firstly, it is the one major attempt the organic movement has made to relate its vision to an urban context, and secondly, that it emphasizes the social dimension of health issues. Peckham failed to influence NHS policy, but its memory persists as an example of a possible alternative to a system heavily reliant on pharmacological relief of illness. And the message has appealed beyond the organic movement to those interested in alternative approaches to health policy and in de-centralized forms of social structure.

Sir Robert McCarrison and his influence

The other major influence on the organic movement's philosophy of health was Sir Robert McCarrison. Like Scott Williamson, he investigated the conditions that make for health and concluded, through his knowledge of the Hunza tribesmen, that food and the way it was produced were the key factors. By the 1920s, McCarrison had influenced Jerome Rodale — who promoted the organic message in the USA, and whose descendents have continued to do so — and was corresponding with Peckham doctors. In 1936 he delivered the Cantor Lectures at the Royal Society of Arts, which were published by Faber under the title *Nutrition and National Health;* and in 1939 he spoke, with Sir Albert Howard, at the launching of the Cheshire doctors' *Medical Testament,* a document which the *British Medical Journal* printed as a special supplement and which gave rise to several months of correspondence. But McCarrison was not given any significant post during the war, despite his worldwide reputation as a nutritionist and his many international honours; nor was he granted a Fellowship of the Royal Society. 'Few doctors,' says Barbara Griggs, 'gave any signs of having been influenced by his work.' One who was, Dr Walter Yellowlees, attributes the lack of interest in McCarrison's work to the discovery of sulphonamide drugs, which proved of great value in the treatment of various bacterial infections. Most doctors, '[b]y their training and attitudes ... not attuned to believing that such a simple measure as dietary reform could prevent many of the diseases which they daily encountered', ignored McCarrison's signpost to health, preferring to study exciting new methods of treating disease.[19]

4. THE ORGANIC ALTERNATIVE: HEALTH AND NUTRITION

Dr Hugh Sinclair

Under the terms of the new NHS, the public was entitled to have these drugs prescribed at need, and doctors found themselves swamped by patients, with no time to give to the subtleties of nutrition, even if they had been interested in doing so. Nutrition was low in status: Lord Horder, chairman of the British Medical Association's Committee on Nutrition, set up in 1947, regretted that nutrition still occupied a 'meagre place' in the medical student's curriculum and that there was no systematic teaching of it. Dr Hugh Sinclair, a Fellow of Magdalen College, Oxford, agreed with Horder and wished to rectify the situation. Sinclair was one of McCarrison's students at Oxford, and became a friend and admirer, inheriting all McCarrison's papers after his death in 1960. As a medical student in the 1930s, Sinclair had become interested in nutrition because he was convinced that 'the most serious problem in medicine arose from alterations in our diet from the processing and sophistication of foods', and that these alterations had led to a dramatic increase in degenerative diseases (e.g. coronary thrombosis, lung cancer, ulcers, appendicitis and allergies). During the war, he worked for the Ministry of Food's advisory team, but in the post-war years found that the interest in nutrition drained away. In 1946 a Laboratory of Human Nutrition was founded at Oxford with the assistance of a grant from the Wellcome Trust, but it terminated in 1955 when it merged with the Department of Biochemistry under Hans Krebs (whose son Sir John Krebs many years later became a target of organicist criticism as Chairman of the Food Standards Agency). This put an end to Sinclair's work in that Department.[20]

In April 1957, Sinclair spoke at London University at a celebration of World Health Day, on the subject of 'Food and Health'. He told his audience that his great-grandfather had been first President of the Board of Agriculture, to which Fisheries and Food had later been added, but that the Ministry which it had now become should be abolished. Since 'the purpose of food is to promote health', the Ministry of Health should, as it had done in the war, dictate nutritional policy, with a Ministry of Food to implement it. Their aim would be 'to produce not dehydrated canned fodder but appetizing wholesome food'. To determine the nature of such food required far more research than the negligible amount currently undertaken. Like Howard, McCarrison and Weston Price, Sinclair believed that traditional societies had much to teach, and that '[m]uch knowledge [was] still to be gained by studying food practices that are hallowed by tradition and established in folk-lore'. He had himself made a close study of the Eskimo diet, adopting it himself for a hundred days. Sinclair's lecture contains one notably curious passage, in which he emphasizes that nutritional experiments on rats and mice are not relevant to humans: 'Man is a different animal'. Curious, because McCarrison's dietary experiments with rats have frequently been cited as evidence for the organic view of diet.[21]

As a scientist, Sinclair had no time for faddism. In a long letter to *The Lancet* in 1956, about the importance of essential fatty acids, he rejected the 'long-haired naturalism' of faddists while pleading for more research into the effects of white flour and manufactured fats.[22] Sinclair later became an authority on essential fatty acids, but at the time his views were dismissed. He suffered academically for his disagreements with Hans Krebs, and it was not until 1972 that he at last achieved his ambition — mentioned in his 1957 lecture — of establishing an institute for the study of human nutrition. This was the Association for the Study of Human Nutrition, to which one of the subscribers was the surgeon Arthur Elliot-Smith, a leading member of the Soil Association's Oxford Group. (It was registered as a charity and its name changed to the International Institute of Human Nutrition: later, the International Nutrition Foundation. Following Sinclair's death in 1990, the Foundation was wound up and the funds used to establish the Hugh Sinclair Unit of Human Nutrition at Reading University. In 1995, Dr Christine Williams became the Unit's first Hugh Sinclair Professor.)

Sinclair contributed to various Soil Association events in the 1960s and '70s: these included a conference at Leicester University in March 1967 organized by the Association's Leicestershire and Rutland Group, at which his paper on degenerative diseases was read for him: the 'Inquest on Health' conference at Ipswich in April 1971, and an international scientific conference at the National Agricultural Centre, Stoneleigh in July 1973, on 'Agriculture, Food Science and Nutrition'. He was also a member of the International Society for Fluoride Research, opposition to fluoridation of the public water supply being another cause dear to the organic movement.

4. THE ORGANIC ALTERNATIVE: HEALTH AND NUTRITION

In January 1957, three months before Sinclair spoke on the occasion of World Health Day, three prominent Soil Association members — Innes Pearse, the dentist Everard Turner and Dr Kenneth Vickery, Medical Officer of Health for Eastbourne — sponsored a Declaration on Health in the *British Dental Journal* and *The Lancet,* in which 440 doctors and dentists put their signatures to a reaffirmation of 'the convictions expressed by the Cheshire Panel Committee regarding the intimate connection between nutrition and health' and arguing for the medical significance of the organic movement's philosophy of wholeness.[23] The Association's call for an 'ecological' approach to health, as against the fragmented analysis of contemporary science, was never likely to affect government policy, but McCarrison's ideas remained influential in the organic movement. Following his death in 1960, Faber re-issued his Cantor Lectures, with a twelve-page postscript by Sinclair on recent advances, and, following one of the Soil Association's Attingham Park conferences, the McCarrison Society was founded in January 1966. 'Why form another Society?' asked its Honorary Secretary Dr S.J. Mount, and answered, simply, that there was a need for one.

The McCarrison Society

Membership of the McCarrison Society for Nutrition in Health, to give it its full title, was open only to members of the medical, dental and veterinary professions. After a year, its ninety-strong membership consisted of, respectively, sixty, twenty-five and five members of those professions. The 1965 Attingham Park conference on the theme of wholefood had aroused strong feelings about the medical profession's lack of interest in the idea of positive health through nutrition, and the McCarrison Society was an attempt to rectify this state of affairs. Mount rejected the criticism of some Soil Association members, that the new Society would draw professional support away from the Association; on the contrary, he argued, it had already drawn the Association to the notice of people who had not previously heard of it. 36 people, of whom 24 were Soil Association members, attended the first meeting, chaired by Perthshire GP Walter Yellowlees, and the Society's aims and strategy were agreed upon. The Society's purpose was 'to enquire as a professional body into the meaning and validity' of three requirements cited by McCarrison as essential for food: 'that it would be grown on healthy soil; that it should be eaten whole; that it should be eaten fresh'. To achieve this purpose, the Society would collect and publish evidence relating to nutrition and health; emphasize wherever possible the importance of nutrition to health; encourage within its means 'such research ecological and otherwise' relevant to its aims; liaise with any other scientific bodies similarly interested; press for review of food and health legislation in the light of its findings, and hold regular meetings with lectures. A typed list of 81 members, probably dating from late 1966, includes the names of various prominent organic figures: Innes Pearse; Aubrey Westlake; Kenneth Rose and Everard Turner, two Leicestershire dentists; Dr Kenneth Vickery; Arthur Elliot-Smith; E. Brodie Carpenter of the Middlesex Group, and four members of the Latto family: Conrad and Douglas Latto, and husband-and-wife Gordon and Barbara Latto.[24]

Douglas Latto spoke to the Society in April 1966 and told members of his friendship with McCarrison, whom he had known at Oxford and often visited. The Society's survival in its early years owed a great deal to Barbara Latto's dedication. She organized a series of successful weekend conferences and became Secretary in 1973. The first of the conferences was held in June 1971, on the topic 'A Strategy for Sound Nutrition', and its details are impressive. Elliot-Smith, the Society's President, spoke on the continued relevance of McCarrison's work, and the Soil Association's recently appointed President, E.F. Schumacher, chaired a discussion, as did veteran organic activist Rolf Gardiner. Speakers included Hugh Sinclair; farmer and Soil Association Council member Hugh Coates; the smallholder Sedley Sweeny; scientist Harry Walters, and a couple of European medical scientists. Margaret Brady's new film on bread was shown. Even allowing for inflation, a full course fee of £2.50 (not including accommodation) sounds remarkably good value for money. In October 1972 the McCarrison Society organized a symposium at the Royal Society of Medicine on a favourite organic topic, 'Diseases Associated with Refined Carbohydrates', chaired by Conrad Latto, which featured a high-quality line-up of speakers, among them Hugh Sinclair; Ian McColl, Professor of Surgery at Guy's Hospital; Denis Burkitt of the Medical Research Council; Dr Hugh Trowell, and, from the University of Bristol, the consultant Kenneth Heaton and the dental scientist Abdul Adatia. By the mid-1970s, the

Society was in a fairly buoyant frame of mind. Finances were sound, thanks to generous donations, and the Society was able at last to fulfil its aim of encouraging research projects when it contributed funds to Kenneth Heaton's investigation at Bristol University into the effects of changing from white to wholemeal bread. The Society had established good relations with the Royal Society of Medicine (RSM), where it regularly held meetings and lectures: in March 1974, Dr Michael Crawford spoke there on the somewhat eugenicist topic 'The Future for Man: Development or Degeneration?', and the RSM's catering department was 'truly co-operative', providing feasts of wholefood.[25]

Nevertheless, looking at the McCarrison Society's position after its first decade, Walter Yellowlees communicated to members what might be termed a desperate optimism. At times they must have felt 'like a faltering David, facing a monstrous Goliath whose breastplate is professional indifference and scepticism, whose sword is blinkered technology, and whose helmet is mind-bending advertisements for worthless food'. The Society's raison d'être was education, but to educate professions and individuals was a daunting task: the Society was small, and lacked premises, staff and wealth. Against it were ranged the financial resources of state-funded university medical education, and the pharmaceutical firms. How then to make the medical and dental professions realise the simple truth that 'expensive technology is called on to remedy faults which industrial man has himself arranged'? — a message not dissimilar to that being communicated at the same period by Ivan Illich. Yellowlees drew comfort from the fact that the McCarrison Society was growing in numbers, and that 'the slings and pebbles of truth ... if aimed aright, cannot be resisted'. One such pebble was the recently published book *Food for Naught* by Ross Hume Hall, which Yellowlees praised for its lucid account of the way in which the demands of trade had triumphed over the needs of health. Hall, a Professor of Biochemistry at McMaster University, Ontario, was one of the speakers at the Society's 1977 conference at Keble College, Oxford.[26]

The McCarrison Society advanced its status as an educational body in the early 1980s when, in conjunction with A.B. Academic Publishers, and with Kenneth Barlow as its first editor, it began publishing *NUTRITION AND HEALTH*. The journal's purpose was 'to provide a broad perspective on the subject of food and its effect on the human body, and to serve as a forum for the communication of recent advances, newer knowledge and conciser reports of research on all aspects of nutrition and foods with particular relevance to the maintenance of health and the prevention of, and recovery from, illness.' With an international board of editors, it was aimed at a wide range of professions — from doctors and nurses to caterers and sociologists — and organizations, from medical schools to community health centres. Its contents were wide-ranging. Hartmut Vogtmann, in an article on the central organic topic of the relationship between soil quality and food quality, argued that staple food grown under organic systems had a number of advantages over its conventionally grown equivalent; Beata Bishop, a disciple of Max Gerson, discussed the role of organic food in curing cancer; Kenneth Barlow discussed the nature of life itself; Michael Crawford and K. Ghebremeskel looked at nutrition and health in relation to food processing; and there was an interesting historical article by V.J. Knapp on fruit and vegetables in the European diet. The journal investigated the possible link between chemical fertilizers, mineral deficiency in the soil, and arthritis, and, in the early 1990s, gave considerable space to the emergence of BSE. The social dimension was not neglected: in the mid-1990s Sean Stitt, Diane Grant and Ciara O'Connell addressed the issue of poverty, suggesting that the inability of many low-income households to afford a healthy diet was serious and widespread, and that the elderly were particularly vulnerable. These articles appeared in 1994 and 1995, and anyone with a knowledge of twentieth-century food history will feel a weary sense of *déjà vu* on reading them. Sixty years on from Sir John Boyd Orr's famous survey *Food, Health and Income* (1936), how much had changed? The McCarrison Society also produced a series of booklets, *The Founders of Modern Nutrition,* under the editorship of the noted food writer and investigative journalist Geoffrey Cannon. We shall look at two of the scientists featured: Dr Hugh Trowell and Surgeon-Captain T.L. CLEAVE, RN.[27]

Two medical scientists: Trowell and Cleave

Like some of the organic movement's most influential figures — Howard, McCarrison, Wrench, St. Barbe Baker — Trowell developed his ideas as a result of his experiences overseas: in his case, over thirty years in East Africa. He was responsible for identifying a disease of malnutrition named kwashiorkor, which in 1952 the World Health Organization described as 'the most serious and widespread nutritional disorder known to medical and nutritional science'. After having endured ridicule and suffered professionally on account of his interest in the disease, Trowell was proved justified, and his 1954 textbook on it was subsequently recognized as a classic of nutritional study. Trowell went on to draw conclusions from his observation of the contrasts between disease patterns in Africa and those in the West, producing what his friend and colleague Denis Burkitt described as his magnum opus, *Non-Infective Disease in Africa* (1960): a work almost entirely ignored in Britain, though highly regarded in Africa.[28] On his return to England in 1959, Trowell trained for ordination and became vicar of a Wiltshire parish and chaplain at Salisbury hospital, further maintaining his links with medicine through chairmanship of the London Medical Group's committee on faith and ethics, and of the British Medical Association's group to study the issue of euthanasia.

He returned to the study of nutrition in the 1970s, working with Burkitt on a book on disease and diet. Trowell was convinced that many common western diseases were unknown in Africa because of the presence of what he termed 'dietary fibre' in the African diet. Full-fibre, starchy foods, he suggested, could protect against heart-disease and diabetes; hence the lower rate of these diseases among monks and vegetarians. Trowell concluded that many illnesses common in western society were in fact man-made, the result of a food technology which produced a diet overladen with fats and sugars but lacking fibre and many nutrients found in wholefood. Obesity, for instance, was very rare among well-nourished Africans. Trowell's research into Western obesity involved the study not just of medical literature, but of paintings, 'to determine when obesity manifested by a double chin became common in Britain' and relate this change to changes in diet.[29] Trowell also studied the increasingly rapid sexual maturation of adolescents, particularly in urban society, hypothesizing that its cause might be the replacement of mother's milk by bovine milk in infant feeding.

First to have been chosen as a 'Founder of Modern Nutrition' was 'Peter' Cleave, as he was known. Cleave died in 1983 but by 1991 was already, in the view of Walter Yellowlees, 'a forgotten prophet'; in his autobiography, published two years later, Yellowlees devoted a chapter to Cleave's ideas. After qualifying as a physician at Bristol Medical School, Cleave joined the Royal Navy, and served as a medical specialist at home and overseas, ending his career as the Royal Navy's director of medical research. He retired in 1962, and it was not until the 1970s that he received recognition for his work on nutrition. His belief in the importance of diet to health was developed early, the result of his eight-year-old sister's death from appendicitis and his teacher Rendle Short's proposition that this illness results from lack of cellulose in the diet. A devotee of Darwin's writing, Cleave built his theory of nutrition on the hypothesis that the human body is inevitably maladapted to the artificial food of civilization. After gathering evidence, he produced his over-arching hypothesis. There was, he reasoned, a simple dietary cause of many degenerative illnesses: what he termed 'saccharine disease', that is, the consumption of ever-increasing amounts of refined carbohydrates — or, to use a better term, fibre-depleted food: our familiar friends white flour and sugar. Cleave identified four types of harm produced by such a diet: constipation, leading to diverticular diseases; diabetes and coronary disease; peptic ulcer, and dental caries. Heaton tells us that Cleave's broad-brush approach, his lofty attitude to peer review, and problems of definition, worked against him. (Cleave thought that the most accurate name for the 'master-disease' was 'Refined-carbohydrate Disease' but that 'Saccharine Disease' was more convenient as the main refined-carbohydrate involved was sugar.)[30]

As one would anticipate, the Soil Association was receptive to Cleave's ideas, with Dr G.E. Breen favourably reviewing his book *Fat Consumption and Coronary Disease,* and *Mother Earth* publishing an article by Cleave and Dr G.D. Campbell on 'The Saccharine Disease' several years before Cleave published his book on the subject. The two men had collaborated on *Diabetes, Coronary Thrombosis and the Saccharine Disease,* which was highly praised by the Medical Research Council. During this period, Cleave's meeting with Denis Burkitt also marked a turn in his fortunes, as Burkitt's connections with various Third World hospitals enabled him to confirm many of Cleave's observations, and his gifts as a communicator (Cleave himself preferred angling to attending medical meetings) were effective in spreading Cleave's ideas and stimulating interest in nutrition. Twelve years after his retirement, Cleave found his ideas at last gaining acceptance. *The Saccharine Disease* (1974) was widely read; in 1976 he was elected to the fellowship of the Royal College of Physicians, and in 1979 the Royal Institute of Public Health and Hygiene awarded him the Harben Gold Medal for outstanding discoveries in the promotion of public health; earlier recipients include Lister, Pasteur and Fleming. Cleave received help from a number of leading Soil Association figures: the surgeon Laurence Knights, who was prominent in the Wessex Group; Walter Yellowlees; Arthur Elliot-Smith and particularly his friend Kenneth Vickery. He prefaced *The Saccharine Disease* with the famous quotation from Horace ('You may drive out Nature with a pitchfork, but she will ever hurry back, to triumph in stealth over your foolish contempt') and appealed to the authority of the Ancient Greeks in his faith that 'correct explanations are nearly always simple explanations'. Provided, he wrote, 'that one keeps strictly within the limits ... set by human evolution ... the danger of oversimplification is incomparably less than that of overcomplication'.[31]

The application of Cleave's principles lay with Britain's Medical Officers of Health, and one of them, Kenneth Vickery, while not uncritical of certain points in Cleave's theory, had been warning of the dangers of refined carbohydrates since the 1950s. Cleave's ideas, if embedded in public health practice, would, Vickery believed, bring about a dramatic reduction in degenerative diseases and the personal suffering they caused; this in turn would relieve the pressure on NHS resources. Nevertheless, much of the responsibility for health improvement lay with the individual, who should observe two basic rules: 'Do not eat any food unless you definitely want it' and 'Avoid eating white flour and white or brown sugar'. Having offered a relatively simple natural diet, Cleave concluded *The Saccharine Disease* by wondering why people would 'take endless trouble over the maintenance of a motor car [when] over the maintenance of that infinitely more delicate mechanism, the human body, they are seldom prepared to take any trouble at all'. Cleave's recognition in the 1970s was swiftly followed by renewed neglect. A 1989 Department of Health report on *Dietary Sugars and Human Disease* made no reference to him; perhaps, Yellowlees suggested, because scientists find it hard to grasp what is simple. Yellowlees also felt that Burkitt had taken over and obscured certain features in Cleave's work. Whatever the reasons, Cleave, like the Peckham doctors and McCarrison, belongs in the company of voices crying in the medical wilderness.[32]

Other medical influences

Scharff, Breen, Badenoch, Yellowlees, the Lattos; Stanton Hicks

Many other doctors and dentists played a part in the organic movement's development. The work of Dr J.W. Scharff, Chief Medical Officer of Health at Singapore until the fall of Malaya in 1942, who had dramatically improved the health of Tamil workers through vegetables grown by Howard's Indore Process, was often cited. Scharff belonged to Cdr Robert Stuart's Compost Club, and was prominent in the Soil Association's Middlesex Group during the 1950s and '60s; another member of that Group, Dr Norman Burman, has recalled his influence. Scharff spoke to the Epsom Group on its foundation in 1963. His reputation survived into the 1980s, when his article 'The Lesson of Malaya' was re-printed in the *Soil Association Quarterly Review*.[33]

Almost entirely forgotten, it seems, is Dr Gerald Breen, despite his membership of Soil Association committees for a quarter of a century: he was on the Advisory Panel from its inception in 1946 and did a long stint as chairman of the Editorial Board. An Irishman who practised medicine in London for almost fifty years, Breen was a specialist in infectious diseases. Following the end of the Second World War, he was appointed epidemiologist to the London County Council, and for the last twenty years of his career was physician superintendent at South Middlesex Hospital. He was a skilled journalist, editing the *Medical Press and Circular* and the *Medical News,* and was for several years medical correspondent of *The Times*. Like many leading organicists, he was artistically inclined, being both a painter and a sculptor. His gifts in these areas led to friendships with many artists, and he was physician to the Chelsea Arts Club. One cannot but wonder why such an interesting and talented man should have been erased from the Soil Association's memory.

Another important figure from the earlier years of our period was A.G. Badenoch of the Scottish Soil and Health group. Badenoch had served in the Royal Army Medical Corps during the First World War, had his back broken and was gassed. Having studied medicine at Aberdeen, he went in 1926 to Africa as a medical missionary and in 1929 to Malaya, in the colonial service, returning to Britain to study for the Diploma in Public Health and then taking a general practice in

Nottingham. Badenoch's own experiences abroad and his knowledge of many people in the colonies led him to conclusions on the relationship between diet and health similar to those of McCarrison: he never came across stomach ulcers or cancers in native populations, whereas in Nottingham these diseases were commonplace. By the late 1940s Badenoch was living and working in Edinburgh, and, with Cdr Robert Stuart and the radiologist Angus Campbell, established the Soil and Health Group and re-cast Sir Albert Howard's journal *Soil and Health* as *Health and the Soil*. In 1949, Badenoch gave the first lecture for the Albert Howard Foundation of Organic Husbandry, *The Minerals in Plant and Animal Nutrition:* the published lecture carries on its front cover the reflection, 'Domini est terra, et plenitudio eius (The Earth is the Lord's, and the fullness thereof)'. According to his son Christopher, Badenoch was a socialist who believed it a moral duty to provide every citizen with health and education, but like so many prominent organicists, he was doubtful about the new NHS, predicting that it would prove very expensive. Badenoch played an active part in the Soil and Health Group from 1948 until around 1954, when he emigrated to Canada. He also contributed to *Mother Earth* during this period, on the topics of bread, sugar and McCarrison's work. The dedication of Badenoch Senior was crucial in establishing a basis for the organic movement in Scotland; his friend Cdr Stuart would carry the torch for a another quarter of a century.[34]

Further north, at Aberfeldy in Perthshire, lived — and still does, at the time of writing — Walter Yellowlees, *A Doctor in the Wilderness,* as he described himself in the title of his autobiography. In any debate on continuity in the organic movement, Yellowlees is a prime example of someone whose career and commitment link the Soil Association's early days to the twenty-first century. The 'wilderness' he refers to is the austere landscape of moor and glen in which he undertook his practice; the intellectual wilderness in which he found himself as a result of his views on health and diet; and what he describes as a 'spiritual darkness' resulting from the retreat of Christianity. Yellowlees spent four years in the Royal Army Medical Corps during the Second World War, which put him off any thought of a hospital career; he wanted closer contact with patients, in their home setting, and took up his practice in Aberfeldy in 1948. Shortly before he was de-mobilized, he read Eve Balfour's *The Living Soil* and 'her introduction to the writings of

McCarrison and Howard opened up a refreshingly new vision of the causes of human disease'. Also, having two farmer brothers gave him insight into the lives of those who work the land, and he identified with them to the extent of using Howard's composting methods to grow garden fruits and vegetables, an experience which convinced him that organic growing worked. He was sensitive to the changes in farming which took place around him in the Upper Tay Valley. When he started his practice, mixed farming prevailed and there was a considerable population; by the late 1980s it had dwindled dramatically, and the small-farm families and their workers had gone, replaced by monoculture and machines. Yellowlees was an early member of the Soil Association and sought to alert his profession to the harmful effects of industrialized food through writing letters to the *British Medical Journal* or *The Lancet* and speaking at conferences. Considering that NHS doctors 'might not like the shrill hostility of some Soil Association members to orthodox medicine', he helped found the McCarrison Society, but, in his view, it made little impact. Nevertheless, and much to his surprise, he was invited to give the 1978 James Mackenzie Lecture, which he delivered at Imperial College, London to the AGM of the Royal College of General Practitioners. The lecture would serve as a good introduction for anyone unfamiliar with the organic approach to health, providing as it does a lucid overview of 'the rising tide of morbidity', the work of McCarrison and Cleave, the lessons to be learned from primitive societies, the harm done by synthetic foods, and the need for administrative changes to health provision and for social changes to restore the mixed, family farm. Faithful to the philosophy of the early organicists, Yellowlees argued that sound nutrition 'is part of a living process, embracing man's relationship with the plant and animal world and .... subject to nature's laws'. The lecture, *Ill Fares the Land,* sold well through the Soil Association's Wholefood shop in London but otherwise sank without trace, never being referred to, as far as Yellowlees was aware, by GP authors writing on preventive medicine. In this respect, it shared the fate of Cleave's work. Yellowlees retired in 1981 and with his wife Sonia ran an art gallery for sixteen years. He also trained as a potter, continued to garden, and remained active in the organic movement; in 1981 he helped found a Scottish branch of the McCarrison Society. His autobiography is essential reading for anyone wishing to study the history of the British organic movement,

4. THE ORGANIC ALTERNATIVE: HEALTH AND NUTRITION

particularly anyone who values a case in favour of the organic approach to health which is built on knowledge and experience.[35]

Also from Scotland, and exhibiting the same qualities as Yellowlees despite subsequently moving to 'the soft south', were the Latto brothers, Gordon, Douglas and Conrad. Their father was Town Clerk of Dundee, and their mother a keen vegetarian: a dietary regime to which all three brothers held during their medical careers. Gordon and Douglas became GPs, the latter specializing in obstetrics and gynaecology, while Conrad became Senior General Surgeon at the Royal Berkshire Hospital. Newman Turner's visitors' book shows that Douglas Latto and his wife visited Goosegreen Farm in May 1952, and Douglas was on the Soil Association Council from 1957 to 1973 (being replaced by his sister-in-law Barbara). Gordon Latto had met Barbara Krebs at a religious conference; they married in 1938, the same year that he set up in general practice in Southend-on-Sea. After the war they moved to Reading, but, choosing not to join the NHS, Gordon set up his main consulting room in London and developed a prestigious private practice, favouring dietary and herbal remedies. Barbara introduced him to the teaching of Max Bircher-Benner, which advocated a diet of raw organically grown vegetables and wholegrain cereals; they became his friends. Gordon Latto was appointed President of the British Vegetarian Society and for ten years was President of the International Vegetarian Society. Dr Kenneth Barlow's daughter Joanna Ray remembers the Lattos as a highly sociable couple, somewhat 'New Age' in their outlook, while Angela Bates recalls Barbara visiting her home in Lincolnshire and carrying out an amateur exorcism on the premises. In 1977, Barbara spoke on the topic of 'Nourishment for Body, Mind and Spirit in a Simpler Way of Life' at a conference on Iona.[36]

Finally in this survey of some leading physicians, it is important to recall the contribution made by a highly distinguished nutritionist, Sir Cedric Stanton Hicks. A New Zealander by birth, Stanton Hicks was Professor of Physiology and Pharmacology at the University of Adelaide, South Australia, from 1926 to 1958, but had close links with the British organic movement. He spent from 1923 to 1926 as a Research Fellow at Trinity College, Cambridge, looking at the chemistry of the thyroid, and through this research came to meet George Scott Williamson and Innes Pearse, who shared his enthusiasm for McCarrison's work on thyroid problems and with whom he would later travel in Europe. During the Second World War he was the director of food supplies for the Commonwealth armed forces. In 1945, at the University of Melbourne, he gave the third Annie B. Cunning Lecture, on the topic 'Soil, Food and Life', which was re-printed with a foreword by Dr G.E. Breen in the second introductory volume of *Mother Earth*. In later years he delivered two other lectures named after other important organic figures: the Sanderson-Wells Lecture of 1950, 'Food and Folly', and the Albert Howard Memorial Lecture of 1958, given at the Soil Association's AGM and concentrating on the Keyline System of fellow-antipodean P. A. Yeomans, who was present at the occasion. Senior Soil Association figures reciprocated his visits, both Eve Balfour and Brigadier A.W. Vickers going to see him when in Australia. Reading Stanton Hicks' work, one is reminded of *Genesis* 6, v.4: 'There were giants in the earth in those days' — or at least, there were intellectual giants writing about the earth. Stanton Hicks was an ecologist years before *Silent Spring*, aware of the way in which Australia's environmental equilibrium had been disturbed and of the dangers posed by the destructive effects of using mechanical inventions to over-ride natural processes. From classical mythology he absorbed the moral of the story of Antaeus, who was destroyed when no longer in contact with the earth; and from history he drew the lesson which Edward Hyams demonstrated in more detail in *Soil and Civilization* (1952): that many organized societies have collapsed through their own destruction of their means of subsistence. Soils were being rendered subservient to industry's demands, but we were ignorant about soil's biologically complex nature, on whose stream of nutrients our health depended. He developed these ideas in his book *Life from the Soil*, co-written with the Australian farmer Col. H.F. White. Stanton Hicks took a close interest in the Soil Association's Haughley Experiment, but considered the alterations to it in the early 1970s disastrous and by 1973 was describing himself as a 'disappointed' member of the Association. Two years earlier, at the launching of the Australian Heritage Society, his tone had seemed closer to despair than to disappointment, as he considered the environmental damage and unsustainability of an economy based on fossil fuels. In 1950, he had begun his Sanderson-Wells Lecture with Blake's mystical stanza 'To see a world in a grain of sand ...', whereas in 1971 he concluded his talk *Ecology and Us* by quoting Shelley's bleak vision in 'Ozymandias'. His final major contribution to the organic movement was *Man and Natural Resources*, published in 1975, the year before he died.[37]

Dental scientists

The role of doctors in the organic movement's development is more significant than that of dentists, but the contribution of dental science nevertheless requires mention.

What McCarrison was to organicist doctors, the American Weston Price was to dentists; indeed, his range of evidence was much wider than McCarrison's, as he had travelled all over the world studying and photographing the jaws, dental arches and teeth of peoples as yet untouched by refined, industrialized diets, and comparing the superior dental health of those who had kept to traditional diets with those who had adopted a western regime. His book *Nutrition and Physical Degeneration*, first published in 1939, was re-issued in 1945 and has remained one of the organic movement's canonical texts. Coming across it by chance in his local library around 1959 or '60, the young Michael Rust was so struck by its arguments that he was drawn to the Soil Association and committed himself to becoming an organic farmer. He was active in the Association's Leicestershire and Rutland Group, in which the dentists Everard Turner and Kenneth Rose were also prominent. When the Price-Pottenger Foundation re-issued the book in the early 1970s, Robert Waller picked it as the Association's Autumn Book Choice, describing it as 'a foundation book' for the organic movement. It retains its status today, Michael Pollan devoting several pages of his book *In Defence of Food* to Price's work.[38]

Another important overseas influence was the New Zealand dentist Guy Chapman, who between the wars had closely studied the effect of western diet on the Maoris, reaching the same conclusions as Weston Price: that, in contrast to their traditional diet it had a thoroughly deleterious effect. He was in touch with Albert Howard as early as 1930, and he founded the New Zealand Humic Compost Club (subsequently re-named the New Zealand Organic Compost Society) five years before the Soil Association was established. From the mid-1940s onwards he spent several years studying the effects of an industrialized diet on native health in Tahiti and Samoa, finding that they were consistent with the effects on the Maoris. Chapman had also undertaken dietary experiments at New Zealand boarding schools; these provided evidence that dietary changes — in particular, the addition of raw foods and compost-grown vegetables — were followed not just by a marked reduction in dental decay but by a decline in infectious diseases such as mumps, measles and colds.[39] Chapman was a broadcaster from 1935 until 1961 and received a huge postbag from people who followed his recommended diets and believed themselves and their children to have benefited as a result.

In Britain, the newly established Soil Association could claim as a founder-member perhaps the country's outstanding dental scientist, Sir Norman Bennett; regrettably, he was already in his mid-seventies and died in 1947. There were, however, two younger members of the dental profession who were active for many years: the above-mentioned Everard Turner and Kenneth Rose. Turner, another Soil Association founder-member, ran a dental practice in Leicester from 1924 till 1968. He was always interested in preventive treatment, followed the development of organic ideas before the war, and found in Eve Balfour's *The Living Soil* the philosophy towards which he had been groping. In 1944 he wrote a letter to the *British Dental Journal (BDJ)* on 'Soil Fertility and Dental Caries', which drew a number of favourable responses, including one from Bennett himself, and in 1946 he and E. Brodie Carpenter had a letter published in the same journal drawing attention to the existence of the Soil Association. The following year, Turner spoke to the British Dental Association's AGM at Bournemouth on 'Nutrition and Dental Health', and arranged for Sir Albert Howard to attend and participate in the ensuing discussion. A Soil Association Council member from 1949 till 1962, Turner frequently contributed to *Mother Earth,* as well as to professional journals and the national press. He was instrumental in producing a 1954 memorandum to the British Dental Association (BDA) on 'The Dental Health of Children': this was in fact a general statement of the Soil Association's philosophy of health and an advertisement for the Haughley Experiment. The Association offered to assist the BDA should it feel disposed to undertake research into the difference in dental health among children fed, on the one hand, on organically grown foodstuffs, and, on the other, on conventionally grown food. Turner also responded in the *BDJ* to criticisms of the

1957 public declaration on health referred to earlier. At the Soil Association's Attingham Park conference in 1959, he and Dr Kenneth Vickery proposed a 'Food and Health Experience Survey', one aspect of which would be a statistical analysis of tooth decay to provide a 'health barometer'. (Nothing seems to have come of this idea.) Turner continued active in the Leicestershire and Rutland group until the early 1970s, and was mentor to Kenneth Rose.[40]

Rose, who served in the Army Dental Corps throughout the Second World War, read Turner's letter in the *BDJ* in 1944 and contacted him on moving to Leicester in 1946. Together they formed the Soil Association Leicester and Rutland Group. Rose was a family dentist who, with his wife Olive, ran a two-acre fruit and vegetable garden. They were also regular visitors to Haughley, where their sons helped with farm work. Kenneth Rose was a founder member of the McCarrison Society and its Treasurer until 1984.

We should also mention Bernard Cooke, who regarded 'Dental Disease as a Yardstick' of a diet's value and a person's general well-being and wrote a long essay of that title for *Mother Earth,* in which he drew attention to wider environmental factors adversely affecting health: air pollution in particular. The essay ranges far more widely than the title indicates; it is a truly 'organic' piece of writing, demonstrating the inter-connectedness of soil, food, environment and waste. Cooke was, in Everard Turner's words, 'a good ecologist' who drew a comparison between the function of dental bacteria and of microbic activity in general. Dental health did not consist in destroying oral microbes, which lived 'in mutually beneficent symbiosis with their human host so long as the latter [was] healthy'. But if the host's diet was deficient in proteins and minerals, the microbes would compete with the host, de-mineralizing the tooth enamel. Cooke's message is thus a variation on a familiar theme: prevent decay by eating a whole-food diet, rather than suppressing the organisms with toothpaste and mouth wash. Cooke contributed a major article on sugar to the Soil Association's journal in 1974 and the following year launched an assault on 'The Failings of Med[i]cine', condemning the medical profession for its neglect of nutrition, its emphasis on specialization, its preference for relieving symptoms over promoting health, and its irrelevant academic training. He quoted the surgeon Laurence Knights, who described his own work as 'a scientific repair service for conditions which, for the greater part, sane living would have prevented', and commented: 'How very true this is of dentistry!'[41]

Health and industrial food

When *Living Earth* merged with *Food Magazine* for a while in the 1990s, some Soil Association members objected to what they perceived as the latter's negative and critical tone. One wonders whether these members had only just joined the Association, for the organic movement has throughout its history directed a barrage of criticism against the food industry (and conventional science, medicine and agriculture), despite the Soil Association's claim to be 'a purely positive Association'. To use a less militaristic phrase, the organic movement has kept a persistently watchful eye on the food industry's excesses. In the first decade of the twenty-first century, when it seems to be assumed that there should be an organic equivalent of every conventional food product, it needs to be borne in mind that for many years the organic movement emphasized the importance, not merely of organically produced ingredients, but of whole food, grown in humus-rich soil and eaten while still fresh. This was the diet to which humans were adapted, and any modification of it, through the sorts of processes described in Chapter 1, would tend to undermine health. That chemical fertilizers might be contributing to the spread of cancer in the western world was a view held by, among others, Edgar Saxon and the Southern Rhodesian farmer J.M. Moubray, one of Howard's early disciples. In the 1980s, Peter Segger considered the evidence on the nutritional effects of artificials to be only circumstantial, but he recognized that there was much debate about the effects on human health of nitrates in the water supply and on animal health of excess nitrogen and phosphates in the soil. The cancer therapist Max Gerson believed that pesticides contributed to the increase in that disease; by the mid-1990s they were being blamed for a decline in sperm count. Genetic damage, birth defects and interference with the immune system have also been identified as possible effects. Exposure to organophosphate pesticides (OPs) appeared to lead to changes in brain function, eye diseases, muscular problems and mental illness. Since such pesticides were introduced in the 1950s, wrote Eileen Fletcher, conditions like dyslexia, allergies, eczema, asthma and hyperactivity had become markedly more common, while MS and ME were no longer rareties. The scientist David Hodges drew attention to research into the effect of agri-chemicals on the health of farmers, which suggested a connection between their use and death from cancer in rural areas, while even the de-mythologizers of 'scares', Christopher Booker and Richard A.E. North, grant the serious health hazards created by OPs. Obesity, which might be imagined a relatively recent concern, had already been noted in the USA fifty years ago, while *Health and Life* magazine reported in 1961 that Cheltenham schools were providing fresh fruit at break-times in order to reduce the risk of children developing the same condition. It did not arouse the level of anxiety that it has come to do over the past ten years or so, scarcely featuring in Soil Association journals.[42]

The issue of food additives and their possible effects attracted much more interest, and is one which can be traced considerably further back than the publication of *E for Additives* in 1984. In 1956, the USA publication *Consumer Reports,* organ of the Consumers Union of the United States, questioned the safety of at least 150 widely used food additives, and two years later, in the same journal, Dr Harold Aaron hoped that Congress might finally be about to pass a bill to protect consumers from chemicals in food, after hearings which had been going on for eight years. While conceding a good deal more than the organic school would have been prepared to on the indispensability of chemicals, Aaron doubted the need for '49 different cake mixes manufactured with competing chemical additives'. These items anticipate the detailed coverage which *Mother Earth* would devote to the topic of additives during 1959 and 1960, and the publication in the latter year of Franklin Bicknell's *Chemicals in Food.* Interest then abated, but surfaced occasionally, for instance in Aubrey Westlake's *Life Threatened* and at Cdr Stuart's Soil Association Week in May 1969, when the food scientist Professor John Hawthorn defended additives and was taken to task by the microbiologist Harry Walters. During the 1970s the topic stayed fairly quiet, though the Sams brothers' *Seed* magazine posited a connection between additives and anti-social schoolchildren, and in 1977 Sidney Alford contributed an article to *The Ecologist* on the need for additives to be controlled. The same year, Sally Bunday founded the Hyperactive Children's Support Group, and her article on additives, which she believed a major factor in hyperactivity, in the *Soil Association Quarterly Review*'s Winter 1982 edition was the precursor of several articles on the topic and of the establishment in 1988 of the Soil Association's Food Additives Campaign Team.[43]

As always, though, the organic movement was up against the power of the corporations and the global economic system, conceding in 1993 that the General Agreement on Tariffs and Trade (GATT) made the increased use of additives more likely. On the other hand, a series of food scares, starting with BSE and salmonella in the late 1980s, and outrage in the mid-1990s at the live export of veal calves, provided the organic movement with more ammunition against what was increasingly condemned as a manipulative and unethical food industry. The more forensic approach of *Food Magazine* gathered evidence of a system whose food — if that was even the right word for it — was less suited to human beings than to the *homo syntheticus* whose advent G.E. Breen had feared way back in 1953.[44]

But the organic movement could not have benefited from the enormities of the industrialized food system without having its own positive alternative to offer. We shall concentrate here on three perennial themes of organic health and nutrition: bread, vegetarianism and herbalism.

Positive alternatives

The wholemeal staff of life

We saw in Chapter 1 how during the 1950s the milling industry at last freed itself from wartime standards of production and was able — as the organic movement interpreted it — to place its desire for profit above concern for the people's health. The de-naturing of bread, a staple of our diet and a food carrying rich cultural and religious significance, symbolized all that was wrong about the food industry's attitude to its products. Lionel Picton, finding only a prophetic utterance adequate to the state of affairs, quoted the biblical lament: 'I will break the staff of your bread ... and ye shall eat, and shall not be satisfied'. The agricultural chemist Dr Hugh Nicol, who would later draw closer to the organicists, considered such an outlook 'vitalistic', harking 'back to the good old days of stone-ground flour', and argued that it was oversimplified; but the value of organically grown, stoneground, wholemeal flour is one of the most strongly held tenets of the organic philosophy, and those who have advocated or, better still, produced such flour are among the movement's most respected figures.[45]

Among the earliest were successive generations of the Allinson family. Dr T.R. Allinson was crusading against roller mills as early as the 1880s, so vigorously that in the end he was struck off the medical register. In 1892 he acquired an interest in a mill at Bethnal Green in East London and had the flour there ground to his strict specifications; the business survived his death in 1918, and his son C.P. Allinson joined the Soil Association and contributed regularly to *Mother Earth* during the 1950s and early '60s, and also to *Health and Life*. Joy Griffith-Jones, in the final issue of *Span*, reported on a visit to the Allinson mills at Castleford in Yorkshire, where members of the Soil Association's Yorkshire Group were shown round by a Mr Frank Cooper, an Allinson's employee for fifty years. 'The end result ... [was] a stoneground flour containing all the health giving essentials of the wheat berry with nothing taken out, and just as important, nothing added'.[46]

For those who wished to produce their own bread, Derek Randal, a young assistant of Newman Turner, offered advice in an early issue of *The Farmer*, insisting that only one hundred per cent wholeness was adequate. Turner was advertising his own wholewheat flour by the end of the 1940s. The Leicestershire and Rutland Group, which attached importance to public education on nutritional questions, realized that such education must be backed by practical examples and so arranged with the owners of the mill and bakery at Rearsby, Mr and Mrs Mapperson, to produce one hundred per cent wholemeal bread, using wheat grown by Friend Sykes. By 1954 the demand was averaging two thousand loaves a week. At Huby, near Leeds, was the biodynamic farmer Maurice Wood, whose flour was entirely free from the addition and subtraction we earlier identified as typical features of industrial foodstuffs. His advertising leaflet proclaimed that 'HUBY FLOUR IS 100% WHOLEMEAL FLOUR *containing every particle of the wheat grain*' [emphasis in the original]; it explained the methods he used and the health benefits of the whole grain, and provided a recipe for baking the Huby Home-made Wholemeal Loaf: 'It's so easy without kneading'.[47]

Probably the most widely known of the various recipes for wholemeal bread was the Grant Loaf, named after its nutritionist inventor Mrs Doris Grant. Dating back to the 1944 publication of her book *Your Daily Bread*, the Grant Loaf was still being praised, in the *Soil Association Quarterly Review*, in the mid-1980s. Grant had been cured of arthropathy (joint pains) when in her twenties by adoption of the Hay Diet, and developed uncompromising views on food which she expressed in a long series of books. She held that white bread should have no place in the diet of those who cared for their health, and that housewives who fed it to their families were irresponsible. Grant is one of those figures who represent continuity in the organic movement: in the 1940s she was an associate of Howard, Picton and Lord Teviot; she was a friend of Newman Turner and spoke at the tenth anniversary dinner of *The Farmer;* and she was on the first Council of the HDRA. In 1973 she updated *Your Daily Bread*, re-titling it *Your Daily Food* and offering it as a means by which 'the housewife' — a figure by now under pressure from 'vociferous advocates of woman's [sic] lib', to use the phrase of Walter Yellowlees in his review of the book — might safeguard her family from the threats to health posed by refined foods and additives. Grant drew attention to the cultural dimension of cookery, agreeing with *The Ecologist's Blueprint for Survival* that it should be regarded as an art rather than a form of drudgery. In particular, there was satisfaction to be had from making home-baked bread, an activity which could go beyond the cultural and touch the mystical. Grant ended her chapter on 'Real Bread' by quoting a passage from a 1956 issue of *The Lady,* in which a certain Pamela Pickering likened the experience of baking to being 'a Stone-Age-Woman-cum-High Priestess': 'a feeling we women are meant to experience now and then' and a return 'to the elemental and basic principles of life of which bread-making is symbolic'. It should be noted, though, that despite repeating such high-flown notions from a somewhat precious magazine, Grant made her recipe feasible both for those who were short of time and those who were short of space.[48]

While Walter Yellowlees saw the McCarrison Society as a David opposing the Goliath of the food industry, Mrs Grant preferred the image of herself and her followers as 'Saint Georginas' fighting 'to slay the Chemical Dragon'. One of her most formidable Saint Georginas was Margaret Brady, a physiologist who became concerned about nutrition when a young mother and who saw the whiteness of an industrially milled loaf not as a sign of purity, but of 'the ghastly pallor of death'. A believer in unprocessed dairy products, raw fruits, nuts and salad, Mrs Brady wrote on nutrition and on children's health, being further enlightened on these topics by Eve Balfour's *The Living Soil*, serving for many years on the Soil Association Council and contributing to *Mother Earth* and later Association journals. In 1961, Grant and Brady contributed to a three-handed discussion in *Mother Earth* on breadmaking, being joined by Cyril Allinson. Brady was in fact noted for her displays of bread-making, and her skills were preserved for posterity in the film she conceived and scripted: 'Our Daily Bread: A Comparison between the food value of Wholewheat and White Bread', which was made with financial support from Allinson. Mrs Brady's scientific training made her aware of the importance of establishing that organically grown food really was superior to conventional: in a debate on the Haughley Experiment at the 1969 AGM of the Soil Association, she was bold enough to suggest that the organic movement did not know that this was so, and urged that the Experiment's organic section needed to be of the highest quality for its products to be properly assessed. She believed that the education of housewives, schools and hospitals, and the effective marketing of wholefoods, were essential features of the Association's work.[49]

One problem of bread-making which caused some debate was the lack of flour made from English wheat: much less than half the flour used in Britain's bread originated from home-grown crops; but an article in the Soil Association journal assured farmers, bakers and consumers that this situation could be improved. Hugh Coates, who farmed near Aylesbury and chaired the Soil Association's Organic Standards Committee, took up the issue a couple of years later, arguing that Britain could save on imports if the government would institute a policy designed to encourage the growing of milling wheat. As so often, vested interests were involved: since the milling and baking groups were based on flour mills at ports, they liked to claim that English wheat made poor-quality bread — a claim which Coates rejected. Coates produced his Springhill bread from his own organically grown wheat, having become interested in milling after attending a baking course in the Black Forest in 1971, during which he learned the merits of freshly milled flour. He imported small-scale mills and became an agent for one particular make of electric mill, and a connoisseur of others. In 1973 he bought a small bakery and proceeded to build up a successful business in wholemeal flour.[50]

The work of Nick and Ana Jones has been considered in Chapter 2; two other successful milling and baking enterprises require mention: the Doves Farm brand of Michael and Clare Marriage, who established their business at Hungerford in Berkshire in 1978, and Andrew Whitley's Village Bakery in north-west England. Whitley's work was noted in *New Farmer and Grower* in 1994 by Bill Starling, at that time Britain's only full-time trader of organic grain, who worked for Gleadalls of Gainsborough, Lincolnshire. Starling described how Whitley had been inspired by E.F. Schumacher, Ivan Illich and John Seymour, and how his experiments in baking led him to buy wheat from Sam Mayall: another instance of continuity between different generations.[51]

Vegetarianism

In this writer's experience, it is quite common to be asked, when you say you are interested in the organic movement: 'Are you a vegetarian?'; and the questioner generally expresses surprise when you say you are not, so you explain that organic farming is based on a cycle of nutrients in which the elements are Soil, Plant, Animal and Man. Of course, it is more complex than this. Iain Tolhurst, one of the leading growers of our time, is a vegan, while the Organic Research Centre at Elm Farm is undertaking research into stockless systems; and, historically, the organic movement has always contained a vegetarian strand. This has never been central, but as the ranks of vegetarians have included Eve Balfour, Edgar Saxon, the Lattos, Doris Grant, Margaret Brady, Juliette de Bairacli Levy and Craig Sams it can hardly be dismissed as irrelevant. Then again, in her detailed study of the organic movement as it was in the early 1990s, Anna Ashmole noted the low profile of vegetarianism. Certainly it has been a persistent subject of debate, and has aroused some strong feelings: Craig Sams recalled that when he first met Patrick Holden, Holden wanted to know if he was 'a bloody vegetarian'. For those who were hoping to steer the organic movement into the mainstream, vegetarianism may have carried unwelcome associations with hippiedom or faddism.[52]

4. THE ORGANIC ALTERNATIVE: HEALTH AND NUTRITION

Although vegetarianism scarcely features in the first twenty years of *Mother Earth,* arguments about its validity can be found in Edgar Saxon's *Health and Life.* H.H. Jones, Secretary of the Vegetarian Society's Manchester branch, wrote to the journal in June 1948 on the wastefulness of flesh foods, arguing that dung was unnecessary as a fertilizer, vegetable compost being just as good; he cited in support the work of the biodynamic gardener Maye Bruce. Philip Oyler disputed this: he had put it to the test forty years earlier and learnt that growing green-stuff for composting wasted land. He also pointed to the example of the Chinese, who, though largely vegetarian, made use of animal manures in their composting. The debate was renewed a year later when Alan Albon, a smallholder who also worked on a 40-acre farm, put the case for mixed farming and suggested that getting rid of animals would lead to increased use of artificial fertilizers. Another smallholder, Roy Bridger, who ran a croft in north-west Scotland and wrote about his own experiment in self-sufficiency long before John Seymour appropriated the phrase, pointed out that livestock were the most efficient means of improving rough grazing and that Weston Price had found some of the healthiest primitive tribes to be flesh-eaters. He expressed his views on the relationship between humans and animals in an eloquent passage worth quoting in full:

> Man and animal have grown up together and are still inseparable. That one species should serve another's needs, that pain is unavoidable, that each individual life comes to an end, these are the uncrossable frontiers of the terms of existence. It could be added that there is a world of difference between the sympathetic disposal of livestock by the perennial husbandman and the reeking mass slaughter-houses of the modern top-heavy densely populated centres of population. What the sane food reformer or lacto-vegetarian is really in revolt against is not true animal husbandry but the unbalanced economy of Metropolis.

This drew a substantial and typically extreme response from Dion Byngham, who had returned to vegetarianism after a lapse. Having claimed that about ninety percent of UK agricultural land was used to grow food for livestock, he fulminated: 'If men cannot exist without an inch of ex-hide encasing each foot, four or five layers of wool covering their skins, and beef, bacon and mutton to fill their bellies, perhaps they ought not to exist at all ... in such a climate. Their very existence under such conditions is a flouting of nature.' This was followed by references to the nation's 'appalling over-population'. A certain T.W. Cheke provided a more balanced analysis of the problems, apparently from a position sympathetic to veganism; and so the controversy rumbled on. When Saxon died in 1956, the journal *The Vegetarian* paid tribute to him. Saxon was a tolerant lacto-vegetarian, who claimed never to have condemned meat-eaters.[53]

There is another fact worth noting from this early period: when Richard St. Barbe Baker launched the environmentalist manifesto the *New Earth Charter* in 1949 — and at the risk of labouring a point, we can note that it preceded *The Ecologist's Blueprint for Survival* by more than twenty years — the occasion on which it was presented was a New Earth Vegetarian Luncheon in London.[54]

Health and Life continued with sporadic coverage of vegetarianism after Saxon's death. In March 1957 it reported on a 'Brains Trust' on the subject, organized by the Reading Vegetarian and Food Reform Society, whose object was 'to advocate, extend and organise vegetarianism and food reform for the benefit of the community'. The meeting was chaired by Dr Gordon Latto, whose brother Douglas was on the panel along with Margaret Brady — looking 'most attractive in the prettiest of pink bonnets', according to the anonymous reporter — and Doris Grant, 'equally attractive in a bonnet of a lovely shade of blue'. One questioner asked whether the country would be over-run with animals if its population turned vegetarian: Douglas Latto's reply concluded that the matter could safely be left to Nature. *Health and Life* did not consider a vegetarian diet essential to good health, and a fair-minded article by Alfred Le Huray in its September 1961 issue suggested that a diet containing protein from animal sources, if containing all the essential nutrients, could result in good health. Nevertheless, the writer, from his own fifty years' experience, believed a carefully balanced vegetarian diet best suited to the needs of *homo sapiens*. This same issue contained a long editorial on the relationship between vegetarianism and Christianity, looking critically at the dangers of spiritual arrogance and divisiveness lurking in all forms of 'isolated virtue' and reminding readers that Edgar Saxon had himself condemned the vegetarian ethic as a negation of the Christian idea of Sacrament. *Health and Life* helped promote meetings of the London Vegetarian Society; one advertisement informed readers of public lectures to be given by Bertrand Allinson, Douglas Latto and the herbalist Claire Loewenfeld.[55]

By the end of the 1960s, the Soil Association was addressing the issue of vegetarianism in its monthly newspaper *Span*. The discussion was essentially an updating of that in *Health and Life* twenty years earlier, though Ronald Lightowler of the London Vegetarian Society was more prepared than H.H. Jones had been, to compromise on the use of animal manure for augmenting soil fertility. Michael Rust, like Philip Oyler, pointed to the valuable lessons offered by oriental agriculture and of the necessity for all wastes to be returned to the soil. Other correspondents aired their anxieties about the likely effects of vegetarianism on agriculture, but Lightowler reassured them that mass conversion to vegetarianism was a distant prospect, dependent on the acceptance of 'a truly spiritual view of life [which saw] sub-human creatures [as] evolving units of consciousness'. The topic surfaced again in the early 1970s, with a letter from an E.G. Cawdrey condemning the 'savagery' of meat-eating and urging vegetarianism because it required less land space to feed people, and one from Geoffrey Rudd of the Vegetarian Society denying the dietary value of meat. A 'non-dogmatic, not over-strict vegetarian', Evelyn Scott Brown, was puzzled by the presence of vegetarians in the Soil Association, seeing the two aims of farming organically and eating non-animal foodstuffs as incompatible. She drew replies from the pseudonymous American 'Lucius Dunius Moderatus', who encouraged her to consume fresh milk and eggs for their protein, and from a vegan Soil Association member, Robert Colby, who supported the Association because its philosophy was far closer to his own than was the philosophy of orthodox farming. Colby had attended the Ewell Technical College course on biological husbandry in July 1972, and was presumably identified as among the sixty per cent of students on that course who were vegetarians. Peggy Goodman, who wrote a regular cookery column for *Span*, felt that the cultural tide was turning towards vegetarianism as housewives, with their 'down to earth plain common sense' looked for cheaper alternatives to expensive meat from badly treated animals. Scientific evidence and philosophical pedigree were on its side, and, although the Soil Association still did not seem to regard vegetarianism as somehow quite respectable, the

Association should remember that it had been itself so regarded until very recently.[56]

Whereas the Soil Association's journal debated vegetarianism in the 1970s, the Sams brothers' 'alternative' magazine *Seed* unequivocally promoted it. Attacking 'The Meat Protein Myth' in its third issue, it put the case for vegetarianism on the various grounds of domestic economy, health benefits, protection of other life-species and fairer worldwide distribution of available protein foods. Negative and positive arguments were adduced: in contrast with the disturbing details of how meat is produced, the anonymous writer presented the attractions — nutritional and economic — of natural foods, and stated that the average consumer in America was taking around twelve per cent more protein than her body required, while two-thirds of the world's population went short of protein. The message was driven home visually by a line-drawing advertisement for the Sams' company Harmony Foods, which contrasted a Viking, about to gnaw on the leg of a dead animal lying at his feet, with a medieval couple standing by a sheaf of corn, the woman demurely holding a posy of flowers. The magazine was soon at odds with the Vegetarian Society, shocked by the latter's endorsement of protein foods made from bacteria feeding on yeast in crude oil. Just as organic cultivation is about far more than not using chemicals, so vegetarianism, argued the editorial of *Seed*'s fifth issue, is about far more than not eating animal products, particularly when the substitute offered is an invention of the chemical industry. 'Vegetarians ... know that there is enough natural protein food available for all the world's hungry if people are allowed to eat basic grains, rather than *inefficiently* [my italics] feeding them to animals for the production of meat'. One unusual article was by Gabriel Pearmain, a vegetarian who spent a month working in a bacon factory and found nothing in his experience there — especially attitudes to hygiene — which encouraged him to abandon his dietary regime. Pearmain's article was backed up with evidence from Professor Curtis Shears and Dr Alice Chase, that fears of a vegetarian diet failing to provide adequate protein and amino acids were quite unfounded. If the worst came to the worst, one could always obtain the latter by growing sprouts in the boot of one's car.[57]

Seed's advocacy of vegetarianism began with practical arguments based on economy and health; but by 1975 one reader — though describing herself as typical of many young people — was presenting an esoteric, mystical case for rejecting meat-eating. Having experienced a 'shift in consciousness', Leah Leneman began to identify with the pain and fear of animals, and realized that she 'could never feel at one with any creature whose flesh I would eat'. A child of the Aquarian Age, 'the age of oneness', Leneman thought it no coincidence 'that almost every New Age religious sect or community is vegetarian'. An advertisement for the Vegan Society next to her letter manifested the same spirit, inviting readers to 'Live in harmony with the compassionate heart of the universe' by going 'the Vegan way so all men can be fed and wide acres be left wild'. Or, one might follow Malcolm Horne and go even further, becoming a fruitarian and 'living off the fruits of the trees and bushes of the earth', which might include nuts but generally meant 'a diet consisting solely of fresh, sweet, juicy fruits'. Mary Heron, who worked at Haughley in the early 1960s, recalled one Soil Association member who pushed fruitarianism to a *reductio ad absurdum* by eating nothing but oranges, a fact fully evident from her skin colour. However, at this point we reach the wilder shores of organicist dietary theory and should re-trace our steps.[58]

It is fair to say that the Soil Association journals were more balanced than *Seed* in their coverage of vegetarianism; the debate surfaced in them periodically during the 1980s and early '90s. One particularly interesting contribution to it came from an anonymous agricultural vet, a non-vegetarian who treated vegetarianism seriously and sympathetically, and examined the question of food production and food shortages both from a planetary perspective and from his own perspective of someone who worked in a rural area where farming was at the heart of community life. The writer saw an analogy between the mixed farming which he would have liked to see revived, which would offer the soil a mixed diet of wastes, and a mixed diet for human beings. In contrast to some of the uncompromising contributors to *Seed*'s correspondence columns, he believed it possible 'to get the best of both worlds, so long as the animals ... have a reasonably stress free life and a humane end.' Such a moderate conclusion, though, was inadequate to relieve the tensions between the organic movement and strict vegetarians. The sticking point was perhaps not so much health, as ethics: certainly, this was what Peter Barclay of the Vegetarian Society argued, when vegetarianism was on the increase in the mid-1980s. While there were good health reasons for avoiding meat, young people in particular were turning vegetarian because of the conditions in which animals were being reared and slaughtered, and because the world's plant food resources were being expended on animal feed while humans starved. The organic movement can respond to the first of these objections by ensuring humane treatment of animals, but the second is harder to deal with — unless, perhaps, one takes the sinister view that it does not matter if people starve to death, because the earth is over-populated anyway. The population issue was subsequently addressed in an anti-vegetarian letter from Alexander Croal, while G.N. Hudson, evidently a vegan, put the vegetarian case against the meat trade with a wealth of gruesome detail. These letters appeared in the *Soil Association Quarterly Review* in September 1985, and seem to belong to a remote age, when two letters, in small print, could fill an entire page and their authors were allowed to develop a line of argument and support it with detailed evidence. One's heart aches ...[59]

As the 1990s opened, the debate — now in at least its fifth decade — continued as Bill Starling, Chairman of British Organic Farmers, objected to the appearance on *Living Earth*'s 'Children's Page' of an advertisement for the Vegetarian Society which, in his view, showed 'a regrettable lack of sensitivity towards organic farming' by ignoring the attention that organic standards paid to animal welfare. Starling considered vegetarianism a luxury, believed that 'Meat in moderation is a natural part of a balanced human diet', and found it ironic that any organic grower should be vegetarian. Starling's wish for a vigorous correspondence was amply granted as vegetarians attacked his views in the journal's next issue. Their letters provide a varied though not comprehensive defence of the view that one can, without any inconsistency, be both a vegetarian and a member of the organic movement, and can be summarized as follows. Human beings are not carnivores (Jon Wynne-Tyson's definitive work on vegetarianism, *Food for a Future,* is cited as an authority) and all dietary essentials can be obtained from sources other than meat, to the consumer's economic benefit. There is nothing ironic in using manure to maintain fertility; doing so and wishing to eat the animals that produce it are quite different matters. Grass is not the primary alternative to meat; a programme of re-forestation would enable the creation of Permaculture systems, 'sustainably provid[ing] nutritious foods for direct human consumption'. From the Elm Farm Research Centre,

Dr Susan Millington wrote to say that organic farming is based on sustainable rotations, and that she believed that stockless rotations could be feasible; Elm Farm was conducting a long-term experiment to investigate this. One correspondent, returning by implication to the message of some of the earliest organic texts — G.V. Poore's *Essays on Rural Hygiene*, first published in 1893, and F.H. King's *Farmers of Forty Centuries*, first published in 1911 — urged the use of carefully husbanded human wastes as a substitute for animal manure. Although the ethical argument for vegetarianism — the callousness of growing crops for animals while humans starved — was missing on this occasion, the spiritual case for it was present: all life is sacred, and humans, the link between the physical and spiritual realms (as in the Elizabethan concept of the Great Chain of Being), should be protective and compassionate towards animals. While the organic treatment of animals was a welcome small step forward, vegetarianism and veganism were 'leaps in seven-league wellingtons'.[60]

Herbalism

Given that the organic approach to cultivation and health places a premium on simplicity, in contrast with the complexity of conventional agriculture and health treatment, it is appropriate that the movement should value the contribution of 'simples', or medicinal plants. Interest in herbalism has always been a feature of the organic approach to health, whether animal or human, and we should note straightaway that at least three outstanding herbalists paid tribute to Sir Albert Howard: Newman Turner was one. Juliette de Bairacli Levy — of whom more below — wrote of her admiration for Howard, who had encouraged her study of Gypsy herbal lore and advised farmers to write to her for cures. She dedicated her book *Herbal Handbook for Everyone* (1966) to his memory. Mrs C.F. [Hilda] Leyel likewise dedicated *Green Medicine* (1952) to the memory of Howard, '*whose life work on plants and the soil ran parallel with, and was complementary to, that of the Society of Herbalists and Culpeper House, to both of which he gave much encouragement and practical help as a Member of the Advisory Committee*' [italics in original]'. One of Howard's basic themes, Mrs Leyel wrote, was that England was a grass country, 'and that wheat, the most important of the grasses, should be grown here in large enough quantities to feed us all.' He had 'a natural understanding about herbal medicine, which he believed to be the right treatment for man and animals.' The Scriptural maxim of the Society of Herbalists was 'The leaves of the tree are ['were', in the original] for the healing of the nations' (*Revelation* 22 v.2). Like so much in the organic movement, one of the main justifications for herbalism was religious. Mrs. Leyel quoted from the Apocryphal book of *Ecclesiasticus* (38 v.4): 'The Lord hath created medicines out of the earth and a wise man will not abhor them'.[61]

Claire Loewenfeld, who established the Chiltern Herb Farms company, regarded herbs as 'little medicine chest[s] of active substances', and, as one would expect of a committed organicist, recommended using them whole. She felt that these 'mysterious concentrated plant[s]' should not be closely analysed, since it was probably better 'to leave nature alone'. Mrs Leyel was similarly holistic, describing the herbs used by herbalists as 'living': by which she meant that they contained 'the enzymes which are the living part of the cells, and the hormones which influence the blood. Nothing is isolated from the plant'. But she warned that poisonous plants are best avoided, and took issue with the preference for them which had resulted from chemists being granted the privilege of dealing with them, thereby assuming a bogus superiority over herbalists. Orthodox medicine did not prescribe herbs in their natural state, and tended to combine them with drugs.[62]

One significant difference between Mrs. Leyel's introduction to her book and Claire Loewenfeld's introduction to hers is the extent to which the latter, published only twelve years later, relates an increasing national interest in herbs to dissatisfaction with the products of food technology, perhaps underlining the rapidity with which the food industry changed during the 1950s. Loewenfeld's positive case for herbs was in part based on a critique of the conventional contemporary diet: an 'interference with natural food substances through artificial aids to cultivation, industrialization and mechanization' which had led to a loss of natural quality and flavour and thence to tasteless, boring food requiring the addition of culinary herbs to render it palatable. As a result, there was a revival of herb gardening, which once had been common practice. Herbs took little from the soil, had natural resistance to disease, and produced their own oils, minerals and other active substances, so it was unwise to force them with chemical fertilizers, which were simply unnecessary. In Loewenfeld's view: 'for no other garden ... is compost making and organic gardening of such importance as for the herb garden'; 'the herbs help the compost and the compost helps the herb' — a perfect, and simple, organic cycle.[63]

Herbalism was fairly slow to make its way into the pages of *Mother Earth*, though Pauline Bulcock, who worked at Haughley soon after the war, recalled that Kathleen Carnley, Eve Balfour's companion, was a skilled herb gardener. There was next to nothing in the journal's first fifteen years, and what there was concentrated on the role of herbs in farmers' leys; however, Margaret Brady did briefly review a book by Mrs Leyel. But 'M.B. and J.J.' — almost certainly Mrs Brady and Jorian Jenks — wrote about their visit to Claire Loewenfeld's herb nurseries near Tring, in the January 1962 issue of *Mother Earth*, and praised the value of herbs as specific remedies and in 'wholesome nutrition and preventive medicine'. What was more, herb growing could still be a cottage industry for those who gardened: though the authors also regretted that production and marketing of herbs had not received a fraction of the funding invested in the development of chemical substitutes for them. Mrs Loewenfeld's own commitment to herbs dated back more than thirty years, to her time as a patient, and subsequently a pupil, of the Swiss nutritionist Dr Bircher Benner, when she learned the value of a raw, vegetarian diet and of herbs in making such a diet more appetizing. During the Second World War, she was called in to try the Bircher Benner treatment on seriously ill evacuees from a London hospital, finding that fresh herbs enhanced both a diet's curative properties and a child's willingness to eat even such an unpopular dish as spinach purée. (Whether her successes outstripped those of the cartoon character Popeye remains to be investigated.) Having little faith in the dried and powdered herbs of commerce, Mrs Loewenfeld also set about seeing how modern equipment could be used to preserve herbs with minimum loss of their natural properties. In October 1962 she provided a list of 'Some Helpful Hints' for culinary benefits. Three years later, she wrote a piece for *Mother Earth* on the role of herbs in organic gardening: practical advice which concluded with the assertion that 'Many more herbs should be grown than most people think necessary'.[64]

The Soil Association's publications may have shown relatively little interest in herbalism during the first twenty years, but the topic was covered in other organicist publications. *Health and Life* advertised training courses run by the British Herbalist Union, discussed the value of nettles and sage, and published a piece on herbs and vegetarianism. The HDRA's newsletter was of course packed with material relating to the virtues of comfrey. Newman Turner's journal *The Farmer* contained a good deal on herbalism. The biodynamic movement had its own authority on herbalism in Maria Geuter, and its own nursery selling biodynamic herbs at Hinstock, near Market Drayton in Shropshire, run by E. and A. Evetts, who described it in the journal *Star and Furrow*.[65]

As with vegetarianism, there seems to have been an increased interest in herbalism from the late 1960s, perhaps a result of the rise of the 'counter-culture'. *Span* reviewed books on herbalism, and its gardening correspondent Brian Furner wrote on the topic. The Association's Leicestershire and Rutland Group invited a herbalist to address it. Joy Griffith-Jones of Haughley contributed a column on herbs to the SA journal, while the *Quarterly Review* even published an item on herbalism and astrology. Such an article, though, was really better suited to the more 'alternative' pages of *Seed* magazine, which published a good deal of material on the various benefits of herbs. Volume 2, No. 4 promoted the Bach flower remedies, which Edgar Saxon had advertised in *Health and Life* back in the 1930s, and recommended herbs in animal treatment; later issues offered herbal cures for natural child care and various herbal remedies. Sue Coppard of WWOOF interviewed Ann Warren-Davis, who had started a herbal treatment clinic in Battersea and whose replies provide a link between the organic pioneers and 1970s occultism. Having inherited a non-organic farm in 1949, Warren-Davis had run into trouble with stock disease and was fortunate enough to come across Newman Turner's books, improving her animals' health through compost farming, and improving it still further through the use of herbal strips. She also found herbal medicine effective in curing her husband's ulcer and gallstones, and her son's scarlet fever; she trained at the National Institute of Medical Herbalists. Her husband, she revealed, was a sculptor who had recently built a wood-henge in their garden, 'which turn[ed] out to have been built 'by chance' at the meeting point of two lay [*sic*] lines — but I don't think any of these things are really chance'. Ann Warren-Davis believed that herbalism was effective because it was 'a natural way of creating a balance and getting the body to do its own work; whereas orthodox practice treat[ed] the symptoms of underlying disturbances': another variation on the theme which, as we have seen, can be found in the organicist approach to cultivation and health. By the late 1970s, Warren-Davis' work was considered sufficiently respectable to feature in the *Soil Association Quarterly Review*.[66]

A popular book of this period was Audrey Wynne Hatfield's *How to Enjoy Your Weeds* (1969), which exhibited a similar faith in nature. Sir Albert Howard had believed that weeds were valuable as indicators of soil condition; for Hatfield they were herbs which, if understood, could not only, through composting, help the growth of plants one wanted to cultivate, but might offer health-giving food or drink. By treating all weeds as enemies to be chemically exterminated, one was missing the chance to benefit from the generosity of nature. Everything had its place and purpose in the ecological pattern.

There had always been a market for herbs: Kathleen Hunter at Callestick near Truro in Cornwall was prominent in the organic movement's early years; in the 1970s there were the Dorwest Growers of Bridport, for instance. Among the younger generation of growers, Geoff Mutton made the case for expanding organic herb production, given the 'general trend towards a greater awareness of healthy and flavoursome food' (which Claire Loewenfeld had discerned twenty years earlier), media coverage of vegetarianism and an increasing interest in alternative medicine. Perhaps Mutton over-estimated the possibilities: in the mid-1990s, *New Farmer and Grower* was debating whether there was any potential for market growth in the sale of organic herbs.[67]

Thus far, we have concentrated on the benefit of herbs to human health; to conclude, we need to say something about the use of herbs for treating animals. This was a particular concern of Newman Turner and of his associate Juliette de Bairacli Levy.

Animal health

Newman Turner and Juliette de Bairacli Levy

Turner summed up his ideas on animal disease in one uncompromising phrase: '[T]here is only one disease of animals and its name is man!'

But he based this view on his own experience, having transformed the health of his animals — they had suffered from abortion, sterility, tuberculosis and mastitis — through adopting 'simple natural cures'. 75 per cent of his animals had been unwell, so he adopted the approach which George Scott Williamson had already taken in regard to human illness and wondered what gave the remaining 25 per cent immunity. He concluded that 'the much-maligned bacteria' were not the primary factor in the cause of a disease, but were 'Nature's chief means of combating it'. Much disease resulted from human interference, so the solution was to return the animals as far as possible to an unperverted state of existence and provide them with the requirements of health which would be available under natural conditions. Deficiencies should be corrected only with natural herbs, which Turner used both for medicinal purposes and to build up the soil fertility in his leys. We should bear in mind that he was a trained, practical farmer, sufficiently respected to have his own column in *Farmers Weekly* during the 1940s, who had to survive economically. Had he obeyed a veterinary surgeon who advised him to slaughter his sterile cattle, he would have been ruined; but through risking his own treatment he cured them and created a pedigree herd. Similarly, he found himself able to cure bovine TB through feeding humus-grown food and providing a diet of mineral-rich herbs. Growing confident in his powers, Turner advertised for other farmers' rejected animals and claimed to be achieving regular cures. His journal *The Farmer* paid considerably more attention to herbalism than *Mother Earth* did during the same period, and he established the Herb-royal Animal Health Association, 'to build a body of wider knowledge and experience in the natural herbal methods pioneered by the author': membership open to breeders of cattle, sheep, horses and goats.[68]

Newman Turner worked very closely with the herbalist Juliette de Bairacli Levy, who put her knowledge at his disposal, and we can conclude this section on herbalism by looking — alas, all too briefly — at the career of a woman who is surely among the most remarkable of the personalities who have graced the organic movement. Levy described her early life in *As Gypsies Wander* (1953). From a well-off background, she studied veterinary science at the Universities of Manchester and Liverpool, but disliked the treatment inflicted on animals in the name of science — especially vivisection, of which (along with concentration camps, Nazi treatment of Jews and Gypsies, and the death penalty) she had a horror. (Some of her relatives died in the gas chambers, and, given her comments on Fascist thugs shouting 'Death to the Jews!', one wonders what she made of the Mosleyite Jorian Jenks being Editorial Secretary to the Soil Association.) She began investigating alternative approaches to animal health; through travel and camping she came to know Gypsies ('true children of Nature' and her professors of medicine) and, later, Bedouin Arabs and American and Mexican Indians, and learned much herbal lore from them, encouraged by her mentors Sir Albert Howard (who had himself regarded the Indian peasant farmers as his most effective teachers) and the Hungarian Professor Edmond Szekely. A passionate dog-lover, she applied herbal methods to treating distemper, and supplemented her veterinary work with herb gathering, especially on Exmoor; this formed the basis of a line of supplements for animals, Natural Rearing Products. After serving in the Women's Land Army as a 'Timber Jill' in the Forest of Dean, she spent time with Gypsies in the New Forest and Surrey, and with the horse-traders of the Yorkshire Dales, where she cured thousands of sheep deemed beyond hope by conventional vets. Her time in the New Forest brought her into contact with the Westlake family, who lived near Fordingbridge. She was a 'gardening friend' of Jean Westlake, daughter of the doctor and ruralist Aubrey Westlake, and in 1953 Aubrey Westlake asked Juliette to speak on behalf of the New Forest Gypsies, whose right to roam freely had been curtailed. Other supporters of the Gypsies' rights included the painter Augustus John, the naturalist Brian Vesey-Fitzgerald, and the writer and broadcaster John Arlott.[69]

Juliette de Bairacli Levy provides a full explanation of her 'almost worshipful' attitude to herbal medicine, and, given that phrase, it is no surprise to learn that it is rooted in a religious philosophy: not from the East, but in the much-maligned Judaeo-Christian tradition. Medical plants are God's creation, 'Teeming in the countryside', and 'it shows disbelief in the power of God to pass them by'. The Bible contains various references to their healing power, and Levy regarded them as a sacred medicine: 'The mystics and the great healers such as that ancient Jewish sect the Essenes, of which many say that Christ was a member, were herbalists'. Humanity cannot excel nature in producing medicines, but industrial civilization has separated us from a life which observes natural law. Levy, like Paracelsus, discovered from her own experience that one needed to learn true medicine from peasants and Gypsies, who were unaffected by modernity and yet, in their way, less naïve than contemporary farmers, who 'habitually permit the modern vet. to inject into or feed to animals, substances about which they know nothing, and often enough of which the vet. knows very little': a process which is 'an insult to the Creator'. That such processes had become so commonplace was due to a combination of the farmer's lack of time and the 'stupendous' vested interests of modern medicine. 'Businessmen who have never owned an animal, fatten like breeding toads upon the ailments of farm stock which need not know sickness at all if they had daily access to the herbs of the fields.' Whereas new commercial products are constantly appearing, nature's remedies and their effectiveness are perennial. Levy described the sense of pleasurable achievement to be gained from gathering herbs and restoring animals to health, and found it extraordinary that farmers, who take so much trouble over growing fodder, should neglect to grow crops for treatment of animal ailments. She praised the American writer Louis Bromfield for recognizing that the good farmer should be — among many other things — a botanist. When Levy described science as 'the ruination of true farming', she was rejecting the agri-business approach described in Chapter 1: the desire to conquer nature through technology. Her own studies, though based on experiment and extensive, systematic knowledge, were of the kind that do not count as proper science in the eyes of those who work in laboratories or stand to profit from the application of the discoveries made therein. Levy claimed that she had experienced no failures with herbs, whether in the treatment of pigeons, camels, dogs, sheep, horses, or her own children. The closer one approached to nature's own methods, the greater the likelihood of success. She had made a study of wild deer, never coming across a hind suffering from mastitis of the udder, or any deer with skin disease; these creatures were perhaps for her what the Hunza tribesmen were for McCarrison.[70]

Juliette de Bairacli Levy's use of medicinal herbs leads us to this chapter's final section, about three people with a specialist interest in animal health: the veterinary surgeon Reginald Hancock, and brother and sister Richard and Rosamund Young, Cotswold farmers.

4. THE ORGANIC ALTERNATIVE: HEALTH AND NUTRITION

Reginald Hancock

Most of Hancock's career falls into an earlier period than that which this book covers, but he was still active in the post-war years and published his memoirs in 1952, and there is a shortage of organicist vets to discuss (we saw earlier in this chapter that the McCarrison Society failed to attract as many vets as it would have liked). An outstanding member of his profession — he was Chief Veterinary Officer to the Royal Society for the Prevention of Cruelty to Animals, and broadcast as 'The Radio Vet' — he was an important and unusual figure in Soil Association history. Although Hancock's autobiography, *Memoirs of a Veterinary Surgeon,* sounds as though it might be either narrowly specialist or the forerunner of James Herriot's comic novels, it is in fact a serious and wide-ranging book, containing a long section about his experiences on the Western Front during the First World War, in the Royal Army Veterinary Corps: a rare perspective for a war writer. It also contains a strong element of social history, or, perhaps, of nostalgia for the world prior to 1914. Hancock contrasts 'the simple healthy routine of living usual to the countryman of the Victorian and Edwardian ages' with the comparative slavery of post-1945 Britain, where the populace received 'from the hands of pallid doctrinaires a maintenance ration just adequate to keep us docile and submissive'; he was similarly unimpressed by the NHS, though he evidently admired many advances in medicine.[71]

Hancock was led towards the organic philosophy by his long-term observation of certain farms whose cattle, which had once been afflicted with tuberculosis or contagious abortion, became free of these diseases. The common denominator in these cases, Hancock, believed, was a change in ownership or tenancy, where the incomer reduced the application of artificial fertilizers in favour of muck and compost. This process, it seemed, vitalized the animals' diet and increased their resistance to pathogens. Hancock had formed this opinion before he became acquainted with Sir Albert Howard's writings. Whereas Howard, McCarrison and Scott Williamson had deplored medical science's concentration on the treatment of disease and its neglect of health, Hancock felt that the medical profession in any case knew 'nothing of the ultimate nature of the phenomena embraced in the term "disease"': hence its insistence on allopathic healing, based on a doctrine of opposites. Hancock did not deny that this approach could work wonders, but he felt that there was also a place for the homoeopathic methods of Hippocrates, rediscovered by the German doctor Samuel Hahnemann in the late eighteenth century. Hancock had successfully used both methods in his treatment of animals. He joined the Soil Association in 1948, soon becoming chairman of its Advisory Panel and supervisor of livestock management at Haughley, believing that the experimental work there held the key to animal health and that the Association, 'by brutal long-term persistence' could 'bring the New Look to farming science'. The veterinary surgeon, he believed, could play a vital role in communicating the Soil Association's message, since farmers generally had confidence in vets. One of Hancock's farmer friends had summed it up by saying that, of all the various officials who visited his farm, Hancock was the only one who ever took his jacket off. The practising vet could present the organic case more convincingly to the farming community than the professional scientist could. Hancock believed that most of his colleagues were thinking along Soil Association lines and were sceptical about many of the edicts promulgated by laboratory workers. Howard had referred to such workers as 'hermits' who lacked any practical experience of farming; Hancock shared his belief that the scientist should observe *what actually happened* in the field. 'The veterinary surgeon ... has seen cattle starving on a lush modern ley, fighting their way through the hedges to get at the despised hedgeside weeds. He, and his father before him, knew that field when it was old meadow, and knew that for generations it had successfully fattened beasts innumerable.' But a modern field, heavily treated with chemical stimulants, would soon be causing bloat in the stomachs of the milch cows which had to graze on it.[72]

Rather unusually for a member of the organic movement, Reginald Hancock looked to physics, rather than biology, to provide his working hypothesis for a study of health and disease. Matter being a whirling mass of particles, undergoing attractions and repulsions of a 'frequency at which the mind faints', it followed that health consisted in the harmonious interaction of vibrant atoms. 'How like an orchestra the total picture becomes! .... One tiny error on the part of one instrument's vibrant contribution is discord, i.e. disease.' Hancock reached the same conclusion as Juliette de Bairacli Levy and Newman Turner: that one should interfere as little as possible with the natural order, for fear of disturbing its harmony. To take the case of bovine tuberculosis, for instance: its germs, existing in the cattle as they do in almost all vertebrate creatures, could generally be regarded as harmless, and could be kept so through the establishment of certain straightforward conditions of hygiene: avoidance of overcrowding and the feeding of a 'proper' (humus-grown) diet.[73]

The Youngs of Kite's Nest

A generation after Hancock's writings, Richard and Rosamund Young of Kite's Nest Farm, Snowshill, on the western edge of the Cotswold hills, were putting such methods into effect. Their mother, Mary Young, came from a long line of farmers; her father had used traditional methods of manuring until forced to take artificials during the Second World War. He was never reconciled to them, mistrusting substances which burnt the trailers they were carried in, and continued to use animal manure after the war. The approach at Kite's Nest appears to have been regarded as traditional rather than self-consciously organic, as Richard Young has said that he had never even heard of organic farming until well into the 1970s. In 1969 he was about to study agriculture at Bangor University when his great-uncle gave him his farm to run, and helped him to get started on it. It was a Cotswold farm which had, until then, remained free of artificials and pesticides; Richard now began to use them, but with a sense of unease, noting that trees died alongside polluted winterbournes. And although he soon became a prize-winning barley producer, he was also uneasy about the wider picture in the Cotswolds, with farmers ploughing up grassland in order to cash in on the rising price of cereals. A turning-point came around 1973/74, when Richard Mayall gave a talk on organic farming to the agricultural discussion group which Young attended. Young went on a farm walk at the Mayall family's Shropshire estate and was very impressed by Sam Mayall, who suggested he join the Soil Association. Richard Young did so, openly declaring himself an organic farmer. But at first he was only negatively organic, ceasing to use the NPK approach but not knowing what to do positively. A television programme led him to approach Barry Wookey, whose methods of cereal growing he then followed. Young went on to play a major part in the work of British Organic

Farmers and the production of *New Farmer and Grower*, for which he wrote a piece on organic veterinary standards.[74]

In the meantime, at Kite's Nest, Mary and Rosamund Young continued to avoid the chemical route, as Mary and her husband Harry had done ever since 1953, and to lay the foundations for what has been, in effect, a long-term experiment in animal welfare: an organic showpiece, not in any touristic sense, but all the more significant for being the Youngs' source of livelihood. The family's success was based on the principles which we have seen expressed by Newman Turner, Juliette de Bairacli Levy and Reginald Hancock: allowing the animals to live as far as possible in tune with nature and express their instinctual behaviour. This is of course a complete rejection of the agri-business philosophy of 'animal machines' whose only interest for the farmer is their 'efficiency': an approach which has resulted in ill-health for the animals thus treated and, quite possibly, for the human consumers of their products. Rosamund Young's loving study of the various creatures at Kite's Nest, summarized in her book *The Secret Life of Cows* (2005) and in DVDs such as *The Calf's World* and *The Secret Life of the Farm*, is the work of a gifted naturalist. In its careful gathering of material it can be considered scientific, while its concern for the individual creatures and their special characteristics demonstrate an artistic perception. The Youngs let the animals make their own decisions on where to go, allowing them out in all weathers, and found the cattle choosing different places at which to drink, or feeding on a balanced diet of old and new grasses. In this way, they gained valuable information on the health of their livestock and on the herbs and other plants which they chose to eat in order to compensate for nutritional deficiencies. The animals were at times pleased to eat from trees and hedges. We noted earlier that Reginald Hancock had observed cattle straining to reach hedgeside weeds; Rosamund Young likewise realized the value of verges and hedgerows, writing a booklet, *Britain's Largest Nature Reserve?* (1991) for the Soil Association's 'Countryside in Crisis' campaign, plus various articles in *Living Earth*. Knowledge of individual animals enabled the identification of unusual behaviour on their part, which was likely to indicate a problem of some sort. If this were the case, human intervention might be required, but in general the Youngs' approach was to leave the animals alone and allow them to play, eat and rest as they saw fit. As JOANNE BOWER,

4. THE ORGANIC ALTERNATIVE: HEALTH AND NUTRITION

of the FARM AND FOOD SOCIETY, put it in her foreword to *The Secret Life of Cows:* 'People watch with amazement a television programme on the social life of elephants — their family groupings, affections and mutual help, their sense of fun — without realising that our own domestic cattle develop very similar lifestyles if given the opportunity'.[75] Such an approach might appear, to the orthodox farmer, a form of sentimentality, or indicate economic negligence; but the Youngs found that their approach brought commercial benefits, producing meat of the highest quality, and that an increasing number of people, unhappy with the suspect offerings of intensively reared livestock, were keen to eat it.

****

From an urban family health club in the 1930s and '40s to a twenty-first century Cotswold farm may seem a long way; but the principles underlying the Pioneer Health Centre and the Youngs' approach to farming at Kite's Nest have a good deal in common. Provide the right conditions; allow the living beings to express themselves freely; observe, but do not intervene unnecessarily; and watch health flourish through the balanced co-existence of vitality and order. How simple — and how frustratingly unrewarding for those who profit from other people's anxieties and ill-health.

5. Commerce and Consumers

The organic movement's success over the past decade or more has been largely identified with an increase in the sale of organic produce: so much so, that when the advent of the 'credit crunch' brought in its train a downturn in sales, the movement's critics were quick to predict its imminent demise and its defenders equally quick to argue that the downturn was only slight and that the committed consumer of organic goods would remain committed. That the validity of the organic philosophy and the movement's success should be measured by the fluctuation of consumer sales would have seemed incredible half a century earlier. Writing in 1959, Jorian Jenks described the organic movement as having 'nothing to sell and offer[ing] few openings for industrial or commercial enterprise'. And as late as 1982, the Soil Association's Agricultural Advisor R.W. Widdowson was insisting — in the face of a younger generation eager to promote organic produce in the market place — that 'The Soil Association has always stood above commercial activity believing that its standards might be corrupted by commercial practice'. Having referred to certain Council members indulging in methods of cultivation which made 'a nonsense of wholeness', he wrote: 'It is to avoid the temptation to condone such practices as these that the Soil Association has (I believe rightly) abnegated commercialism for itself'. Angela Bates, another of the senior generation, and one who would resign her office of Vice-President in protest at the approach which her younger colleagues were taking, was more ambivalent, giving it as her opinion that 'there is always likely to be, and perhaps ideally should be, some tension between our commercial and philosophical aims'. Mrs Bates in fact had her own commercial interest: she was a director of an animal feeds firm called Vitrition Ltd which in 1979 led her into some controversy. Where the Soil Association was concerned, she subsequently came to feel that involvement in commercial activities was incompatible with its charitable status.[1]

Angela Bates' reference to the Soil Association's commercial aims is worth noting, given that the Association's original statement of its aims said nothing about commerce, emphasizing instead the need to promote research and spread knowledge of matters relating to the soil and health. Eve Balfour objected to advertisements in *Mother Earth,* and until 1951 the journal did not carry any. In 1959, the year that Jenks described the organic movement as having nothing to sell, only about six or seven per cent of an issue of the journal was devoted to advertising, and those pages were kept to the front and back, not intruding into the rest of the material. But the April 1959 issue of *Mother Earth* contained within it the seeds of a change of approach which would grow, albeit slowly at first, in significance, as C. Donald Wilson demanded action to promote the marketing of organically grown food.[2] This chapter will look at the organic movement's often ambivalent attitude to commercialism, and at some of the initiatives which its members developed in order to promote the sale of organic produce.

Newman Turner and the Whole Food Society

Although the Soil Association in its early years kept strictly to its role as a charity concerned with research and education, that did not mean its individual members were required to be similarly austere. Given that many of them were farmers and growers, this would have been simply impossible; and, if you believed that organic produce was likely to be more nutritionally valuable than conventionally grown — even though this had not been proved — then you would of course want to provide it for those who wished to consume it. Newman Turner was a key figure in the early attempts at creating a market for organic food.

Unlike *Mother Earth,* Newman Turner's journal *The Farmer* had no inhibitions about commercial advertising. A typical issue such as that of Spring 1948 might promote Hill's Pedigree Seeds, Bersee Wheat from Elsoms of Spalding, the Sarum Igloo Aluminium Poultry House, the Watton 3-in-1/Combine Plough, various food reform/vegetarian guest houses, the B.K. Stone Mill, Lusty's Natural Product Company ('Give Nature a Chance!'), and Garlisol, a deodorized garlic-based antiseptic. The Small Ads columns included a directory of health food stores and health practitioners, including naturopaths. Particularly intriguing was the advertisement placed by a Jane Topping of Worthing, who offered 'instructive reconditioning'.

As if running a farm and a magazine were not enough for him, Newman Turner ran a co-operative enterprise from December 1946 onwards, whose aim was to put producers and potential consumers of organic produce in touch with each other. In the autumn of 1948 it was formally established as the PRODUCER CONSUMER WHOLE FOOD SOCIETY LTD its inaugural meeting being held at Friends House in London. Turner himself was President, and his colleague Derek Randal Vice-President. On the Committee were some of the organic movement's most notable members at that time: Doris Grant, F.C. King, Robert Henriques, Dr Cyril Pink, Stanley Williams and the Berkshire farmer Hugh Corley. Randal described the Whole Food Society's relations with the Soil Association as 'very cordial'. The Association 'was concerned primarily with research and showing comparisons between organic and inorganic methods', while the Whole Food Society was 'a group of people *already convinced* [my italics] of the value of organic methods and the great need for food grown by these methods.' The Society was intended as an alternative to the National Health Service. It had emerged as a response to the 'tremendous demand for properly grown food', which far exceeded the available supply. 'The aim is to start at the bottom of the ladder of health and provide service to help people to get whole food, so that they can ensure their own wholeness, or good health; rather than use outside technical assistance and attempts to cure disease externally by the ineffective methods in widespread use to-day'. The Society looked to balance the interests of both producers and consumers, bringing them together in a reciprocally beneficial arrangement. It would 'ensure adequate rewards for the primary producer using [organic] methods' and 'facilitate the direct supply of this whole food fresh to the consumer'. And, long before the Soil Association struggled to come up with a suitable certification symbol, the Whole Food Society was experiencing similar problems in its attempt to design a Whole Food Mark. As soon as organic or whole food is being offered for sale, some sort of safeguard is required, and the Society looked for a Mark to 'appear only on produce complying with Society's rigid definition of Whole Food'. But also, as soon as bureaucratic control is necessary, complexities are spawned and demands upon the time and energy required to deal with them increase; and so it was that, more than two and a half years after the Society's formal establishment, a meeting was held which involved not merely the Whole Food Society but the Albert Howard Foundation, the Biodynamic Agricultural Association and the Soil Association, at which it was agreed that a Mark was indeed essential and that the bodies now concerned should elect a committee to deal with the matter. Those present included Maye Bruce, Louise Howard, Dr C.A. Mier, Peggy Goodman and Richard St. Barbe Baker, in addition to Turner, Randal and Henriques.[3]

The *Whole Food* newsletter devoted the front page of its August 1951 issue to the Mark, and one sees in this editorial that Newman Turner and his associates were wrestling with the sorts of problems that later generations of organic entrepreneurs would also face. Given an increasing demand for food which had not been sprayed with poisons or otherwise 'manipulated', it was inevitable that commercial interests would unscrupulously take advantage of the situation by advertising their foods with such terms as 'organically grown' or 'free from chemicals': phrases which had no precise definition. The Whole Food Society deprecated the use of such terms by companies not specializing in the supply of whole foods and looked to the rapid establishment of a Mark with an authority which would be a sign of a producer's integrity. It was therefore vital that producers and consumers co-operated and contributed financially to ensure that the Mark would be nationally launched. Such a project was 'an essential instrument for re-establishing mutual confidence in everyday transactions, in a sphere where that confidence [had] ... been seriously undermined'.[4]

And that, it seems, was that: another organic project scuppered by lack of funding. It turned out that most growers did not want a Mark, and the Soil Association lacked the financial resources to tackle the problem effectively. But we should not concentrate on the signifier to the exclusion of the signified. Newman Turner and Derek Randal played a valuable role, symbol or no symbol, in bringing producers and consumers together. Their *Produce Bulletin,* published every two months until the *Whole Food* news-sheet was launched in the winter of 1948–49, provided Society members with the names of producers and a list of the produce they had available; consumers were therefore able to contact directly the producers nearest to them and arrange direct supplies. The early organic movement's familiar bugbear, the superfluous middle-man, would therefore be denied the opportunity to tamper with — i.e. process — the food 'or make a profit out of the community's need'.[5] Later, *Whole Food* provided a county-by-county list of available produce, with the suppliers listed by number. Information about whole food suppliers was eventually incorporated into *The Farmer*.

Newman Turner's work with the Whole Food Society seems to have been almost entirely forgotten, but it anticipated problems (of setting standards) and methods (direct sales, co-operatives) which would feature when a later generation of enthusiasts again sought — this time with greater success — to establish a market for organic produce. And Turner himself shrewdly anticipated the fact that the relationship between producers and consumers, which he had tried to keep in balance, would change. Writing in 1955, as the Age of Affluence dawned, he predicted that: 'The end of austerity for the consumer means the beginning of austerity for the farmer.'[6]

The Soil Association edges towards consumerism

When in 1959 Donald Wilson made his call for a market in organically grown produce, he was responding to a question frequently asked by visitors to the Soil Association's London office: 'Where can I buy organic food?' Until February that year, Peake's Farms of Colchester had been supplying high-quality organic produce to London outlets, but the demands on Mr Peake's energies, physical and nervous, had broken his health. There was, Wilson believed, little financial inducement to grow organically produced food and sell it specifically as such. How might such inducement be created? Perhaps potential consumers themselves were partly to blame, being too apathetic to seek out and pay extra for one hundred percent organic food. Such 'humbug' on the part of Soil Association members meant that organic producers found it easier to sell their goods in the conventional wholesale market, where their distinctiveness was lost. Those producers who tried to deliver directly found themselves undertaking long journeys, and several organic delivery rounds had failed because of the time and expense involved. There was also the problem of standards: who would define them?, and how much produce would meet the one hundred percent organic standard? Wilson was convinced that the Soil Association must involve itself in issues of handling, transporting, storing and marketing organic foodstuffs, asking rhetorically: 'If we cannot help to establish organic food distribution in London, have we any right to have our headquarters there?'[7]

The Wholefood shop

The result of Wilson's soul-searching was an enterprise which proved long-lasting and would provide an important link between different generations of organicists: the Wholefood shop. Mary Langman, one of its directors, described it as 'the first retail outlet in London for the produce of organic agriculture'; a claim which may be exaggerated, given the existence of Roy Wilson's Iceni shop in the 1930s. Nevertheless, Wholefood's importance to the post-war organic movement is unquestionable. Working with Earl Kitchener, Donald Wilson established the Organic Food Society in 1959. He was fully backed by the Soil Association, who granted him a year's sabbatical. The Society's retail premises opened in Baker Street in July 1960. The journalist and nutritionist Barbara Griggs, writing shortly before they began doing business, saw it as part of a craze for 'getting back to nature', of which the Health Juice Bar in Knightsbridge was another indication. (If so, its message was nothing new, as Edgar Saxon's inter-war journals *The Healthy Life* and *Health and Life* demonstrate.) Wholefood's aims in fact chiefly concerned the practicalities of organic economics. They were twofold: the first aim was that which Newman Turner's Whole Food Society had set out to achieve: bringing organic growers into contact with those who wanted to buy organic produce. The second was to demonstrate that, if organically grown produce was offered to the public, 'there would be a demand not only from the dedicated believers in organic principles but from gourmets and others who appreciate quality and flavour; that this demand would grow, and that a profitable business could result'. Writing in the mid-1970s, Mary Langman considered that the first objective had been fully achieved; the profitability was always more of a problem, and Wholefood's survival owed a great deal to the generosity of Henry Kitchener. The shop lost a lot of money in its first three years, and in 1963 LILIAN SCHOFIELD was appointed General Manager. Robert Clayton, a lawyer and property developer, was brought on board to provide business expertise, but his involvement proved disadvantageous; Miss Schofield's contribution was much more significant.[8]

Lilian Schofield was a trained pharmacist and an experienced manager, having run a busy branch of Boot's the Chemists. A keen gardener, she joined the Albert Howard Foundation of Organic Husbandry after hearing Sir Albert speak; as we have seen, the Foundation later merged with the Soil Association. When orthodox medicine proved incapable of dealing with an attack of ulcerative colitis, she turned to nature cure and eventually recovered at the Kingston Clinic in Edinburgh.[9] These experiences engendered in her an antipathy towards chemicalized medicine which made it impossible for her, in all conscience, to return to her job at Boot's. The opportunity to work at Wholefood was ideal for her. The shop was discovering that insufficient organic produce was available, so Schofield set about tracking down additional sources, finding them not just in Britain but in California, Israel, France and Spain. She also adopted a pragmatic 'next-best' approach, stocking certain goods which, although not strictly organic, could be guaranteed chemical-free. But fresh fruit and vegetables, dried fruit, and wines, all had to be fully organic.

By early 1965, Wholefood was at last beginning to break even, an achievement celebrated in a *Times* article on the first of February 1965 featuring a photo of 'the energetic and adventurous' Lilian Schofield at her dairy shelves. The anonymous correspondent described the shop as having 'a window like a harvest festival', and emphasized its difference from the usual range of 'health food' businesses, which tended to cater chiefly for vegetarians and those who wanted lose weight; among its many products Wholefood provided ham, chicken and 'some of the best sausages in London'. If prices were higher than average, so was the quality, and many local residents supported the shop for that reason. And then there were the celebrities — actors such as Michael Redgrave, Albert Finney and Rupert Davies — who patronized it. The article's tone was remarkably free from the sorts of jocular sneers so common to coverage of organic foods, and this would have encouraged Lilian Schofield and Mary Langman, who reminded readers of *Mother Earth* that the shop's purpose was 'to prove that organic husbandry was not just an ideal for cranks'. A similarly positive article on Wholefood appeared in *The Guardian* in the autumn of 1967, offering a fair-minded explanation of the higher cost of the food it sold. This same piece, by Shirley Lewis, also referred to another magnet for showbiz people: Crank's Salad Table in Carnaby Street, founded in 1961 by David and Kay Canter. Crank's developed close links with the Soil Association, hosting a meeting there in 1969 at which the surgeon Laurence Knights was chief speaker. By the early 1970s it would be providing wholefood meals on PanAmerican flights out of London.[10]

A full account of the vicissitudes which Wholefood endured would require a chapter in its own right. It would be a tale of various additional enterprises — salad bar, restaurant, butcher's — undermined, it seems, by various forms of incompetence on the part of Robert Clayton and the builders he employed. Through all these upheavals, the food shop itself survived and Mary Langman established an impressive book section containing major works on organic thought and practice. In the summer of 1983 the shop moved to nearby Paddington Street and, at around the same time, Wholefood Trust Limited, a new charitable foundation, was set up, all the shares in Wholefood Limited being transferred to it. The Trust's aim was 'to carry on research in the fields of nutrition and organic growing, with diffusion of information to the public, and to research into methods of storing, processing and distributing foodstuffs in ways that maintain maximum nutritive value'. Profits from the shop would be ploughed back into the Trust, whose Council of Management was chaired by Lord Kitchener; among the Council's other members were Yehudi Menuhin, who had been involved since 1960, and the long-term servant of the organic movement Mrs Elizabeth Murray.[11]

Lilian Schofield died in June 1989, and a memorial meeting was held that September at St. Marylebone parish church. The roll-call of people invited and of those who wrote in response to the news of her death, makes instructive reading for anyone interested in tracing the threads of the organic network, for here indeed is a convergence of different generations. Eve Balfour, Doris Grant, Lawrence Hills, Walter Yellowlees, Joanne Bower and Hugh Sinclair all paid tribute by letter to Schofield's work; and among the younger generation invited to the service were Hugh Coates, Clare Marriage of Doves Farm, Lawrence Woodward, Peter Segger, Patrick Holden — who began his career as a commercial farmer by selling his carrots to

Wholefood — and the Sams brothers, Craig and Gregory. We can now turn to consider these two important figures in the development of organic entrepreneurialism.

The Sams brothers and Seed magazine

From Nebraska, Craig and Gregory Sams had been made aware at an early age of the relationship between food and health, as their father had been cured of serious illness by adopting a macrobiotic diet; he had suffered from post-traumatic stress and internal bleeding after serving in the forces during the Second World War and was restored to health by a Japanese doctor in Hollywood. Having come to Britain in the mid-1960s, the Sams brothers established Seed Restaurant in West London in 1967. This 'legendary hip' watering-hole, with its 'groovy clientele' was, in Gregory Sams' view, the first restaurant in the UK knowingly to promote organic foods. It offered 'a mostly vegetarian diet based on wholegrain and organic foods and free of additives', believing such fare to be 'the essential foundation for a sustainable future in a world running out of resources, with a growing population and increasing degenerative disease'. The restaurant's success led to the establishment of what is claimed as Britain's first natural foods store, Ceres Grain Shop. As other retailers joined the growing market, they formed the customer base for Harmony Foods, who imported wholegrain rice during the 1970s and made peanut butter and no-sugar-added jams. They were first with a whole range of foods, among them sourdough bread, organic baked beans and brewed soya sauces, and they subsequently enjoyed great commercial success with Gregory's Real Eat company, producing Vegeburgers, and with Craig's Whole Earth empire and Green & Black's chocolate and ice cream.[12]

The Sams brothers' interest in organic and whole foods brought them into contact with the Soil Association, which Gregory joined as early as February 1968. It seemed to him like a club for gentleman farmers, but it was the only body with any interest in wholefoods and organic produce, and it could offer advice and experience. Its Wholefood shop was a possible outlet for Harmony Foods products. Bill Vickers, the Association's General Secretary, told Lilian Schofield to have nothing to do with these hippies, as Wholefood was a mainstream business; but the Sams brothers overcame the obstacles and placed their products in the shop. In the early 1970s, Gregory Sams was invited to sit on the Soil Association's Standards Committee, but he found the farming knowledge way above his head and did not stay long. Nevertheless, he formed a good relationship with Lilian Schofield, whom he much admired, with Mary Langman and with David Stickland. Stickland, though in general dismissive of what he termed the 'beard-and-sandals' generation, was, like the Sams brothers, entrepreneurial in outlook, and considered their work to have been vital in helping establish a market for organic produce.[13] Also around this time, Gregory and Sam, with their father's support, founded *Seed* magazine. Its survival was aided by advertisements for Jordans products (muesli in particular) and Aspalls (apple juice, cider and vinegar). Gregory Sams believes it to have been prescient about various health and environmental issues, but once his father withdrew his support it ran into debt and was unable to benefit from the interest in organics which increased markedly during the 1980s. Given Craig Sams' later prominence in the organic establishment, it is worth spending some time examining the curious blend of elements that went to make up *Seed*.

Seed magazine: hippiedom and entrepreneurialism

Like the other journals studied in this book, *Seed* impresses by its range of material and the devotion that went into its making. Its general tone and style reflect the influence of the so-called 'underground' press of the late 1960s and early '70s, and to that extent *Seed* could be considered something quite new in the organic movement; but a case could also be made for it having an affinity with Edgar Saxon's journals of the 1920s and '30s: the many advertisements for alternative health treatments, honest foods and vegetarian restaurants; the articles on natural living, spiritual awareness and the need for social transformation; the advice on organic gardening and nutritive cookery. In fact, the C.W. Daniel Company, which published several of Saxon's books and the classic organicist text *The Wheel of Health* by G. T. Wrench, advertised in *Seed* the books of Edward Bach, whose flower remedies Daniel had been promoting since the 1930s. *Seed* featured two articles on Bach and the healing power of his medical methods. Also like *Health and Life, Seed* looked for a more joyful form of living than that which industrial society — ugly and disease-ridden — offered. 'Natural people are happy people,' announced its first editorial. Such people demonstrated their higher level of awareness by eating natural foods, campaigning against pollution and wanting their children 'to grow up in a world that is not stifled by technology and its wastes'. They were 'the ecologists, the conservationists, the vegetarians, macrobiotics, the consumer advocates, the family planners.' (The birth control had to be 'natural', though, observing lunar fertility periods. Any similarity to the Roman Catholic Church's 'rhythm method' was no doubt coincidental.) Unfortunately, the public image of these natural people was unappealingly mirthless. Of course, the issues were serious; but *Seed* believed that natural living was compatible with high spirits, and it aimed to combine serious efforts with humour and satire. In this respect, it was closer in spirit to the later journal *VOLE* (see Chapter 6) than it was to *The Ecologist*. It chose as its original sub-title 'The Journal of Organic Living', though its definition of 'organic living' was somewhat broader than observation of the Rule of Return; it included 'crying unashamedly at something which strikes you as sad', 'going a full day without saying a word' and 'drinking a cup of ginseng tea instead of three pints of beer'. Other examples of the organic lifestyle included the spiritual discipline of studying 'the Bible, Koran or other religious book' to discover 'its real meaning'; a wholesale rejection of the mass media (which sat uncomfortably with the magazine's interest in actors and pop stars), and, more predictably, a refusal to eat any food containing additives and preservatives. Bach flower remedies were not the only link with the early organic movement that could be found in *Seed*: McCarrison, the Men of the Trees and Juliette de Bairacli Levy all received mentions, while a resident of Aberfeldy in Scotland wrote about the work being undertaken there, with Walter Yellowlees actively involved, to turn the Upper Tay Valley into the equivalent of what we would now term a 'transition town'. Lawrence Hills, W. E. Shewell-Cooper and John Seymour all contributed.[14]

Seed also contained many elements of what would come to be known as 'New Age' thinking, and was often distinctively pagan in tone. There were various references to astrology (both Western and Chinese), including an advertisement for The Temple of Bread, 'An Aquarian Age Natural Foods Restaurant' in Notting Hill, and guidance on astrological food chemistry; articles on Geodesic domes, nature spirits,

Druidry, Saturnalia, dowsing and witches' knowledge of medicine. There was an interview with a dervish sheikh; and Dr Arabella Melville, in an issue featuring on its front cover a picture of nudists in Piccadilly Circus, blamed St Paul for the sexual inhibitions of those who attended rock festivals. One article, on flying saucers, is of particular interest: not so much for its content as for its author, Dr E. Graham Howe, a Harley Street psychologist who had written several books for Faber and Faber in the 1930s and '40s and been much admired by Henry Miller. Howe had studied deeply in Eastern religious thought, and Eastern and esoteric philosophy of course featured in *Seed,* with Noel Freeman contributing several articles on the Buddhist tradition. An anonymous review of J.G. Bennett's *The Masters of Wisdom* showed that the connection between the organic movement and Gurdjieff's system still existed.[15]

Readers irritated by *Seed's* tendency towards mysticism were rebuked in an editorial which praised those 'romantics' who wanted to resist the seemingly inevitable drift to an Orwellian society. A bit of romantic mysticism harmed nobody, and made those who adopted it much happier. 'If a group of mystics like the Findhorn Trust ... consult elves to grow organic crops, and are successful in doing so, is that bad?'[16] The magazine in fact instituted a regular feature on mysticism, 'Messages from a Star', urging its readers to reject the complexities of technology and exploitation of the natural world, relying instead on the inner resources of a refreshed spirit and the outer resources of God's generosity as revealed in nature.

What constituted 'natural living' might well have struck some readers as extreme. A 'Suds-Bashing Special', for instance, concentrated on 'The Case Against Washing', the need to wash being 'greatly exaggerated' by companies like Unilever. Short-sighted readers were more than once urged to discard their glasses, improving their sight through diet, sunshine and exercises. The magazine offered advice on natural child-rearing and recommended various familiar features of alternative medicine: homoeopathy, comfrey and other herbs. Those wishing to practise 'natural' agriculture were offered the example of the Japanese farmer Masanobu Fukuoka (whose methods were indeed highly productive).[17]

Natural living also implied a strict dietary regime. *Seed* opened its assault on conventional foodstuffs by quoting an American nutritionist who considered milk to be 'the greatest single cause of diseases in our human bodies' and continued its campaign through providing evidence of the benefits of a macrobiotic diet and through satirizing and exposing the bogus claims of the processed food industry. An early issue reviewed a powerful indictment of that industry, *Food Pollution: The Violation of Our Inner Ecology* by Judith Van Allen. Volume 2, No.3 concentrated on the case for natural foods, examining in detail the social impact of changed eating habits and the effects of industrial food production on the soil, on animals, and on the human beings who ate the food. *Seed* argued that food quality affected not just people's physical state, but their consciousness: 'The way we perceive the world is to a large part governed by the condition of our nervous system.' Proper diet was the basis of social understanding: 'How can we speak of social justice when we eat more food than is required? How is it possible to consider peace until we have experienced it on a cellular level?' The magazine drew attention to evidence that additives might adversely affect children's behaviour, making them anti-social and unable to concentrate. Rather more controversially, it argued that a macrobiotic diet increased the harmony between individuals and their environment, perhaps thereby helping the development of a sixth sense. Whereas in its early issues *Seed* saw health as chiefly dependent on correct diet, its perspective broadened after a couple of years as it began to give more consideration to the social dimension of healthy living. One might say that it began to develop a more truly ecological approach to the problem. 'It's not enough just to eat well if you live in a basement flat close to a superhighway. The pigfat won't get you, but the fumes will.' The magazine therefore covered topics such as alternative technology, solar power, urban farming and gardening and the prospect of a world without petrol.[18]

Seed certainly gave the impression of a fairly radical opposition to technological capitalism. One of its cartoons, in a style reminiscent of Robert Crumb, offered a nightmarish vision of a supermarket whose checkout operator was Death himself; while another satirical offering de-constructed advertising slogans: for instance, 'For regularity, try our super PUFFED BREAKFAST BRAN' translated as *'Thank God you people are obsessed with your bowels, or all this left-over bran from our milling operation would end up as cow fodder'*. More profoundly, a page headed 'Grains of Truth' quoted a Japanese proverb: 'In the hum of the market there is money, but under the cherry tree there is rest'. In practice, though, it seems that the hum of the market exerted a greater appeal than did the cherry tree, as one can discern in *Seed* some germs of the consumerism that would later dominate the organic movement. Perhaps the most fascinating blend of early organicist philosophy and the entrepreneurial spirit can be found in a half-page on 'Natural Foods: The New Consciousness' in Vol.2, No.3. *Seed* placed the idea of a natural order at the heart of its philosophy, as the organic pioneers had done, but used it as the key to creating 'organic' businesses. 'There is a natural order. If enough individuals purify their bodies and minds through natural eating and natural living, then all mankind will begin achieving greater harmony with this natural order.' Business and technology would then follow the lead of 'new consciousness people' and start providing natural foods and, in time, other environmentally friendly goods. And so (just as democratic socialism would abolish capitalism from within the system) organic consumers would, through their purchasing power, transform agri-business and the food industry. That was the theory, and it would come to dominate the strategy of influential sections of the organic movement.[19]

Given that Craig Sams would one day sell Green & Black's to Cadbury's, the *Seed* editorial in Vol.2, No.8 takes on considerable retrospective significance. Cadbury-Schweppes had just entered the health food market, joining other large corporations such as sugar producers Booker-McConnell, Alliance Breweries and Imperial Tobacco. In a fit of chronic optimism, the editorial decided that this was 'not necessarily an alarming situation', since these organizations embodied 'prestige and stability' and could use their resources to educate 'the masses of ordinary people' out of their nutritional illiteracy. Whoever wrote the editorial was indulging in either a subtle form of irony or, more probably, a sanguine fantasy, when he or she envisaged Cadbury and company extolling the virtues of organic foodstuffs. *Seed*'s 'alternative' stance was thoroughly pragmatic, placing faith in the established channels of media power as a means of spreading its message. Hence the prominence in its pages of 'celebrities' from the world of pop music and the media: another aspect of the organic movement which has flourished, like the green bay tree, since the mid-1990s. *Seed*'s first two issues featured interviews with singer Paul Jones and actor Terence Stamp; the latter showed his relaxed approach by announcing: 'I believe you should eat what you like as long as it makes you feel groovy'. (Sir Robert McCarrison couldn't have put it better.) They were followed over the next few years by Paul McCartney, BBC radio disc-jockey 'Emperor' Rosko, Spike Milligan and Keith ('Captain Beaky') Michell. Quite early on, a number of readers voiced their complaints about the attention paid to pop stars; the magazine's response was one which has grown familiar in an age when the organic message is so frequently spread by television celebrities: that media stars influence the population at large. 'If admiration of a musical celebrity can influence [the population] to eat properly, then let music be their passport to health and happiness.' After all, John Lennon and Yoko Ono, John Peel and Marc Bolan had patronized the Seed macrobiotic restaurant, and Gregory Sams had provided natural foods at various festivals of the late 1960s/early '70s, including the Isle of Wight festival in 1970, and Glastonbury; so there was a link between the world of pop music — or at least, certain parts of it — and the wholefood movement.[20]

David Stickland and Organic Farmers and Growers

As we have seen, the Sams brothers' hard work was respected by David Stickland, a man who shared with them an entrepreneurial outlook but little else. Stickland is one of the most controversial figures in the history of the British organic movement, and the hostility between him and various members of the Seventies Generation persisted, on both sides, for many years: civil wars are traditionally the most bitter. Yet, as Angela Bates, one of his supporters, has pointed out, on the face of it he and the Seventies Generation had a good deal in common, given their shared desire to create a market for organic produce.[21] Rightly or wrongly — and we might find ourselves in a legal minefield if we tried to determine which — the Seventies Generation objected to the methods Stickland employed in pursuit of this aim, which they believed debased organic standards. Stickland in return regarded what he considered to be their purist, hippie approach, as unrealistic for practical farmers, who, as businessmen, were looking to make a profitable living. (But it was the Seventies Generation's discovery that its members could survive only by selling their produce which made it so keen to develop a market, while few in the organic movement have displayed greater business acumen than the one-time 'hippie' Craig Sams.)

Stickland's love of farming stemmed from the fact that his parents ran a school with about 180 acres of land, on which they kept dairy cows. Someone gave them one of Newman Turner's books as a Christmas present, and they converted to organic; Turner's ideas seemed to make sense, the cows' manure helping enrich the chalk land on which the Sticklands farmed. David Stickland's support for organic methods was always based on a simple, aesthetic philosophy: 'It's a nice way of farming'. He spent time as a trainee on a dairy farm in Oxfordshire, then joined Dunn's Farm Seeds and during his time there met Sir George Stapledon, who was one of the directors. Further business experience followed, with East Kent Packers at Faversham, a large (non-organic) co-operative dealing in fruit. Stickland was an inspector and then worked on the marketing side: this was during the second half of the 1950s. But he wanted to farm again, and in 1961 emigrated to New Zealand, where he spent a year as a sheep farmer and then took up organic dairying, obtaining blood, bone and fish from a nearby abattoir. This was a period of 'slavery', from which he turned to organic vegetable growing, as one of just four organic farmers in New Zealand at that time. He survived financially because there was a demand for organic produce: most vegetable growing was controlled by the Chinese, who drenched their tasteless crops with pesticides (a far cry from the Chinese peasants whom F.H. King had described in *Farmers of Forty Centuries*). But the agricultural colleges provided no assistance, and he suffered from flooding, so he travelled again, this time to the more go-ahead Canadian province of British Columbia. Here Stickland worked for farm co-operatives and established a journal, *Canadian Agriculture and Horticulture,* in the face of implacable opposition from the agricultural chemical companies. This editorial experience served him well when he returned to England in the early 1970s and made contact with the Soil Association, which was seeking someone to replace Robert Waller as its journal's editor. Following meetings with Eve Balfour and E.F. Schumacher, Stickland was appointed. Waller wrote to him warning that he would be able to endure the General Secretary, Brigadier Bill Vickers, for only two years. Stickland initially dismissed this prediction, but subsequently came to realize it was spot-on, finding Vickers 'dreadful'.[22]

Stickland began his stint as editor by rejecting Waller's approach as old-fashioned and too philosophical, and re-constituting the journal as a monthly publication more relevant to those in the business of producing and marketing organic goods. He did not think highly of *Span,* and closed it down. He was not particularly interested in nutrition (which many have regarded as the Soil Association's central concern); above all, he wanted to help organic farmers obtain a good premium, but he felt that the Association, as a charity, did not at that time want to know about marketing. Although Stickland liked Schumacher personally, he considered him 'wrong about almost everything': an attitude which certainly distinguishes Stickland from the Seventies Generation. And he had more time for Eve Balfour than certain members of that generation did, finding her approach always practical. Drawing a fine distinction, Stickland did not consider her particularly autocratic, but felt she had a clear sense of what needed to be done. Balfour in fact gave her blessing for him to set up Organic Farmers and Growers (OFG).

David Stickland's admirers praise the energy he showed in pushing for farmers' interests, and this was soon evident in the Soil Association journal: his second editorial regretted that the Association was failing to attract farmers. (In Stickland's view, there were about three hundred 'worthwhile' farmers in the Association at this time.) Anticipating what would later be achieved by the Organic Growers Association and British Organic Farmers, Stickland told readers that the Association was considering forming regional Farmers and Growers Groups, as existing Association groups tended to dwell on topics such as gardening and health foods rather than enabling farmers to discuss their methods and problems with each other. One such Group in fact already existed: the Northern Organic Farmers' Group, chaired by Robert Milnes-Coates, whom Stickland came to know well and greatly respect. The Association's Organic Marketing Company (OMC) would be closely linked with these groups. The OMC, wholly owned by the Soil Association but with its own registered trademark, was an offshoot of the Association's Organic Marketing Committee, chaired by Hugh Coates, its aim being 'to encourage the growing of organic produce in accordance with standards defined by the Association and to assist in the distribution and marketing of such produce'. (We see here that the Soil Association was concerned about marketing some years before the Seventies Generation turned its attention to that issue, though of course it did not achieve anything like the success which would mark their endeavours during the 1980s.) In 1973 the OMC ran a pilot scheme for the production and marketing of organically grown wheat. By the summer of 1974, the OMC was proving beyond the Soil Association's capacity to support without additional help, and the Organic Marketing Committee recommended the establishment of an independent and financially self-supporting co-operative to develop organic production and marketing, owned and financed by those who joined it. It was hoped that this body would demonstrate that organic methods could be economically attractive and thereby help draw farmers into the organic movement. The Soil Association Council agreed to the establishment of Organic Farmers and Growers Ltd, an organization which — ironically, in the light of later events — was thought likely to help raise the Association's profile in agricultural circles. (Hugh Coates gave OFG some money to help start it and was a director for many years.) Early in 1975, Stickland began contacting farmer members of the Soil Association, at which point the story becomes obscured by controversy. All one can say for certain is that Stickland left the Soil Association and that OFG became his own separate project, albeit with the support and participation of some, like Hugh Coates and Michael Rust, who remained in the Association. There was considerable bad feeling towards him, but he was committed to supporting organic farmers and from 1978 was invited to contribute a regular column to the *Soil Association Quarterly Review*'s 'Farming Observations' section.[23]

Stickland initially recruited seventeen farmers for OFG; by 1978 he had 65 farmers and market gardeners as members, ranging from smallholders to some with over a thousand acres, and including tomato and vegetable growers, arable farmers and dairy specialists. He had obtained a government grant, but had to sink a good deal of his own money into the project; he also received support from the Blythe-Currie sisters, who owned an import-export business. Stickland felt that one of the main problems which organic farmers faced was, quite simply, a sense of isolation, and he spent much time visiting his members, advising them by telephone and producing a newsletter. There were no textbooks in those days, much misinformation circulated, and organic farmers were regarded with derision or hostility. One of

Stickland's main reasons for setting up the International Institute of Biological Husbandry (see Chapter 7) was that he was tired of being called an idiot and a crank, and wanted to give biological methods (he preferred the concept 'biological' to 'organic') some respectability. Stickland found what he termed the 'sandal-y' tendency in the organic movement a further disadvantage, giving quite the wrong impression to his target population of experienced, business-minded farmers. Certainly it was Stickland's more conventional, less ascetic persona which would appeal to the Oxfordshire farmer Charles Peers, when he heard Stickland speak: here was a man, Peers thought, whose years in different branches of the farming industry enabled him to appreciate commercial realities, and who was happy to enjoy his pipe and a pint of beer. Peers committed himself to organic methods and subsequently spent several years as a director of OFG. Nevertheless, and as Stickland accepted, OFG in its early days owed a good deal to the 'sandal-y' types; at a time when the supermarkets regarded organics as a non-starter, Infinity Foods and Whole Earth were good customers for OFG produce, providing thousands of pounds' worth of business. OFG also found a market with Lilian Schofield at Wholefood. Grain products were its chief commodity, though it worked to develop the marketing of stock, fruit and vegetables.[24]

If Stickland had raised hackles by the way he took over OFG, the question of standards was to render him even more unpopular, particularly with the Seventies Generation, though also with senior figures in the organic movement like Mary Langman. The 'purists' — to use a term which Stickland employed with negative intent — perceived a dilution of standards in the establishment of the OFG2 grade, which permitted, in one year only, use of 50 units of nitrogen, 30 of phosphate and 30 of potash (the dreaded 'NPK'), and of an MCPA (herbicide) spray. For Stickland, OFG2 standards were an environmentally sound compromise suitable for farmers grading up to organic or whose fields were as yet insufficiently fertile. (It is worth recalling that earlier figures such as Laurence Easterbrook and Jorian Jenks accepted that use of chemical fertilizers might be necessary to give the land an initial boost, though that was before the days of commercial standards.) OFG2, in Stickland's view, was 'more sensible' than pure organic, and gave better tonnage without harming the soil, whereas for the 'purists' it represented an opportunistic muddying of the waters. In 1984,

Stickland announced that the problem was solved: OFG's alternative grade was re-named 'Conservation', which eliminated any reference to 'organic' or 'biological'. This ingenious way of sidestepping the issue did not prevent him, though, remaining a bogeyman for members of the Soil Association and OGA/BOF. Stickland later sold the designation 'Conservation Grade' to Bill Jordan, whose muesli, cereals and granola business is one of the success stories of health food marketing.[25]

As the move towards unified British standards got under way in 1987, the Soil Association was compelled to deal with OFG, whose OFG1 grade was the other main organic standard in the UK, though, according to the Association, plainly inferior to its own. But when the Soil Association journal became *Living Earth* in 1988, no reference to OFG appeared in its pages, despite the fact that, as Stickland wrote to point out, it was the UK's largest commercial organic farming organization. The Soil Association, evidently quite certain of its position on the moral high ground, responded with a piece comparing the Association's standards favourably with OFG's. Exactly how this was supposed to contribute to harmonious relations between the two bodies is unclear, and was unclear at the time to D.B. Grier, who found the article 'depressingly parochial'. Not that the animus was confined to one side: recalling the debates which took place in the late 1980s as the various organic bodies struggled to establish national standards for organic products, Stickland wondered how the UKROFS chairman Professor Colin Spedding found sufficient reserves of patience to tolerate the high-principled approach of figures such as Patrick Holden and Nic Lampkin. But Spedding, whom Stickland much admired for his skills as a mediator, managed to elicit sufficient agreement from the different parties to ensure that the organic movement's standards became part of government policy. Stickland was heavily involved in the work, being, he claimed, a member of all the committees which were created; but he was, he believed, the only one of those involved who thought that the whole business went too far: indeed, that minutely detailed discussion of chemical processes was 'an utter waste of time'. Looking back on it many years later, he considered that standards should apply only to the growing and selling of organic crops and that what happened later should not concern the various organic bodies.[26]

OFG is one of the longest-lived organic projects, and one of which Stickland, despite his detractors, felt proud. In his view, its achievements were 'to find and provide a market for a quality processed organic product', to offer continuity and some regulation in the market-place — providing millers, for instance with a continuous supply of organic grain — and ensuring that farmers received a worthwhile premium for their goods. It provided farmers and buyers with greater certainty and ensured that, once the public became aware of organic produce, the farmers were there, ready to provide it. When there was no UK market for a product, OFG sold to Germany, the Netherlands and Denmark in particular. Stickland took satisfaction in having led an organization from its outset and watching it grow. He remained an object of hostility for many in the organic movement, but others — even those who feel that he did to some extent prioritize commercial considerations over strict integrity of standards — recognize him as someone who was full of energy and keen to start new initiatives for the sake of the organic cause. Richard Thompson, whose father Michael was a founder of the Soil Association's Yorkshire Group, was a director of OFG for many years; like Charles Peers, he felt that Stickland was shrewd in his assessment of what might make commercial farmers consider converting to organic methods. Other supporters include Deidre and Michael Rust, Marcus Ridsdill-Smith, Angela Bates and David Hodges, who worked closely with Stickland for the International Institute of Biological Husbandry. Although Hodges grew distant from Stickland, he nevertheless felt that Stickland had played a valuable role in stirring up the latent energies of the organic movement.[27]

Organic standards

It is understandable that the Soil Association should have felt protective about its standards: it had taken a lot of time and trouble developing them. In the spring of 1967, Douglas Campbell and the Kent farmer and Association Council member Simon Harris visited Lawrence Hills to discuss the setting of standards for wholefoods. These were eventually established in 1972, and the Association distributed a set of them (featuring the new symbol), and a list of permitted substances, the following year. The symbol would be awarded only to products which were one hundred percent organic, its aim being to apply a stricter definition than currently existed to the much misused word 'organic':

in other words, to deal with the same problem that Newman Turner had faced twenty years earlier. By the autumn of 1973 the Association was inviting applications from those wanting to use the symbol; Aspalls apple juice company at nearby Stowmarket was the first business to register. From the outset there were complaints that the Association's approach was 'impractical' for smaller-scale commercial producers (the same complaint continues to be made today), and a letter from a West Sussex grower eloquently expressed the frustrations of his lot; but David Stickland, in a detailed and helpful reply, urged him to support the Soil Association symbol as a likely means of improving the prospects for people in his position. One had to make a start somewhere.[28]

During the remainder of the 1970s members of the Association's Organic Standards Committee put in much work on refining and extending the range of standards and on establishing the symbol as 'synonymous with complete organic integrity'.[29] The Committee was chaired by Alan Brockman, who was succeeded by Robert Brighton. Livestock standards were produced in 1977, guidelines for processors, packers and millers drawn up, and the permitted list revised. Standards for dairy producers followed. The Committee also appointed its first inspector, Graham Shepperd.

Graham Shepperd, standards inspector

Shepperd came from a farming background and, after qualifying as a physiotherapist, in 1960 he returned to the land. His mother had always been interested in health issues, and he began to question orthodox veterinary use of antibiotics, bringing up calves without them and investigating the potential of homoeopathy. Unwilling to follow the accelerating East Anglian trend towards arable monoculture, he took on a hill farm in Somerset in 1965, specializing in livestock and sheep. He had by this time already come into contact with the Soil Association, and, like so many in the organic movement, was impressed by the writings of Friend Sykes and Newman Turner. He was fortunate in that Eric Clarke ran an active Soil Association Group in West Somerset; he also forged links with the Biodynamic Agricultural Association, attending its conferences and visiting Alan Brockman at his farm near Canterbury; he found Brockman's work particularly inspiring. Among his neighbours, though, he suffered the isolation that David Stickland had identified as one of the organic farmer's chief problems, and he tended to keep quiet about the methods he was using.[30]

Having spent a couple of years as secretary to the Soil Association's Standards Committee, he was appointed Inspector and for a further two years devoted himself to getting to know all — literally all — the organic holdings which existed at that time, from Scotland's Black Isle to Cornwall. In Shepperd's view, the drive to create standards was not a response to the market, because scarcely any market existed at that time, but rather to the sense that the Soil Association had a duty to ensure that organic foods were produced to a high quality. He regretted the split with OFG and the way in which the Association subsequently wrote off Stickland's work; Shepperd found him approachable and felt him to be genuinely concerned about what he believed to be the Soil Association's neglect of its farming members. (British Organic Farmers emerged from a similar feeling.)

Shepperd is another of those figures who provides an interesting link between two different periods in Soil Association history. He was influenced by some of the early organicist writers, and found *Mother Earth* under both Jorian Jenks and Robert Waller an outstanding publication, presenting a broad spectrum of ideas. The spiritual dimension to the organic philosophy, notably Steiner's esoteric system of Anthroposophy, appealed to him. But he also sympathized with the Seventies Generation and their efforts to create a market for organic produce, both through creating co-operatives and through persuading supermarkets to take an interest. As an inspector, he enjoyed an excellent view of what was happening, being especially impressed by Iain Tolhurst's holding in Cornwall. He also admired the work put in by Peter Segger and Patrick Holden, and found Elm Farm's soil analysis service and studies of crop rotations valuable. The older generation, he felt, saw the Soil Association as 'a nice cosy club' and were insulated from the rigours of the marketplace. Like John Butler and Peter Mansfield, though, Shepperd disliked the confrontational atmosphere which developed in the early 1980s.

Peter Segger, organic entrepreneur

Thanks to the determination of Peter Segger and his associates, the drive to promote the Soil Association symbol as the guarantor of 'food you can trust' — as their slogan had it: Edgar Saxon's 'honest food' in a new guise — gained considerable impetus in the mid-1980s. Segger's strategy was as follows: 'from now on the Soil Association stands for ... *The Promotion* of the *Production and Consumption* of *Organic Food* at every level' [emphasis in original]. The public should be educated as to the value of organic produce and the organic movement should then ensure that such produce was available, correctly graded and attractively presented: 'the days of shrivelling and decrepit vegetables rotting in the corner of wholefood shops had to be finished'. The Organic Growers Association was playing a vital role in this change, and in late 1983 Segger established another initiative, of which he was Chairman, Organic Farm Foods (OFF), based in Clapham, for which Ginny Mayall worked. It bought direct from farmers and growers in many parts of the UK, distributed to London restaurants and shops, and provided a cash-and-carry service for individuals, groups and wholesalers. Segger's venture attracted media attention and was part of a broader trend: in addition to Brighton's Infinity Foods co-operative (established in the early 1970s and one of the longest-surviving organic projects ever undertaken in the UK), there were distribution networks springing up in South Wales, Bristol, Somerset and Herefordshire, with plans for other schemes in several major industrial cities. In Segger's words: 'Never in the history of the organic movement ha[d] there been such excitement, enthusiasm and activity in the production and distribution of organically produced food'. Key to it all, in his view, was the symbol, representing the interconnected wholeness of soil, plant, animal and man. (David Stickland related that the Association originally opted for a design portraying a cockerel standing on the back of a cow, to which he strongly objected. If this is so, then the Soil Association surely has at least one reason for gratitude to him.) The organic movement had already, and perceptively, been anticipating the day when organic standards would become a matter of government concern, and from 1981 to 1983 Lawrence Woodward chaired the British Organic Standards Committee in order to arrive at the standards upon which the different parties could agree: in Richard Young's view, it was Woodward who ensured that these standards were imbued with real strength. At the 1983 Royal Show, the new standards were made public; that same summer, Graham Shepperd was appointed Symbol Inspector Co-ordinator. Once the processes of administration, inspection and monitoring had been fine-tuned, a national campaign, with a distinctly imperialist feel to it, was launched. 'We must aim,' wrote Peter Segger, 'to make sure that shops only buy Symbol products, unless there is no choice available. We must ensure that all producers apply for the Symbol and put it on all their packaging ... We must ensure that all wholesalers only buy Symbol produce wherever possible ...'[31]

Government and supermarket involvement

Events moved very rapidly during the mid-1980s, so that by 1986 the Symbol Scheme was performing a wider function than merely guaranteeing a producer's commitment to organic principles and practice. Increasing public demand for organic produce was a matter of national interest, and the Scheme was becoming an arbiter of genuine organic produce and, in effect, the inspector of the organic food industry. As ever, there existed the threat of unscrupulous businesses cashing in, so policing the entire food chain, including wholesalers and importers, became essential. The organic movement was about to take a momentous step: the Soil Association was by now involved in discussions with the Ministry of Agriculture, ADAS, the NFU and the Trading Standards Office, and, at an international level, was helping IFOAM advise the EEC on a directive for organic standards in Europe. It was clear that the British government would very soon become involved, and Patrick Holden, at that time co-ordinator of BOF, was determined that the organic movement should present a united front to officialdom and agree on a common set of standards. In practice, as an article by him in the *Soil Association Quarterly Review* made explicit, this meant that the Soil Association's own standards must prevail. The point was re-emphasized in a *Review* editorial in June 1987, the month before the government established UKROFS: 'organic' must be defined positively, as a systems approach, rather than negatively, as no more than a refusal to use chemicals and artificial fertilizers. The first UKROFS board meeting was held in November 1987. Professor Colin Spedding, who chaired the discussions, saw his role as facilitating an agreement between the different parties — which included the Soil Association, Elm Farm Research Centre, Organic Farmers and Growers, the HDRA and the Biodynamic Agricultural Association — on a consistent set of standards; he was not concerned with the arguments as such, but with helping the various factions to co-operate. This was not always easy. He found Lawrence

Woodward's input helpful, but regarded Patrick Holden as someone who would have been prepared to jeopardize the entire process for a minor point of principle. No doubt it was Holden's uncompromising approach which ensured that the UKROFS standards, published in 1989, closely followed those of the Soil Association, which, for organic producers, were 'the Bible'. By this time, Europe-wide standards were on the way. IFOAM had been working on the development of standards and definitions for organic farming since 1976: both it and other organic farming organizations, as well as consumer groups, began to put pressure on the EEC to develop a regulatory framework, and the result was Regulation 2092/91, published in 1991 and brought into force the following year.[32]

The breakthrough at government level was made possible not just by the energy and publicity skills of the Seventies Generation, but by the fact that during the 1980s the supermarkets were starting to stock organic produce. In June 1971, the recently appointed Soil Association President E.F. Schumacher had urged the Association to incarnate its values in the world of commerce, and fourteen years later it looked as though this might be happening. In the autumn of 1985, *New Farmer and Grower* reported that all the major supermarkets were showing an interest in organic produce. Safeway had taken the lead, first stocking it in 1981; in 1985 it began to stock organic produce in all 124 branches of its stores. By 1988, 2.7 per cent of Safeway's fresh produce was organic, and in 1990 Sir Alistair Grant, Chairman of Safeway, visited the HDRA at Ryton and sponsored that year's National Organic Wine Fair. During the late 1980s and early '90s Safeway also sponsored the Organic Food and Farming Centre, organized by Gaye Donaldson, at the Royal Show. Tesco began to get in on the act as well, in 1991 making a point, for a fortnight, of selling organic vegetables at the same price as conventionally grown; this experiment led to a fifty percent increase in sales of organic vegetables *after* their prices had been raised again.[33]

Responses to the new commercialism

Its critics

Given the organic movement's traditional mistrust of big business, the profit-skimming middleman and the centralized state, its preference for the local and regional, and its environmentalist perspective, one would expect to find some soul-searching and debate about the involvement with government standards and supermarkets; and so there was. Where UKROFS was concerned, Lawrence Woodward saw the situation as one which all successful radical movements had to face: whether 'to become involved with the establishment and risk being taken over, swamped and diverted, or to stay outside and risk being irrelevant to the mainstream.' Such irrelevance might have been 'glorious', but at this point, just after UKROFS had been established, Woodward tended to favour being 'a vital part of society', as long as the Register guaranteed organic principles. Some years later, though, Woodward was forced to conclude that his fears had proved more perceptive than his optimism. Writing in the Tenth Anniversary Issue of *New Farmer and Grower,* he conceded that although the organic movement had been obliged to enter the mainstream in order to demonstrate a sense of responsibility, 'the mainstream is changing us more than we are changing it'. The recent 'successes' had been 'based upon accommodating the essential character of conventional approaches to food production' and, in effect, 'assisting in "greening" the *status quo,* not changing it'. Julian Rose expressed similar concerns, noting that many organic producers now found themselves swamped by bureaucratic pressures and could expect no magic shield if they entered the mass-market fray. From West Wales, where they had been farming since the mid-1980s, Colin Johnson and Arabella Melville wrote a forcefully argued piece on why they preferred to be non-approved organic producers. They pointed out what they felt to be the naivety of the Soil Association's strategy: the organic movement had meekly handed over its standards to UKROFS, simply because it wanted to be accepted as part of the established order, and producers were now sinking under regulations. But Johnson and Melville had taken up organic farming 'as part of a life philosophy, and ... ha[d] always known that [they were] *de facto* up against the Establishment... The organic movement, on the other hand, appear[ed] to see itself as part of the *status quo,* but with a different agricultural method.' (The weight of their criticism has not diminished with time.) Johnson and Melville intended to continue farming as before, cut out the middle man and sell to people who shared their philosophy. Francis Blake responded to their attack on behalf of the Soil Association, arguing that UKROFS had been necessary, since self-regulation did not work, and that it had helped the organic movement gain recognition and respect. The Association was helping small producers through Community Supported Agriculture schemes, and everyone there 'embrace[d] the deep and complex philosophy which underpins organic farming'. (The point was, though, whether that philosophy could ever be compatible with the values of the existing social and economic system. Like the organic pioneers, though presumably from a different political perspective, Johnson and Melville thought not.)[34]

Johnson and Melville were among many in the organic movement who were wary of working with the supermarkets. Anne Spearing, a final-year horticultural student, saw Organic Farm Foods (OFF) as 'hell-bent' on supplying supermarkets, and wondered how this sat with the clause in the recently published *Charter for Agriculture* about encouraging de-centralized markets. Spearing's letter drew support from David Urwin, who uncompromisingly condemned 'the whole business of supermarkets' as being 'out of harmony with ecological principles': centralized, technological and ignoring the food miles issue. Defending his enterprise, Peter Segger described OFF as 'a combination of idealistic and pragmatic marketing', the vast majority of whose sales were to wholefood shops. It was necessary to accept that supermarkets were an important *option* for growers: after all, they sold 68 percent of fresh foods, and OFF was working towards a 'store within a store' principle. The Organic Growers Association, with which OFF was very closely associated, did not tell growers where they could and could not sell, and both bodies aimed to encourage a greater diversity of local outlets. Segger's hopes for OFF were to prove unfounded, as the business collapsed in November 1986 owing £96,000.[35]

Segger's response did not satisfy Urwin, who wrote again to explain why. His argument is worth looking at in some detail, as it eloquently puts a case which became marginalized by the organic movement's later embracing of consumerism. This was not merely an issue of 'business opportunities': a philosophy was involved. Was OFF requiring supermarkets to display material explaining the Soil Association's ideas? If not, then growers' co-operatives might own or rent shops in order to promote the Association's philosophy. Urwin feared that the pragmatic approach was, 'as ever', obscuring the movement's ethics. It was all very well to bandy about words like 'holistic' and 'ecology', but they implied social and economic change, not just agricultural.

> The better quality of life that many search for is a reaction to consumerism, excessive economic growth, over-affluence, centralization ... Supermarkets belong to this ... world, and selling through them is implicit acceptance of a system we should be fighting. The holistic approach to life demands meaningful, satisfying labour for all — not the drudgery of shelf-filling and checking-out in multiple stores ... A lot of farmers and growers will argue that they are in the business of producing healthy food, not trying to change the world. But if they imagine that chemical-residue-free food is the complete answer to the health of the world, they have ... failed to see that we all have far greater responsibilities.[36]

With economic recession beginning to bite in 1990–91, Julian Rose similarly reminded organic producers of the movement's wider aims, pointing out that supermarkets were based on the very ideas — standardization, increased quantity for the sake of profit — that the organic movement should oppose. The whole idea of a 'green consumer' was a contradiction in terms. Instead, the organic movement needed to foster 'agricultural models that have a built-in social, ecological dimension in their distribution as well as their farming practices', and to counteract the 'consumer is king' materialist dreams of the Thatcher era. Particularly interesting, in view of the Soil Association's subsequent swoon into the arms of consumerism and celebrity culture, are Patrick Holden's remarks to a 1991 seminar on marketing. Among the various adverse effects of supermarkets, he noted their tendency to reduce crop diversity, the increased distance between producers and consumers, the absence of environmental values in their economic practice, and their threat to the organic movement's early ideals. Supermarkets should be just one of a range of potential outlets, among which Community Supported Agriculture was especially important.[37]

As the 1990s passed, some commentators began to fear that, in cosying up to the supermarkets, the organic movement might share the fate of the Lady of Riga, who, according to the limerick, went for a ride on a tiger and ended up inside it. Towards the end of the period we cover in this book, *Living Earth* published a well-informed, three-page article by Hugh Raven and Tim Lang of the SAFE (Sustainable Agriculture, Food and Environment) Alliance on 'The Shops We Love to Hate', in which the authors argued that supermarkets had a detrimental effect on local economies, public health and the environment. Despite some admirable achievements, the supermarkets required greater public control: Raven and Lang recommended, among other policies, the reflection of true transport costs in product prices, a reduction in packaging, the opening of wholesale fresh food markets to the public, and experiments with direct marketing, box schemes and Local Exchange Trading Systems. The Parliamentary Select Committee which was investigating out-of-town shopping needed to look at the shape of the whole food retailing system. In *NFG,* Caroline Dumonteil was more confrontational about the supermarkets' concentration of retail power, looking at the issue from the growers' viewpoint. Growers were locked into a production-line system to which they had to respond at a moment's notice, the risks and costs being pushed down the line to suppliers. The logic of the supermarkets' approach implied monoculture, restriction of variety and control of natural factors: a mass-supply logic incompatible with organic values. Dumonteil clearly had no doubt that the excitement ten years earlier about Safeway's interest in organics had been misplaced: ' ... the inexorable supermarket machine steamrollers the supplier into the same subservient mould as the rest of the growers, while their produce is lodged precariously on the shelves at unaffordable prices to ordinary people just to give the store a good image'.[38]

Although Patrick Holden insisted that the interests of producers and consumers were *'absolutely inseparable'* [italics in the original] — in this he differed from Newman Turner, quoted earlier — there does seem to have been something of a divide within the organic movement between producers and retailers.[39] This is of course a statement which requires considerable refinement and provides a topic for much further study: David Stickland and Peter Segger, for instance, combined farming and growing experience (though Stickland's was considerably more extensive than Segger's) with knowledge of marketing and a strongly commercial mentality. And, as we have seen, the 'Seventies Generation' became interested in marketing of necessity, in order to survive. But their interest in ensuring that they could sell their own produce should be distinguished from that of those who were interested in selling other people's, and one finds in *New Farmer and Grower* the persistent wariness — which has a good organicist pedigree — of the peasant towards the trader, of which we have just seen some examples. Catherine Dumonteil's 'peasant' perspective makes an instructive contrast with the approach of CHARLOTTE MITCHELL, a trader who, like her good friend Craig Sams, had no background in food production and, also like Sams, came to prominence in the Soil Association in the early 1990s.

The growing influence of retail

Charlotte Mitchell had suffered from multiple sclerosis when an art student, and came across one of Cherry Hills' books on diet and nutrition: this led her into the wholefood movement, and in 1975 she joined the staff of Real Foods in Edinburgh. Subsequently, she became its Trading Director. To find suppliers, she attended many food fairs, and this is how she met Patrick Holden, Lawrence Woodward and Craig Sams, during the early 1980s. In the mid-1980s she produced *The Organic Wine Guide* with Iain Wright, and contributed to the *Green Consumer Guide* (apparently not seeing any contradiction in the title, as Julian Rose would have done). High-level positions in the Soil Association followed: Treasurer in 1991 and Chair for seven years from 1992, during which period Craig Sams was Treasurer. Mitchell's view of the Association was surprisingly non-holistic, in that she felt it placed too much emphasis on farming and too little on food, as if the two had ever been regarded as separate in organicist thought. (Perhaps worse, to someone of an entrepreneurial spirit, its name was not 'sexy'.) The implication seems to be that the interests of retailers should take priority over those of producers. Mitchell saw supermarkets simply as a fact of life: if the organic sector was to grow, then organic products simply *had* to be sold in them. (Mitchell herself helped to establish organic foods in Waitrose in the final years of the twentieth century.) And while the presence in the organic movement of such large-scale organizations was worrying, they nevertheless ensured that the amount of organic produce available would expand, even if only for reasons of commercial profit, and this would in turn ensure the improved quality of ever-greater areas of soil.[40]

This book highlights the elements of continuity in organic history between 1945 and 1995, but if there have been changes, then one of the most significant is surely to be found in the area of attitudes to retail.

Differing perspectives are neatly encapsulated in a written exchange between Craig Sams, Tim Lobstein of the Food Commission and Lawrence Woodward in 1993, on 'Organic Junk Food'. An oxymoron? Not for Sams, by then taking his first steps towards substantial wealth with Green & Black's. '[W]hen Mars bars go organic,' he wrote, 'we'll know we've won the battle.' Lobstein and Woodward vehemently rejected this assertion, and it indeed requires a powerful capacity for fantasy to conceive of the organic pioneers endorsing it. They would more probably have felt that the battle would be won when Mars bars of any kind were no longer manufactured.[41]

The early 1990s saw some debate as to whether the organic movement should settle, commercially speaking, for being a niche market. The boom of the late 1980s had come to an end, and organic producers — along with many conventional farmers — were facing economic problems: some were on the verge of bankruptcy. Should the organic movement retrench and opt for supplying a small section of the fresh food market? Jonathon Porritt, addressing the BOF/OGA conference at Cirencester in January 1991, was quite clear that organic farming was about far more than supplying dedicated organic consumers: its environmental benefits made it an essential part of a 'blueprint for survival'. The government's attempt to marginalize it indicated a refusal to devise a comprehensive agricultural response to the problems which the 1990s would bring. But however important the movement might have been for ensuring a stable national and planetary future, individual producers were urgently concerned about the prospects for their own financial stability; many of them were beginning to wonder whether, as Nigel Dudley put it, the organic movement had 'made a fundamentally wrong decision in putting so much effort into the handful of large retail chains that control seventy per cent of the food retailing market'. The German organic movement, which was more successful than the British, had decided to avoid dealing with major retailers, and there were good reasons for this. Such businesses were unreliable — in 1990, Marks & Spencer had suddenly abandoned its interest in organic food — inflexible and wasteful: as much as thirty percent of organic produce might be discarded for being the wrong size or shape.[42] There were also the environmental considerations. Dudley accepted — grudgingly, it would seem — that there might have been a case for compromise if the tactics proved successful, but in fact they had not. The organic movement should have been concentrating on its most committed supporters who, by definition, were those consumers least likely to shop in supermarkets.

Alternatives to the retail system

Co-operatives

What, then, were the possible alternatives to conventional retail systems, for those who wished to avoid them? As we saw earlier, Newman Turner had addressed this problem by setting up his Whole Food Co-operative in the 1940s; such an initiative, working 'from the ground up', fitted in well with the organic philosophy. But co-ops, whether in organic or conventional agriculture, have always proved difficult to sustain, a state of affairs for which the farmer's strong sense of individualism has traditionally been held largely responsible. Helen Browning, Soil Association Chairperson, expressed the same thought more negatively by blaming farmers for a lack of producer discipline. Organic farmers and growers had certainly tried hard to make co-operatives work: indeed, Patrick Holden, in the mid-1980s, regarded them as the key to marketing success. By the start of 1985, growers in West Wales had founded an Organic Growers Co-operative, while early in 1986 Iain Tolhurst founded Cornish Organic Growers (COG). COG was initially seen as a possible blueprint for other such ventures, but it folded after only a year: perhaps through lack of 'producer discipline', as the rules were felt by some to be too restrictive. In 1986 another West Country initiative followed: this was Somerset Organic Producers (SOP), in which John Blake (Chairman), Laura Davis (Secretary), Dick Vernon-Dier (Manager) and Lawrence Hasson of the Organic Advisory Service played leading roles. SOP was one of three co-operatives which by 1988 had received legal and financial assistance from the government agency Food for Britain: West Wales Organic Growers and Eastern Counties Organic Producers (ECOP) were the others. SOP attracted members from Devon to as far east as Sussex, and took a professional approach to marketing. As it began to increase its sales to supermarkets, the issue of 'stricter discipline' raised its head, the multiples being very demanding. SOP fared better than

COG had done, surviving until March 1992; its chief problems were transportation and failing to attract new members. ECOP, with Rick and Pam Bowers and Grahame Hughes as its leading spirits, drew its members from an even broader area than SOP did: from Suffolk up to the East of Scotland. Thanks to a somewhat unlikely link with Geest Industries, who acted as its marketing agents, ECOP enabled its members to sell to Waitrose, Asda and Budgens, and offered technical advice and help with transportation. In May 1991, a specialist dairy co-operative of 25 founding members, British Organic Milk Producers, was established, whose Chairman was the biodynamic Sussex farmer Dirk Bauer. Bauer and his business partner Michael Duveen were responsible for the Busses Farm brand of yogurt, and took a two-pronged approach to selling their product, simultaneously supplying supermarkets and expanding their local market. In their view, a regular demand for large orders was essential, otherwise distribution costs pushed profit margins down to almost nothing.[43]

Farmers' markets and box schemes

Farmers' markets were another possibility. The first regular organic food market in Britain was launched in April 1992 at London's Spitalfields; the entrepreneur Elizabeth Taylor, whose idea it was, had been inspired by the success of similar markets in France. It was far from being a local event, though: stallholders included Arthur Hollins from Shropshire and Brian Tustian from Northamptonshire. Box schemes, on the other hand, had — at least, in those days — greater environmental credibility, given their local scale of operation; they belonged recognizably to the world of Miss Kathleen Talbot's Village Produce Association, which featured in H.J. Massingham's book *The Small Farmer* (1947). The pioneers of this approach were Tim and Jan Deane, at Christow, on the edge of Dartmoor.

The idea of by-passing markets altogether had first occurred to Tim Deane in 1982, when he was working as a stockman and tractor-driver on someone else's farm. After moving to Christow in 1984, he and Jan began growing organic vegetables to sell, which they did through local shops and sometimes in Bristol or even London if they could arrange transport; but the time and effort involved tended to cancel out the small gains. They therefore welcomed being involved in SOP, which enabled them to grow a larger quantity of vegetables and made them more money. But by 1990, smaller producers were finding themselves increasingly disadvantaged, and, around this time, Tim Deane experienced a moment of illumination in Sainsbury's, which 'shone a light on the whole murky business of marketing organic vegetables'. Deane came across what he recognized as his own white cabbages, which he had raised with such care and delivered to the co-operative fifty miles away. They had then travelled to Lampeter and on to the Sainsbury's depot at Warrington, in order to end up in a store ten miles from where they were grown. The Deanes had sold these cabbages at 9p a pound, and a week later they were on sale, clear-wrapped, at 40p a pound. What sense was there in aping the ways of the conventional market? Some other approach was needed, and it was the Somerset grower Charles Dowding, another SOP member, who provided the clue to what this approach might be. Dowding was selling some of his produce through boxes, purely for pleasure, but did not think the idea had any profitable potential. The Deanes, however, were facing serious financial difficulties and therefore willing to experiment with something more environmentally sound than having anything to do with Sainsbury's. In 1991 they invited local people to subscribe to a scheme for vegetable boxes; eighteen responded, and by the end of that season, through word-of-mouth recommendation, Northwood Boxes had 45 customers, all within a three-mile radius. SOP's closure in March 1992 lent an extra urgency to the Deanes' endeavours, and that summer they increased the range of crops they grew, organized an efficient packing-house and enlisted the help of Exeter Friends of the Earth in delivering to households in the city. They had 120 customers and the farm's profitability doubled.

> This was a revelation to us — to be appreciated ... for our work and the quality of what we grew and the way that we grew it. The farm became part of the community around it, serving a tangible need, and, if our customers were to be believed, promoting family health and rekindling pleasure and anticipation in cooking and eating ... As well as that we gave them contact with the source of their food, the land and the grower, and for many that was rain in the desert of bland

anonymity that rules the retail trade. As time went on we were also able to express something of what organic farming means ...

The Deanes also started to enjoy something like financial security: a predictable income, with cash in hand. It was truly a virtuous organic circle. They aroused the interest of Channel Four television, who made a programme about the box scheme, and in 1993 were able to sever all connection with the wholesale market, packing nearly two hundred boxes each week. Jan Deane took the success story out to a wide audience, with the Soil Association's help, speaking at various conferences.[44]

****

So in the mid-1990s, with the rise of farmers' markets and box schemes, the organic movement had found a way, compatible with its philosophy, to ensure the economic survival of the small growers whom it had always championed. The mutation of the idea from its local basis to the 'regional' range of Guy Watson's Riverford Organics lies outside our chosen period. Let us bid farewell to the vexed issue of organic marketing while it is in one of its more upbeat phases.[45]

6. Ecology, Environmentalism and Self-Sufficiency

Rachel Carson's *Silent Spring,* first published in Britain early in 1963, looms large in the history of environmentalism — a milestone which blots out any view of the landscape through which one has just passed. It sometimes seems that nobody had heard of ecology or been aware of environmental damage until a modest marine biologist scooped the publicity pool and drew everyone's attention to these subjects. And lo!, the environmental movement sprang 'in warlike armour drest / Golden, all radiant', like Pallas Athene from the head of Zeus.

Organicist ecology before Silent Spring

Without wishing in any way to deny the impact and significance of *Silent Spring,* the point nevertheless needs emphasizing that the organic movement had been promoting an ecological perspective and expressing concern about environmental damage years before Rachel Carson wrote her book. Richard St. Barbe Baker founded the Men of the Trees in 1922 to combat soil erosion in Africa. R.G. Stapledon was writing in the mid-1930s on the natural and social ecology of rural Britain; *The Rape of the Earth* (1939) by G.V. Jacks and R.O. Whyte was a worldwide survey of lost soil fertility. Dr Kenneth Barlow and Philip Mairet, key figures in the early organic movement, had both sat at the feet of the ecologist Sir Patrick Geddes. In 1947, Mairet gave a talk at the Summer School of the Men of the Trees entitled *The Ecological Basis of Civilization.* He began by drawing a familiar organicist distinction between economy and ecology, arguing that Western civilization had thought too little about the sources of life-material. The destruction of trees in parts of the British Empire had seemed economically justified at the time, but this had been a short-term approach, a form of 'ecological sin' which drew an overdraft on the future instead of investing in it.

Mairet did not hope for the end of civilization, but believed that its salvation lay in somehow developing a culture which would relate to Nature symbiotically, not parasitically. This culture would look to improve the quality of life of other beings, not forcing them to human will but allowing them to develop according to their own nature and excellence.[1]

Richard St. Barbe Baker identified 1947 as 'one of the most significant years in the history of the Men of the Trees for it was at the General Meeting in the Chelsea Physic Garden that they assumed world leadership in earth-wide regeneration, launching the new Earth Charter which was translated into most languages.' Baker drew up the Charter with his former forestry pupil Henry Finlayson and it was presented in 1949. Three years later, Baker led the First Sahara University Expedition, whose object was to carry out an ecological survey which would estimate the speed at which the Sahara was advancing towards food-producing lands on its southern perimeter. Baker's account of the work to try and reclaim the Sahara through afforestation programmes can be read in *Sahara Challenge* (1954); the project was under way twenty years before *The Ecologist's Blueprint for Survival* appeared and demonstrates the organic movement's commitment to ecological projects during a period, the 1950s, which seems often to be regarded as merely a prelude to the more exciting events of the subsequent decades. Baker's autobiography, *My Life My Trees* was published, coincidentally, in the years that *The Ecologist* first appeared: 1970. It was re-printed in a new edition by the Findhorn Community in 1979, with an introduction by the Community's co-founder Peter Caddy, demonstrating Baker's links with a younger generation of earth-healers. He was a frequent visitor and offered valuable advice on how to transform 'a barren stretch of sandy coastline into a wonderful garden possessing rich soils'.[2]

The concept of ecology appeared frequently in the pages of *Mother Earth* in the years BC (Before Carson); indeed, it appears in the first essay of the first introductory issue, 'Why It Happened', by Eve Balfour, as she urges that 'a steady increase in our knowledge and understanding of Nature's biological laws and the significance of the complex interplay between living organisms must be brought about. The investigation and interpretation of this interplay has its own science called ecology ...' In the Spring 1948 issue, Jorian Jenks related the principles of ecology to the practice of mixed husbandry, arguing that 'the more that the farm ... is built round the machine the less likely it is to conform to a balanced ecological pattern. Yet such a pattern is ... indispensable as a basis for biological efficiency ... [E]cology can confirm and enrich traditional practice ...' The same issue carried an article by St. Barbe Baker on 'Applied Ecology in Forestry', while that autumn there was a piece on the impairment of hill lands through failure to study their ecology. One particularly interesting item from the early years is a long article entitled 'Agricultural Ecology as a Branch of Science': interesting because its author, Dr Frank Fraser Darling, would make a considerable impact with his 1969 Reith Lectures, *Wilderness and Plenty*, a canonical text for a younger generation of environmentalists. The organic pioneers were promoting his work nearly twenty years earlier. Fraser Darling defined ecology as 'the science of causes and consequences', though he wondered if perhaps it was more accurately 'an attitude of mind' than a science: the attitude of someone who could predict where certain agricultural practices were likely to lead if they interfered with the natural order. Later in the 1950s, Rolf Gardiner reviewed Fraser Darling's *West Highland Survey* under the heading 'An Ecological Masterpiece'. For H.E. Lobstein, the ecological approach to agriculture was the antithesis of the economic approach: they were divided by a 'Green Curtain'. While the economist's 'rational' approach to profitable agriculture might bring benefits in the short term, its neglect of ecological principles would, in the longer term, prove destructive. Lobstein quoted Edward Hyams' parable of the apparently unprofitable honeysuckle whose destruction brought in its wake the disappearance of bees and hence of food. Such was the Soil Association's commitment to the concept of ecology that in April 1954 it established an Ecological Research Foundation, whose purpose was to take over the Association's interests in the farms at Haughley and, rather more ambitiously, to find out 'the ecological consequences of all human activities'. The project proved short-lived, but the will existed.[3]

The fact is, that during the 1950s there was considerable interest in ecology and its application to farming. In 1956, E.M. Nicholson, Director-General of the Nature Conservancy, addressed the British Association for the Advancement of Science on that topic, discussing the need for research into soil erosion and into the effects of farmers' attempts to eradicate wild places and pests. Farmers and foresters,

Nicholson argued, are themselves predators, and ecologists should study their activities in relation to the balance of nature. Jorian Jenks envisaged the organic movement playing a leading role in establishing new sciences of agricultural and horticultural ecology: much more demanding sciences than the simplistic, monocultural approach of orthodox agriculture. Only through close observation and much trial-and-error could the organic husbandman achieve the ecological balance which would 'enable him to take full advantage of natural affinities and antagonisms'; an area that the HDRA was investigating. Such an approach would help to control pests without the *ad lib* application of pesticides. The Soil Association was particularly excited by a letter on 'The Relationship of Man and Nature' published — astonishingly, as we might think today — by the Royal Bank of Canada and stressing the essential need to respect the ecological framework without which human life could not survive. Here was a rare instance of economics in sympathy with ecology.[4]

The organic movement and the new environmentalism

We can see, then, that an interest in ecology was not something which emerged in the organic movement as a response to Rachel Carson's best-seller; it was integral to the organic philosophy from the start. The organicists understandably gave an enthusiastic welcome to *Silent Spring*: Dr Reginald Milton reviewed the US edition before the British one was published, summarizing its thesis as follows: 'That attempts to control the natural processes of life by the use of chemicals ... must be carried out with great caution, because the eradication of one component of natural life can have far-reaching effects upon a much wider sample of the natural kingdom.' Milton noted that one American reviewer had identified 'organic gardeners, anti-fluoride leaguers, those who cling to the philosophy of vital principle, pseudo-scientists, and [food] faddists' as probable enthusiasts for Carson's ideas. As so often, the self-styled rationalists resorted to abuse, and a writer in *New Scientist* followed suit prior to *Silent Spring*'s British publication with an attack which relied a great deal on lofty contempt. Protected by the pseudonym 'Geminus', the reviewer anticipated a positive reception for Carson's message among devotees of wholemeal bread, nudists, and 'tweedy gentlewomen' living in 'picturesque towns in Sussex', but 'earnestly hope[d] that Miss Carson's book [would] not make the wider impact she has sought'. It is a pleasure to savour the extent to which 'Geminus''s earnest hopes have been so thoroughly dashed.[5]

For those familiar with the organic movement's history prior to 1962, *Silent Spring* seems less like the herald of a new era than it does a confirmation of what the movement had been saying for twenty years or more. Like the organicists, Carson argued that monoculture was a dangerous rejection of nature's variety and its checks and balances; that pesticides and herbicides were potentially dangerous and organophosphates probably carcinogenic; that vested interests promoted these poisons through aggressive marketing; that agricultural chemicals were a threat to earthworms, whose value to the soil Darwin had established (Sir Albert Howard had written an introduction to Faber's 1945 re-issue of Darwin's studies of the earthworm); that chemical controls were self-defeating because pests developed resistance to them; that biological controls must be experimented with; and that scientists should develop a sense of humility in the face of complex natural systems and, particularly, of the ecology of the soil, a subject which scientists had neglected. 'The very nature of the world of the soil has been largely ignored,' Carson wrote, though she did not examine the work undertaken by those who were exceptions to her generalization: Waksman, Howard and Albrecht, for instance. She shared their philosophy, though, concluding the book with some thoughts of the kind which the organicists had been expressing for many years. The 'barrage' of agricultural chemicals was '[a]s crude a weapon as the cave man's club', ignoring the delicacy of the fabric of life; the notion of 'control of nature' was 'conceived in arrogance' and led to tampering with vast forces which were not understood.[6]

Silent Spring's impact gave a powerful boost to the organic case, and although the Soil Association did not greatly benefit from it in terms of increased membership, it was an Association member who helped promote it in the British press. John Davy, Science Correspondent of *The Observer*, was given a full front-page spread in that newspaper's *Weekend Review* on 17 February 1963 to analyse the use of pesticides in Britain. Davy was not an alarmist: '[T]here is no immediate cause for clamorous anxieties about the hazards to humans in Britain — but there is certainly no cause for complacency. There can be no question of giving potent poisons the benefit of the doubt ...' He found the fact that insects were developing resistance the most sinister feature of pesticide use and so recommended study of biological methods in order to establish 'new and *lasting* balances ...between crops, pests, parasites and predators'. On the following page, *Silent Spring* was favourably reviewed by the Cambridge zoologist W.H. Thorpe; and Thorpe it was who wrote Carson's obituary for *Mother Earth,* following her death in the spring of 1964. He praised her for having 'given the conservationist and the naturalist the courage and the enthusiasm to build on the foundations which she laid'.[7]

Robert Waller and Michael Allaby

To Robert Waller and Michael Allaby, who both started working at the Soil Association early in 1964, Carson's boost to ecological awareness provided the Association with a fine opportunity to take a leading role in the cause of environmentalism. Allaby arrived at the Association by chance, but *Silent Spring* had strongly affected him and so he sympathized with the Association's aims. Waller's interest in ecology stemmed from his years as a producer of radio programmes on agriculture for the BBC's West of England service; he had met Sir George Stapledon and been deeply influenced by his ideas on 'human ecology'. In 1964, Faber published Stapledon's writings on that topic, which Waller had edited and to which he contributed a long introduction. By 'human ecology' Stapledon meant a point of view which took into account both man's biological nature and his self-created cultural environment. Stapledon had drafted his ideas in the late 1940s, but Waller believed them prophetic of the sorts of changes which 1960s environmentalism was beginning to work towards. In his role as editor of *Mother Earth,* Waller rapidly ensured the publication of some weighty articles on ecology, not by tweedy Sussex ladies but by respected scientists like Geoffrey Hull, who was a consultant biologist to the OECD, and Professor Lindsay Robb. Victor Bonham-Carter wrote on 'The Ecology of Rural Planning'. Hull made the point that even from a purely economic view the study of ecology made sense: a contaminated environment cost the national economy a great deal of money to clear up. An 'élite corps of ecologists' was required, scientists who had studied broadly and deeply in a wide range of disciplines.

(Readers will find this prospect comforting, inspiring or rather sinister according to their temperament.) Another prominent scientist who highlighted the dangers of ignoring ecological factors in land use was Professor G.W. Dimbleby, in a lecture to the British Association.[8]

Kenneth Mellanby and Edward Goldsmith

The organic movement's most distinguished British ally in the world of ecological science was probably Dr Kenneth Mellanby, who from 1955 to 1961 had been the head of Rothamsted's entomology section. He then became the first Director of Monks Wood Experimental Station in Huntingdonshire, chief research station of the Nature Conservancy. Monks Wood apparently boasted the largest concentration of ecologists in Europe. Mellanby spent thirteen years there, investigating the effects of industrial agriculture on the countryside, and for some years worked closely with the Soil Association, serving on its Council in the early 1970s: Michael Allaby was very friendly with him. His book *Pesticides and Pollution* (1967) appeared in the prestigious Collins 'New Naturalist' series. According to Anne Chisholm, in her 1972 study of ecologists *Philosophers of the Earth,* Monks Wood was a cautious institution; its employees were civil servants who did not wish to be linked to extremists and cranks, which makes Mellanby's membership of the Soil Association all the more significant. In the late 1960s and early '70s his pro-environmental activities featured frequently in *Span:* he spoke on water pollution, chaired environmental conferences, addressed the Association's Wessex Group, gave the 1971 Essex Hall Lecture on world pollution and reviewed a variety of books. In the end, though, Mellanby's scientific detachment caused him to take a more equivocal attitude to the organic movement. By 1975, in his book *Can Britain Feed Itself?* — Lt.-Col. G.P. Pollitt had asked the same question thirty years earlier and concluded that the answer was 'yes' if chemical fertilizers were used — Mellanby referred to 'the so-called "Organic movement"' and 'so-called "Ecologists"', with their 'emotional antipathy' to chemical pesticides and fertilizers. Evidently this was the civil servant in him, separating himself from the more extreme or cranky fringes. He still admired the work of the more scientific, 'excellent' organic farmers and believed that it might be feasible for British agriculture to go organic. Eve Balfour reviewed the book in the *Soil Association Quarterly Review* and was heartened by its conclusions, brushing aside Mellanby's jibes.[9]

There was no love lost, though, between Mellanby and Edward Goldsmith. Mellanby had contributed a letter to the first issue of *The Ecologist* on the environmental dangers of artificial fertilizers, but in the late 1970s Goldsmith subjected him to a brutal verbal assault entitled 'What Makes Kenny Run?' (The title refers to Budd Schulberg's 1941 novel *What Makes Sammy Run?* about an excessively aggressive writer who reaches the top in Hollywood through stabbing people in the back.) In it, Goldsmith condemned Mellanby for apparently using his eminent position to downplay the destructive effects of industrial agriculture. Mellanby responded by saying that his chief concern was 'The Truth', and that his supposedly contradictory statements were in fact the result of his attempts to give a full picture of highly complex evidence. All dogmas, 'even in what are called "basic ecological principles"', must be supported by facts. Goldsmith retorted that basic ecological principles should no more be considered dogmas than should be the Second Law of Thermodynamics. In the correspondence columns, Jon Tinker, the Director of Earthscan, expressed his dislike of Goldsmith's 'fundamentalist abuse' of Mellanby; while disagreeing with Mellanby's views on DDT and chemical fertilizers, he felt that Mellanby had a better claim than Goldsmith to be termed an ecologist. Brigadier Vickers of the Soil Association, however, praised Goldsmith's attack on the former Soil Association Council member, considering that Mellanby's views on chemical fertilizers were 'in defiance of all known facts'. By the early 1980s, Mellanby seemed to have lost enthusiasm for organic farming, believing that it had little to offer to the conflict between food production and wildlife conservation. Reviewing his book *Farming and Wildlife*, Angela Bates — who knew Mellanby well, having often travelled with him to meetings at Haughley — believed his attitude to chemical pesticides verged on the complacent. But her overall assessment of the book was fair-minded and she did not condemn him for heresy.[10]

The Soil Association may have seemed old-fashioned, narrow and élitist to many in the environmental movement of the late 1960s, but in Edward Goldsmith's view it contained some of that movement's most interesting people. He regarded the headquarters at Haughley as 'a veritable "brains trust" of farmers, scholars and activists': among them, he was particularly impressed by the scientists Professor Lindsay Robb and Harry Walters; by Eve Balfour, who had put the 'brains trust' together, and by Robert Waller, who, like Michael Allaby and Lawrence Hills, became one of the Assistant Editors of *The Ecologist*. Early issues of the journal were in fact printed on the same presses that produced Soil Association publications, before Goldsmith re-located it to Wadebridge in Cornwell. Since Waller and Allaby had been involving the Soil Association with the emerging environmental movement, it was fully to be expected that Goldsmith's project would attract them. Waller wrote what Goldsmith described as a 'brilliant' article for *The Ecologist*'s second issue, in August 1970, entitled 'The Diseases of Civilisation: The Declining Health of Urban Man'. Another factor which helped transfer the loyalties of Waller and Allaby to *The Ecologist* was, ironically, the appointment of E.F. Schumacher as Soil Association President from the beginning of 1971. Within three years, the success of *Small is Beautiful* would elevate Schumacher, formerly a brilliant economist at the National Coal Board (NCB), to the position of environmentalist guru: a role which would help kill him. As the Soil Association's new President, facing one of the more serious of its periodic financial crises, he insisted that it should limit its scope, not try to address all aspects of environmentalism, and concentrate chiefly on the cultivation of the soil. This might be seen as a narrowing of its holistic vision; nevertheless, this was what happened, and it was welcomed by David Stickland, Waller's successor as editor of the journal. There was further irony in that it was Waller and Allaby who had persuaded Schumacher to play a more active role in the Association.[11]

E.F. Schumacher

A point worth considering in relation to the vexed issue of continuity in the organic movement is that Schumacher had joined the Soil Association as early as 1951. During the war, he had worked as an agricultural labourer, a formative experience for him in that it set him thinking — like the early organicists and their 'orthodox' opponents — about the post-war future of British agriculture. At that time, though, his conclusions were more in tune with those of the orthodox school: he rejected the idea of a system of smallholdings, favouring instead State intervention on a large scale, with perhaps a third of agricultural land being nationalized. Like the early organicists, he wanted to stop the drift from the land, but believed that only the State 'could make agriculture into a career'. Schumacher's feeling for the land, discovered during his wartime period as an agricultural labourer, expressed itself in a love of gardening when he settled in Surrey in 1950. He had started working at the NCB and joined its gardening club and the Soil Association. Evidently he joined the Association for the advice on gardening which it offered, but his daughter and biographer Barbara Wood tells us that as a result of his membership '[h]is eyes were opened to a whole new way of thinking'. In this, he must have differed from many, perhaps most, of the Association's members, who would have joined it because they were already in sympathy with its philosophy. Although a prolific writer, Schumacher did not contribute to *Mother Earth* until Robert Waller published an abridged version of a paper he gave to a conference on factory farming which Ruth Harrison organized. This was on the problem of world food shortages and the role that Intermediate Technology could play in solving it, and it showed how far Schumacher had travelled from the views he espoused twenty years earlier: he was by now a wholehearted advocate of the organicist view that measuring productivity by output per acre was of far greater importance than the conventional economist's standard of output per man. There was, he maintained, no positive correlation whatever between them, and, for feeding the world, only the former was significant. Large-scale, capital-intensive, labour-saving techniques disadvantaged the poor, whereas human-scale technology could grant them independence and enable them to increase the productivity of their land. Schumacher is another figure whose importance tends to obscure what existed before him, but his ideas are part of the same train of thought that the early organicists were articulating two decades or more earlier; L.T.C. Rolt, better known as an industrial historian and the biographer of Brunel and Telford, had been advocating small-scale agricultural technology in 1947.[12]

Considered dispassionately, *Small is Beautiful* is an unlikely commercial success: a collection of essays and talks on 'seemingly disjointed subjects', as the blurb on its dust-jacket admitted, many of them first published or given several years previously and mostly on aspects of economics. The title itself was not Schumacher's, being suggested by the publisher

Anthony Blond; but it resonated with a public distrust of powerful, faceless organizations, whether commercial, bureaucratic or military. And it appealed to the organic movement, with its characteristic preference for the smallholder over the large-scale farmer, for the local and regional over the centralized, and for the responsible entrepreneur and craftsman over the capitalist combine. We shall consider the religious dimension to Schumacher's thought in Chapter 9; here, we can note one aspect of his philosophy which identified it closely with the organic movement: his sense of the limits to human activity imposed by the finitude of certain natural resources. Modern man, in his misconceived battle with nature, laboured under the delusion of unlimited powers, but was destroying 'the irreplaceable capital ... without which he can do nothing' (a point which Sir Albert Howard had made thirty years earlier). Schumacher's observations on fuel consumption, first published three and a half years before the oil crisis of 1973–74, were particularly far-sighted, anticipating the contemporary organic movement's interest in 'peak oil' by around 35 years. Schumacher based the case for organic agriculture, though, not so much on industrial agriculture's dependence on non-renewable resources, as on the environmental and social damage which industrial agriculture caused: the harm it inflicted on 'human ecology'. In Schumacher's view, agriculture's second task, after producing food, was 'to humanise and ennoble man's wider habitat', and it could fulfil this function only through obeying the law of return, through diversified cultivation and through de-centralizing its distribution system (or, as we would now say, by addressing the 'food miles' issue). Organic husbandry was thus an important means by which the environment could become healthy and beautiful once more.[13]

Curiously, the fact that Schumacher was Soil Association President when he gained worldwide fame did not boost the Association's fortunes, which caused considerable soul-searching among its senior members. Schumacher's fame may even have been counter-productive, since the demands it made on his energies contributed to his sudden death in 1977, at the age of 66, which robbed the Association of a brilliant and personable figurehead and led to a period of instability as first Eve Balfour and then the rather ineffectual Lord O'Hagan succeeded Schumacher. Nevertheless, Schumacher was a powerful spokesman for the organic movement in general and helped draw into it some important younger figures.

Resurgence

The organic movement was also connected to the new environmentalism through the journal *Resurgence,* founded in 1966, which Schumacher helped found and for which he wrote frequently. Its first editor, John Papworth, in his original statement of intent, described it as a 'peace publication': an expression of democratic grass-roots resistance to the power structures which were leading the world towards a 'monstrous biological anticlimax'. Papworth was a Soil Association member, and *Span* commended *Resurgence,* saying that it 'aim[ed] to translate many of our aims into political terms and to foster a new ecological approach to world problems'. Two months later, *Span* reported that *Resurgence* was accumulating debts and urged Soil Association members to make donations, since it would be a great loss if it were to close. It did not, of course, and became a close ally of the organic movement and a forum for the expression of organicist ideas. During its first ten years, Michael Allaby wrote for it on 'The Politics of Nutrition' and, following the winter of 1973–74, on the question of whether Britain could feed itself in the event of a long-term energy 'crunch'; while Robert Waller, in an article on 'Agriculture and Community', attacked agri-business for disregarding 'the social requirements of rural life and the balance of town and country', a problem he was observing at first hand in rural Norfolk.[14]

Self-sufficiency

John Seymour

Another frequent contributor to *Resurgence* was John Seymour, like Schumacher a guru to the 1970s 'eco-activists'. Like Schumacher too, he provided a link between the organic pioneers and the younger generation. Born in 1914, he had already lived a varied and nomadic life by the time he joined the King's African Rifles at the start of the Second World War: his experience included farm labouring, in Essex and South Africa; sheep farming; snoek fishing, and copper mining. He had also studied agriculture at Wye College and Leeds University, failing to complete either course; he disliked academic study and preferred to mix with craftsmen or, in Africa, the bushmen. During the war he was stationed in Ethiopia, Ceylon and Burma; in Ceylon he came to love what he saw of the peasant life, particularly the way in which feast days were linked to work on the land. On returning to England he worked briefly for the Ministry of Agriculture, finding employment for prisoners of war, but found his niche broadcasting, writing, and recording the histories of country folk and gypsies. In 1954 he married Sally Medworth, a talented potter and artist, and they settled at a remote property, The Broom, near Orford in Suffolk. Seymour recounted the story of their journey towards self-sufficiency in *The Fat of the Land* (1961), which Jorian Jenks reviewed in *Mother Earth,* describing the Seymours as 'Peasants by Choice' and identifying the strength of the peasant economy as 'its ability to cock a snook at [the] money economy'. A year or two after this review appeared, the Seymours moved to Fachongle Isaf in Pembrokeshire, a smallholding which was to become a centre of environmentalist thought and activity following the success of their books on self-sufficiency in the 1970s. Here again, the 1973–74 fuel crisis proved influential, demonstrating the value of independence in an economy vulnerable to any shortage of fossil fuels. The ideal of what Jorian Jenks, more than a decade before the television sitcom, described as the 'good life', was in fact an eminently practical, albeit demanding, response to the threat of a collapsing infrastructure; but it also appealed to those who wanted to escape a polluted environment and to practise a form of cultivation which respected nature rather than regarding it as an enemy to be quelled. Seymour's *Complete Book of Self-Sufficiency* (1976) was so successful that it had an unpredictable commercial result: the emergence of its publishers Dorling Kindersley as a force to be reckoned with, enabling Peter Kindersley to become one of the twenty-first century's growing band of 'organic millionaires'. And of course, John Seymour's income from writing ensured that he did not have to rely on his smallholding's produce in order to survive, while Sally's work enabled him to undertake the writing.[15]

Practical Self-Sufficiency/Home Farm

Seymour's success rapidly spawned a magazine for those who intended to take his ideas seriously: this was *PRACTICAL SELF-SUFFICIENCY,* established by KATIE THEAR and her husband DAVID, and run as a cottage industry from their smallholding near Saffron Walden in Essex. First appearing in 1975, it incorporated *Goat World* in 1982, changed its name to *HOME FARM* in 1983 without changing its format, and incorporated the newsletter of the Small Farmers Association in 1986. Also that year, it started to describe itself, in defiance of *New Farmer and Grower,* as 'Britain's Organic Farming and Growing Magazine'. Katie Thear had attended the Soil Association's 1975 course at Ewell on organic husbandry, and been particularly impressed by Dr Anthony Deavin's ideas.

At this point, readers will be anticipating more lists demonstrating the range of topics covered in *Practical Self-Sufficiency/Home Farm* and revealing more filaments of the organic network; and they will not be disappointed. During the 1980s the magazine featured articles on the following aspects of the self-sufficient life: animals and the law, arc welding, bantam hens, biological pest control, the black economy, chainsaws, chinchilla ranching, computers, coppicing, drystone-walling, field beans, financial planning, foaling, foxes (and how to deal with them), geese, green manures, herb borders, log-splitting, marketing for commercial horticulture, meat (natural), milk-machines, security, sheep and sheepdogs, sows (breeding of), tanning, water systems, weather forecasting, wheat-growing, wholefood cookery, wind power and zoonoses. This is in fact a very selective list. Although the magazine provided a wealth of practical advice, it could not guarantee a happy ending for those who followed it: Geoff Connolly's article 'We Failed!' recounted his personal experience of having to return to urban living.[16]

Various figures from the organic movement appeared in its pages. The Small Farmers Association, with which it was closely connected, had as its Vice-Principal Sir Richard Body, and Broad Leys Publishing, which published *Practical Self-Sufficiency/Home Farm,* also published Body's book *Red or Green for Farmers* (1987). Body contributed articles on agricultural policy during the mid-1980s. There were links with the HDRA: Lawrence Hills wrote on comfrey, Alan Gear on the HDRA's search for a new site and Sue Stickland on seed and potting composts. W.E. Shewell-Cooper also contributed, on the varied topics of farm composting and drying flowers for winter. Kite's Nest Farm was featured; Elm Farm Research Centre advertised its soil analysis service; from the Soil Association, Francis Blake advised on organic egg production; and Penny Strange promoted Permaculture. *Practical Self-Sufficiency/Home Farm* is typical of organicist publications during the period covered by this book, in its depth of commitment and range of content.[17]

John Seymour and environmentalism

To return to John Seymour: the term 'self-sufficiency' can connote an idea of retreat, remoteness, parochialism or even selfishness, but Seymour continued to travel widely in his later years, undertaking during the mid-1980s an itinerary of 'many tens of thousands of miles' across four continents in the company of Herbert Girardet for the BBC television series *Far From Paradise: The Story of Human Impact on the Environment*. This was a study with a long historical perspective, which looked at post-1945 developments in the light of what happened to ancient civilizations which had destroyed their forests and soils. The book of the series contains the same sort of photographic evidence — though in colour — which Jacks and Whyte had provided in *The Rape of the Earth* more than forty years earlier: gully erosion, forests turning to desert, dust storms, as well as more specifically contemporary images of crop-spraying, burning hedges and forests, and the abstract landscape patterns of industrial farming. The answer to such evils was 'the organic alternative', or biological farming, with its attention to many factors: 'the rate of unemployment and crime in the cities, the probable oil and natural gas reserves of this planet, the nutritional value of crops, the probable effects on soil degradation, erosion, disease, soil biology ...' In short, only the organic perspective of human ecology was adequate for assessing the impact of a particular agricultural technique or philosophy.[18]

Seymour and Girardet dedicated the book 'To the founders of the organic movement especially Lady Eve Balfour, and to those who are carrying on their work'. Girardet's final contribution to it was an essay called 'The Amplification of Man'. The essay does not mention one of the other founders of the organic movement, Philip Mairet, but nevertheless provides an interesting link with an essay which Mairet wrote at the beginning of our period, 'A Civilization of Technics'. Mairet — a close friend of the historian Lewis Mumford, who was to become a guru to American environmentalists in the 1960s and

'70s — prophetically saw the ecological damage which a technological society would inflict on the natural world. Although not opposed to technology as such — in fact, admiring many of its achievements — Mairet believed '[t]he age of machine power ... to have modified the status of Man in Nature'. Western society was now disposing of collective powers far greater than ever before, with the result 'that the technical means we propose to employ are making societies not more, but less, able to provide for their primary biological needs. Already before the war, it was brought forcibly to our notice that the technical hypertrophy of our culture was depriving it of contact with its living sources and tending, indeed, to dry them up'. Nonetheless, Mairet believed that our technological civilization could, in principle (as the work of the Tennessee Valley Authority showed in practice) help 'restore, root and branch, the natural, terrestrial organism upon which it is grafted'. His cautious optimism was to prove unfounded. Forty years after he wrote, Girardet powerfully summarized the effects of 'amplified man' during the intervening period, which had been 'to turn land covered with lush vegetation into wasteland [rather] than to do the opposite'. The results of this powerful technology, Girardet pointed out, were now there for all to see; 'progressive' scientists could no longer afford to ignore the findings of ecology, 'the science of planetary housekeeping'. There was no real conflict between ecology and economics: the former was simply '*long-term* economy', since running down planetary resources was 'bad economics'. The effects of such a policy were suicidal: as John Seymour wrote in the opening essay of *Far From Paradise,* mankind was 'destroying the community of which he [was] a part: the living community of the soil'.[19]

Sedley Sweeny and smallholdings

Less well-known than Seymour, but equally remarkable, was another ex-soldier whose ecological perspective led him to active participation in the organic movement, to write on the art of managing a smallholding and to run courses on self-sufficiency. This was Major Sedley Sweeny, who retired from the army in 1954, apprenticed himself to a Breconshire farmer for eighteen months and bought a 286-acre farm with hill rights on the Brecon Beacons and a flock of five hundred ewes. He came to know many hill farmers; one of them lent him a book by Sir Albert

Howard (or, in another version, by Friend Sykes) and he decided to go organic, rejecting the efforts of a Ministry of Agriculture advisor who tried to persuade him otherwise. He spent ten successful years at Pen Twyn, adding beef and dairy cattle and poultry to his sheep and finding that organic methods helped the health of his livestock: 'my vet's bills were much lower than many of my neighbours'.' From 1966 to 1968, Sweeny managed a Tibetan orphanage for the Save the Children Fund, then returned to the UK and spent a year as the Director of the Tibet Society in London, through which he met Ruth Harrison. Sweeny went back to Wales and in 1972 founded the Tibetan Farm School near Tal-y-Bont: Schumacher was its President. Ironically, the School foundered because the Dalai Lama, who visited it in 1974, refused to send any more students unless more modern methods were introduced. Sweeny's next project was to convert his own ten-acre smallholding nearby into a Smallholders' Training Centre, for British youngsters who wanted to 'return' to the land; its aim was to teach them that smallholding required both skill and patience. Sweeny has recorded laconically that 'These requirements proved too much, and we had to close down'. He subsequently made many visits to India and Nepal, undertaking rural development work and preaching organic self-sufficiency. Oxford University Press commissioned him to write a book, which was published in 1985 as *The Challenge of Smallholding*. He was active in the Soil Association, serving as a Council member and as chairman of its Smallholders Committee, and participating in its conferences. He knew members both of the older generation (Dinah Williams and Sam Mayall), and of the younger: Peter Segger interviewed him for the *Soil Association Quarterly Review*'s 'Focus on Members' column in 1983. He was also a life member of the HDRA. Sweeny eventually returned to his native Canada and, in his late eighties, was actively involved in the Self-Sufficiency Co-operative on Cortes Island near Vancouver and in supporting the Tibetan Ecoforestry Training Partnership, teaching Tibetan refugees and their Indian neighbours how to re-establish a sustainable environment. Sweeny is another of those figures who connect the earlier organic generation to later generations; as we have seen, he was familiar early on with the work of the pioneers; he knew Eve Balfour and read *The Living Soil;* he admired Stapledon's work on human ecology and was still in touch with Robert Waller in the mid-1990s. Like Seymour, Sweeny committed himself to organic husbandry and local self-sufficiency because he believed they were the way in which the environment could be restored after its 'terrible misuse' in the post-war decades. The alternative was that the human race would commit suicide.[20]

****

In a discussion with Sedley Sweeny at the May 1964 Attingham Park conference, Sam Mayall had started by saying that 'The Soil Association must be vitally concerned with everything that affects our total environment'; but he added the caveat that its foundation was agricultural husbandry.[21] As we saw above, Schumacher took the same view, and it is possible that his insistence on the Association restricting the breadth of its concerns helped reinforce its 'gentleman-farmer' image, when, on the face of it, the Association should have been playing a major role in the new environmentalism. This is not to say that the organic movement failed to contribute its ecological philosophy to the wider environmental movement, making the case for organic husbandry as a valuable means by which environmental damage could be reduced. But it was done more through publications other than the Soil Association journal: chiefly through *Resurgence*, *The Ecologist*, and *Seed*. There was also a fairly short-lived but very interesting magazine called *Vole*. We shall look (as always, necessarily briefly) first at the wealth of material on organicist concerns which made its way into the first ten years or so of *The Ecologist*, and then at *Vole*. *Seed* has already been discussed in Chapter 5.

Environmentalist journals

The Ecologist

Improbable though it might seem to those familiar with *The Ecologist* as a magazine prepared to grant space to the daughters of 'organic millionaires' who wish to promote their sex-toys businesses, Edward Goldsmith's editorial in its first issue was an uncompromising attack on western civilization's 'accepted values', with more than a hint of 'eco-fascism' about it. Agriculture entered the picture as early as the third paragraph, and — despite the fact that Goldsmith was referring to traditional, not industrial, methods of cultivation — as a harmful development, enabling man to increase his numbers 'beyond ecological requirements' and 'set out on his career as a parasite'. Goldsmith was concerned chiefly with the issue of supposed over-population, but referred to more central organicist themes and issues: the finitude of resources; the need to respect the delicate mechanisms of the biosphere; the dangers of monoculture and loss of bio-diversity; the tendency of medicine to treat the symptoms rather than the causes of disease; the dangers of reductionist science; the preference for quantity over quality in agricultural production.[22]

During its first year, *The Ecologist* featured several writers prominent in the organic movement. Lawrence Hills had his own regular column, 'Down to Earth' and in the journal's first issue provided a lively review of the organic movement's long-term bête noire Dr Magnus Pyke. Professor Lindsay Robb appeared in the first issue, asking whether a merger of agriculture and medicine was required: the answer of course was 'yes', just as Eve Balfour and Sir Albert Howard had thought a quarter of a century earlier. Michael Allaby wrote on a variety of topics, including the Conservation Corps (which the Soil Association had formed, in conjunction with the Council for Nature, in 1968), and the disappearance of hedges. Joanne Bower, of the Farm and Food Society, looked at the use of antibiotics in factory farming. Two contributors linked this new publication with the early organic movement: Roy Bridger, the crofter who in the 1950s had written for *Health and Life,* and L.B. Powell, who in the late 1940s had produced Country Living Books, which had promoted self-sufficiency skills and the writings of Lord Portsmouth (Powell's co-director), Ronald Duncan, Laurence Easterbrook and H.J. Massingham; and also those of James (Jim) Worthington, who later wrote regularly on poultry for the Soil Association journal. Bridger wrote on pesticides and Powell on the destruction of Britain's soil fertility. And there should be nothing surprising about such connections. It is now forty years since the first issue of *The Ecologist* appeared, but little more than half that time had elapsed between the post-war years and 1970. Once the Soil Association's journal had shed the philosophical dimension which the editorships of both Jorian Jenks and Robert Waller had provided, *The Ecologist* would be a natural alternative forum for those interested in the organic movement's broader implications.[23]

Agriculture played a significant role in the document which brought *The Ecologist* to wide public notice: *A Blueprint for Survival,* published in January 1972. The *Blueprint* was supported by a very impressive array of scientists and academics. Among them were Professor Derek Bryce-Smith of Reading University; the naturalist Peter Scott; C.H. Waddington, Professsor of Animal Genetics at Edinburgh; Dr E.J. Mishan, whose 1967 book *The Costs of Economic Growth* provided evidence for the 'limits' philosophy of the environmental movement; Sir Peter Medawar of the Medical Research Council; Sir Julian Huxley, whose links with the organic movement can be traced back to his involvement with the Economic Reform Club and Institute thirty years earlier, and who wrote a book on the Tennessee Valley Authority, a project much admired by some of the early organicists; and John Hawthorn, Professor of Food Science at the University of Strathclyde, who attended a couple of Cdr Stuart's Soil Association Weeks at Chirnside. The list also included a couple of names we have already come across: Professor G.W. Dimbleby and Sir Frank Fraser Darling. Michael Allaby, one of the *Blueprint*'s authors, recalled being taken aback by the extraordinary amount of media interest which it generated: it was far greater than anything which the Soil Association had ever managed, though much of what it said would have been familiar to supporters of the organic movement, being a re-casting of the sorts of ideas outlined by Mairet and St. Barbe Baker twenty years earlier. Given that Western humanity should be aiming to minimize ecological disruption if it wanted to have a stable future, changes in agricultural practice were essential; and the changes which the *Blueprint* recommended were those which *Mother Earth a*nd the HDRA's newsletter had been recommending since their inception: reduction in pesticide use in favour of integrated control systems; chemical fertilizers to be phased out and replaced by organic manures, ley systems and diversified farming methods, and so forth. This was, the *Blueprint* argued, simply a return to traditional husbandry, but a change from 'flow fertility' to 'cyclic fertility' aiming — as biodynamic farming does — to create as closed a cycle on the farm as possible. In calling for a much more efficient disposal of domestic sewage, the *Blueprint* was repeating a message whose origins can be traced back to Hugo's *Les Misérables* (1861) and, in Britain, to G.V. Poore's proto-organicist text *Rural Hygiene* (1893). As we shall see in Chapter 7, municipal composting for the sake of fertility was a cause to which the organic movement had been devoted since the 1940s. Then there was the need to deal with the threat of famine: a threat to which Massingham and Hyams had drawn attention twenty years before the *Blueprint*. (The spectre of famine, either of quantity or of quality, has always haunted the organic movement.) For *The Ecologist,* population control was one essential means of dealing with it; the other was to ensure increased food production, which could be achieved, not by extending agriculture into marginal lands but by improving existing farming. This did not mean, of course, adopting industrial methods, but preserving and restoring soil structure. The *Blueprint* argued that predictions of doubling food production within ten years through technological inputs were the fantasies of '"experts" who fail to take into account basic ecological, physical and biological principles'.[24]

The *Blueprint* looked for the creation of 'the stable society', defined as 'one that ... can be sustained indefinitely while giving optimum satisfaction to its members'. Its features included 'the invention, promotion and application of alternative technologies which are energy and materials conservative': for example, Intermediate Technology; and de-centralization of polity and economy, with the formation of self-regulating and self-supporting communities. In 1974, the Centre for Alternative Technology was established at Machynlleth in mid-Wales, and aroused the support of the Seventies Generation of organic activists, whose attempts at establishing communes reflected an interest in the second of the *Blueprint*'s proposed features of the new order. Without wishing to ignore the differences between the *Blueprint*'s essentially scientific and secularist approach to such issues, and the markedly religious approach of some of the organic pioneers, we should nevertheless note that the latter, too, had sought a stable society, rooted in the regional and local, ecologically sustainable and valuing community: particularly the community provided by the traditions of the Church and the agricultural cycle. Massingham praised surviving examples of such communities in his post-war travel writings, and it is no surprise to find him featured in *The Ecologist* in the mid-1970s. Given his opposition to the whole concept of planning, though, one can only speculate on what he would have made of a 'blueprint' for creating a devolved society of local communities, had he lived into his mid-eighties.[25]

6. ECOLOGY, ENVIRONMENTALISM AND SELF-SUFFICIENCY

The specifically organicist presence on *The Ecologist*'s board of assistant editors — Allaby, Hills, Papworth and Waller — lasted through the first ten years (Allaby stood down in 1979) and by the early 1980s many articles had appeared in the journal of a kind which would have been quite at home in *Mother Earth*. The nutritionist Ross Hume Hall contributed articles on typically organic issues, such as the adulteration of food's nutritional value by processing; the true contents of a pizza; the connection between diet and cancer, and — a topic which has caused concern in Britain in the recent past — the quality of hospital food: in this instance, in the United States. Joanne Bower covered the battle within the EEC to ban the use of hormones in livestock production and, with Anthony Etté, put the sort of case for the religious dimension to good husbandry with which Massingham's generation would have been very familiar. The Arizona soil scientist Gary Nabham wrote on the value of traditional crops as an aspect of cultural heritage, arguing that their destruction caused not just loss of genetic diversity, but social impoverishment. Laurence Roche wrote on the relationship between forestry and community, and Lawrence Hills on the capacity of 'forest farming' to supply food crops. Providing another link with the organic pioneers, Goldsmith interviewed the ninety-year-old Richard St. Barbe Baker; while in the mid-1970s Dr Kenneth Barlow trawled the national press to find 'Information for Survival'. Baker struck a rare note of optimism, seeing hope in the intense feelings which young people expressed about trees and the future of humanity. In general, the tone was one of pervasive gloom, with the cartoonist Richard Willson providing a succession of dystopian images. For the Tenth Anniversary Issue, Goldsmith composed an editorial headed 'Preparing for the Collapse'; though one might always argue, as the journalist Malcolm Muggeridge used to, that the most nightmarish prospect of all was to imagine contemporary Western civilization continuing indefinitely, and Goldsmith did display a sort of optimism in outlining the putatively beneficial effects which ecological apocalypse might bring in its train. But *The Ecologist* was — perhaps realistically — a darker publication than *Resurgence,* the Soil Association's publications, the HDRA newsletter, *Seed* and, particularly, Richard Boston's *Vole,* founded in 1977.[26]

Vole

Boston was a *Guardian* journalist and expert on beer and comedy who had made a cameo appearance in Jacques Tati's anti-technological satire *Playtime* (1967). *Vole*'s first editorial identified the journal as 'environmentalist' and 'conservationist'; 'in favour of decentralisation and small units' and opposed to consumerism and unlimited economic growth. It was interested in many organicist causes, including alternative technology, self-sufficiency, good cheese and good bread; like the organic movement it deplored the irresponsible squandering of land and energy, and Britain's dependence on finite sources of fossil fuels. Boston celebrated playfulness rather than prophecy and implicitly distinguished himself from *The Ecologist* in announcing: 'The Vole [no italics *sic*] does not claim to have a message that is going to save the world: indeed, it instinctively distrusts those who make such promises.' This did not prevent two of *The Ecologist*'s assistant editors from writing for *Vole:* Michael Allaby was a frequent contributor, on topics such as the domination of the food system by an ever-smaller number of companies, West Country mackerel fishing, Jacques Cousteau and BBC TV Bristol's Natural History Unit. He reviewed books, including one by his *Ecologist* colleague Lawrence Hills, who also contributed frequently: on English apples, for instance, leafmould and other matters of concern to gardeners. Allaby wrote a profile of Hills in the journal's second issue.[27]

There was not just a strong organic presence in *Vole,* but one which showed itself aware of the movement's history. The first issue contained a feature by Jeremy Bugler on the Mayalls' farm which referred to Friend Sykes' *Humus and the Farmer;* Edward Abelson wrote about H.J. Massingham, and D.B.C. Reed contributed pieces on Distributism, land reformer Henry George and the economic reform ideas of Major C.H. Douglas, whose theory of Social Credit had been integral to the organic movement's development in the 1930s. Christopher Hall wrote on Henry Williamson and Gillian Darley on the Whiteway Colony in the Cotswolds. Colin Ward and John Seymour wrote for *Vole* and, from among the younger generation, the nutritionist Gail Duff provided recipes and reviewed books on herbalism; tomato-growers Douglas and Penny Blair were featured, and John Butter, who ran the Cowley Wood Conservation

Centre in North Devon, wrote with a certain scepticism about self-sufficiency. Marion Shoard, whose *The Theft of the Countryside* provided so much ammunition for the organic movement in its criticism of agri-business, wrote several pieces for *Vole,* and Richard North, one of the magazine's assistant editors, reviewed her book. Another difference between *Vole* and *The Ecologist,* apart from their contrasting tones, was in their treatment of Kenneth Mellanby: whereas the latter journal bitterly attacked him, *Vole* made him the subject of a respectful and encouraging profile by Joanna Kilmartin. The magazine took an interest in home food production, water pollution, city farms, rural industries, protection of forests, the dangers of sugar, and the dark side of milk production. It monitored developments in the nuclear and oil industries, in medical merchandising, transport and road policies, and urban development. Despite its rejection of all political parties, it showed some sympathy for the Ecology Party, reporting on its 1980 conference and praising Jonathon Porritt's insistence that society's problems must be tackled from a completely fresh standpoint. (This was long before Porritt decided that capitalism could be given a human face.) *Vole* exhibited a love of particular places: it celebrated, and highlighted threats to, Britain's natural beauty and rural communities. It revelled in the traditional and enjoyed lambasting all that was ugly and intrusive in contemporary life: it favoured heavy horses, pub pastimes, Amberley Wild Brooks, brass bands, village schools, bee-keeping, dovecotes and wheelwrights, and disliked motorway service stations, television and micro-technology. One prescient issue — Volume 2, No. 2, in November 1978 — was devoted to the value of slowness, and although it did not develop the idea of the twenty-first century's 'slow food' movement, it implicitly referred to the philosophy of two early influences on the organic movement, the art historian Ananda Coomaraswamy and Eric Gill, in its call to blur the division between work and leisure. In short, *Vole* was full of articles reminiscent of those which Massingham and Ronald Duncan produced in the late 1940s, seeing no contradiction between such celebration of the past and involvement in the new environmentalism. Like all independent publications, it struggled to survive, and its attempt to campaign entertainingly for the environmentalist cause came to an end in the early 1980s.[28]

The 1980s: *The Ecologist* and Jonathon Porritt

The Ecologist and *Resurgence* survived, and continued to put the case for organic agriculture and an organicist philosophy of the environment. *The Ecologist* did so primarily from a scientific, and *Resurgence* from a spiritual and artistic perspective; the range of both journals was worldwide.

Although in the forefront of contemporary environmentalism, *The Ecologist* was happy to accept that an older generation of thinkers and activists had something to offer. James Fitzwilliams recommended environmentalists to take 'A Leaf Out of St. Barbe's Book' in an article on Richard St. Barbe Baker; his essay was nicely placed next to one by Vandana Shiva on the Chipko movement, which was struggling courageously to protect India's forests. Later the same year, 1987, a whole issue of the journal was dedicated to Baker's spiritual heirs. Brian Keeble wrote a long essay on Eric Gill, and Grover Foley contributed two articles on the American polymath Lewis Mumford, 'philosopher of the earth'. Mumford had been a friend of Philip Mairet and Kenneth Barlow, who had both, like him, been disciples of the ecologist Sir Patrick Geddes. Certain other long-term members of the organic movement appeared in *The Ecologist* during the 1980s: Doris Grant condemned fluoride as 'Poison in Our Midst'; Sedley Sweeny wrote about his experiences in India, and A.H. Walters summarized a speech on 'Nitrate and Cancer' which Dr Geoffrey Taylor had given to the McCarrison Society in 1982, not long before his death. John Papworth, Robert Waller and Lawrence Hills remained as assistant editors throughout the 1980s. Hills drew attention to an issue which would become increasingly important: the threat of seed patenting by corporations and the need to preserve vegetable varieties.[29]

The Ecologist expressed perennial organicist concerns. One particularly interesting article, by Bharat Dogra, looked at traditional Indian agriculture. Going back beyond Sir Albert Howard's period in India, Dogra recounted how in 1889 the British government sent the agricultural chemist J.A. Voelcker to study farming practices in India. Voelcker's conclusions were not what the government expected to hear: like Howard a couple of decades later, he considered that Indian cultivators at their best were at least as good as, and in some cases better than, English farmers and that it would be an easier task to suggest improvements for English agriculture than for Indian. Traditional agriculture was actually the way forward, Edward Goldsmith argued in an article on small farmers in Sri Lanka. In Britain, there was increasing evidence that monoculture led to soil erosion and that the situation was severe enough to warrant attention: a view all the more significant for being proposed by a scientist at the Silsoe agricultural college. Pesticides remained a cause for concern: Francis Chaboussou argued that they in fact increased the number of pests, while Peter Snell drew attention to the contamination of food by residues and the British government's attempts to play down the problem. The effects of chemicals on food quality were also examined by Hartmut Vogtmann, a good friend of the British organic movement. And a new danger, genetic manipulation, was beginning to emerge, as ever more complex and profitable 'solutions' were developed to deal with the problems which industrial agriculture created. But simpler solutions were emerging too. Although, as we saw in Chapter 2, Permaculture owed a good deal to the Keyline system of P. A. Yeomans, *The Ecologist* welcomed it as a new response 'to the ecological crisis brought about by de-forestation, air and water pollution, desertification and erosion', and Penny Strange outlined in some detail its practice and principles, a couple of years after Bill Mollison was awarded his 'Alternative Nobel Prize'.[30]

Looking back at *The Ecologist*'s first issue on the journal's twentieth anniversary, Edward Goldsmith praised the perceptiveness of its various contributors while regretting that the intervening two decades had only served to provide more evidence that their various anxieties had been fully justified. Michael Allaby's predictions about the Green Revolution, and Lindsay Robb's about the effects of food production methods on health, had been proved correct, Goldsmith believed: he cited salmonella in eggs, listeria in cheese and Mad Cow Disease, in support of Robb's thesis.[31]

By 1990, organic cultivation was established as one of the practical strategies which might be used to reduce environmental damage and pollution. Jonathon Porritt's manifesto for 'the politics of ecology', *Seeing Green* (1984), showed little awareness of organic history: he described *Small is Beautiful,* published a mere eleven years earlier, as a 'golden oldie'. In quoting Lester Brown of the Worldwatch Institute his starting-point was the same as that of Jacks and Whyte 45 years earlier: soil erosion and disappearing forests; and his remedies for the ills of industrial agriculture were mixed, rotational farming and Permaculture. Like Peter Bunyard and Fern Morgan-Grenville in their book *The Green Alternative* (1987), he believed that organic methods were likely to prove more economic as the cost of fuel and chemicals increased. The 'Age of Oil' was passing, and the future would see a return to 'the principles of good husbandry'. It is probably coincidental that in 1945 H.J. Massingham had written of *The Wisdom of the Fields* and that Porritt headed his section on husbandry and Permaculture 'The Wisdom of the Land'; but if so, it is a coincidence which reveals an affinity of spirit with his organicist forebears.[32]

Satish Kumar, organic history and a Green Books manifesto

Resurgence — the 'soulmate' of *The Ecologist,* as it described itself — began to expand its activities during the late 1980s and to take a greater interest in organic history. In 1986, its editor SATISH KUMAR undertook a pilgrimage round Britain during which he visited Eve Balfour, then almost 88 years old, at her Suffolk home; they made chappatis together. Kumar admired her gifts as a gardener, and mused on the parallels between soil and soul: how might wasteful negative emotions be transformed into energy for the soul, in the way that the waste products of cultivation are transformed by organic methods into fertility for the soil? By chance, the present author sent, on spec, an article about Massingham and Hyams to *Resurgence* only a couple of months later; Kumar, his interest in organic history stimulated by his visit to Lady Eve, responded by commissioning an anthology of early writings under the title — which he specified — *The Organic Tradition*. This volume appeared in a series of 'Organic Classics' along with an anthology of Massingham's writings edited by Edward Abelson, and re-prints of Walter Rose's *Good Neighbours* (introduction by Brian Keeble) and of L.T.C. Rolt's *High Horse Riderless* (with an introduction by John Seymour). The Organic Classics were launched at a day conference in Bristol in September 1988 on 'The Predicament of the British Countryside: An Organic Perspective', which celebrated the centenary of Massingham's birth and was jointly organized by Green Books — *Resurgence*'s new publishing venture, which Kumar had started the previous year — the Schumacher Society and the

Soil Association. 'H.J. Massingham,' said the booking form for the event, 'was a great exponent of a comprehensive philosophy of life based on organic, ecological and conservationist principles. Therefore the conference organised to commemorate his centenary will serve as a catalyst to all shades of the modern Green and environmental movement.' The conference provided a pleasing conjunction of older and younger supporters of the organic movement: John Seymour travelled over from Ireland, while Robert Waller needed only to take the train from Bath. The impressive array of speakers included Sir Richard Body, Jonathon Porritt, Julian Rose, Richard Young, Clare Marriage, Katie Thear and Gareth Rowlands. Thus the early organic movement overlapped with the world of New Age philosophy, prominent in both the articles and the Small Ads columns of *Resurgence*. There is another striking example of continuity to be found in the pages of *Resurgence* around this period: the presence of Dion Byngham as a book reviewer, writing about Henry Williamson and P. L. Travers. Byngham had been involved with the organic movement for more than half a century, having written regularly for Saxon's *Health and Life* in the 1930s; Travers, who lived on until 1996, had been on the editorial board of the *New English Weekly* and contributed to Lymington's *New Pioneer*.[33]

Green Books has of course not merely survived, but flourished, over the past twenty years, and promoted the cause of organic and sustainable cultivation; in 1989 it published Lawrence Hills' autobiography. One of its earliest publications deserves mention here, as a good example of how the organic movement's concerns blended into late-twentieth-century environmentalism.

The Countryside We Want : A Manifesto for the Year 2000 (1987), although edited by Charlie Pye-Smith and Christopher Hall, was the brainchild of organic farmer and Labour peer Peter Melchett, who in 1984 created 'The 1999 Committee' with the sole purpose of producing this book. It was, in effect, a vindication of the organic husbandry school's view — as related in the Prologue — of where the new, post-war approach to agriculture was likely to lead. As the Committee's writers summed it up, it had, during forty years, damaged both the physical environment and the social ecology of the countryside, eroding both the soil and rural community life. Food production methods degraded animals and the national diet. The practice of dumping food overseas created serious problems for Third World economies.

Much of the problem stemmed from the single-minded pursuit of productivity without regard to other factors; Melchett's alternative to this policy was to bring together 'a number of isolated interests to produce a coherent set of policies for the countryside of the future': a 'holistic' approach which Melchett described as breaking new ground. Breaking new ground in terms of differing from government policy, certainly, but not in terms of organicist thought. Of course, there were significant differences between the 1999 Committee and the Kinship in Husbandry forty years earlier (though both were convened by aristocratic landowners); but there were plenty of similarities, too, in their concern for small farmers and a well-populated countryside, their dislike of the Forestry Commission and of wasteful farming practices, their recognition of the link between food production methods and the people's health, and their belief that mixed farming beautified the environment and protected its ecological stability. Beyond the Kinship, a major influence on it, lay Stapledon's major, comprehensive analysis of farming, environment and social ecology, *The Land: Now and To-Morrow* (1935); and Stapledon's biographer Robert Waller filled his copy of *The Countryside We Want* with enthusiastic marginalia (as well as some less enthusiastic comments about Oliver Walston which, alas, cannot be repeated here).[34]

The Soil Association and 1990s environmentalism

Although focusing on Britain much more than *The Ecologist* did, the Soil Association remained aware of international developments affecting the environment. In 1991 Colin Hines drew attention to the new GATT agreement, 'a gleaming new engine for accelerating free trade', and imagined readers asking what it had to do with organic agriculture. In outlining its probable dangers, Hines was maintaining the organic movement's traditional mistrust of free trade; those familiar with the movement's earlier history will know that it paid considerable attention to the workings of the international finance system. The homogenization of trade standards for which GATT was working would mean a lowering of environmental standards, with any country attempting to implement tougher legislation being likely to suffer retaliatory trade sanctions. Hines called for policies of 'green protectionism' to bolster regional self-reliance in food and ensure it was produced with minimum environmental degradation. Tim Lang drove home the point the following year, shortly before the 'Earth Summit' held at Rio de Janeiro in June 1992. (The issue of *Living Earth* in which his article, 'What Future for the Earth?' appeared, featured on its cover a typically repellent — though not inaccurate — Ralph Steadman cartoon portraying humanity as a spoilt baby defecating over the entire globe.) GATT officials had finally — after twenty-one years of deliberation — addressed environmental issues in a document which argued 'that increased trade brings the wealth needed to protect the environment'. Lang itemized examples from all over the world of how 'food imperialism' was imposing the model of Western intensive farming on societies which still practised traditional methods. The result was an increase in national debt which led directly to environmental degradation: in Latin America and the Philippines tropical forests were being felled to pay off the debt. How, Lang wondered, could the 'wealth' brought by increased trade recover the lost forests? The Soil Association set about establishing a Responsible Forestry Programme, to provide a certification service for producers of 'sustainable' timber, thereby 'encouraging good forest management instead of destructive logging'.[35]

The Association was also busy making the case for organic farming as beneficial to Britain's environment. In February 1994 it hosted a seminar on 'Organic Farming and the Environment' at the House of Commons, which included presentations by broadcaster and dairy farmer John Humphrys and the naturalist David Bellamy. Soon afterwards, Patrick Holden was involved in discussions with the Minister of Agriculture (Mrs Gillian Shephard), the Secretary of State for the Environment (the organic movement's old friend John Selwyn Gummer) and the European Commissioner for the Environment. At the end of 1994, the Soil Association issued its policy statement *Subsidies Without Set-Aside,* 'outlining proposals for the full integration of agricultural and environmental objectives within future CAP reforms'. This document was greeted with 'widespread acclaim', and MAFF 'immediately undertook to ...': no, not begin to implement its suggestions, but 'prepare a co-ordinated response to its content'.[36] However, the Association had enjoyed some success the previous year when UKROFS adopted its environment and conservation standards as part of their national organic standards. Protection of the environment was now — officially, as it had always been unofficially — an integral part of organic agriculture.

We can conclude with a reference to the Lady Eve Balfour Memorial Lecture which broadcaster, CPRE President and 'embryo organic farmer' Jonathan Dimbleby gave almost at the end of our period, in October 1994. Dimbleby believed that a switch from farm price support to direct payments for environmental management would encourage a variety of farming methods and shift the emphasis from food production to other beneficial 'products' such as landscape and wildlife protection and an enrichment of communal life in the countryside. Organic farming would be the best means of achieving these ends. Dimbleby concluded by quoting from Schumacher, who had 'placed the concept of organic agriculture within the totality of what is at issue for all those millions of us who care about the environment.' Schumacher had referred to the 'battle with nature' and pointed out 'that if we win that battle we will be on the losing side'.[37]

7. The Role of Science

Lord Taverne's brainstorm

A common criticism of the organic movement in the twenty-first century and over the years is its supposedly 'unscientific' nature: the phrase 'muck and magic/mystery/mysticism' has been directed at it since time immemorial. One of the movement's most hostile recent critics has been Lord Taverne (the one-time Labour MP Dick Taverne), a barrister and 'Distinguished Supporter' of the British Humanist Association, and therefore admirably qualified to pronounce on matters of agricultural and nutritional science. He is a spokesman for Sense About Science, an organization which claims to have as its chief purpose the promotion of an evidence-based approach to scientific issues, and has made the case against the organic movement in the following fashion:

> ... [it] has murky origins; its basic principle is founded on a scientific howler; it is governed by rules that have no rhyme or reason; it is steeped in mysticism and pseudo-science; and, whenever it seeks to make a scientific case for itself, the science is shown to be flawed.[1]

Taverne follows up these measured, evidence-based assertions by informing his readers that the Soil Association was established in 1945 — which is not the case — and that its original inspiration came from Rudolf Steiner, which is not the case either. As far as 'murky origins' are concerned — Taverne is presumably referring to the radical Right influences prominent in the organic movement's early years — one might consider inviting him to examine his own glass house before throwing stones at the organic movement; but perhaps the former SDP enthusiast is content to belong to an organization with its roots in the

Revolutionary Communist Party and the journal *Living Marxism*. Readers of Taverne's outburst who have any critical sense at all will be wary of sweeping claims that organic standards have '*no* [my emphasis] rhyme or reason' — has Taverne studied the long process by which they were established and continue to be modified? — and that *all* evidence for the organic case is *always* wrong. Such statements are not about evidence; nor is his assertion that 'The organic farming lobby ... do not believe in the scientific method'. On the contrary, the agricultural scientific establishment has shown minimal interest in encouraging research into organic methods: the reason being that powerful vested interests will discourage any sort of research which might render their products unnecessary. Although the organic movement has weakened its own case in this regard by identifying itself so closely with consumerism that it has created its own vested interests, 'the power of the multi-million pound organic farming lobby' to which Taverne refers, remains insignificant in comparison with the resources of its opponents.[2]

Taverne's attack on the organic movement is so simplistic, inaccurate and intemperate that one inevitably suspects motives which have little to do with the pure light of Reason. It raises questions about the nature of science and about what should count as scientific knowledge; about whether scientific method should be considered the only valid means of gaining knowledge, and whether the knowledge so gained should be prioritized over other kinds; and about the extent to which Taverne's 'science' is bound up with corporate interests. Taverne prefers abusive language to evidence, so this chapter's main aim will be to provide further evidence that during the period 1945-95 there was always present a strongly scientific strain to the organic movement. 'Further' evidence, because much is provided in other chapters. Just to re-cap: Taverne claims that the Soil Association's establishment was inspired by Steiner. It was in fact inspired chiefly by the work of two internationally respected scientists, Sir Albert Howard and Sir Robert McCarrison, and Eve Balfour's initial response to her discovery of their work was to instigate the Haughley Experiment, of which more below. In Chapter 4 we saw that medical scientists such as Stanton Hicks, Sinclair, Burkitt and Cleave made important contributions to organicist nutritional theory, and saw in Chapter 6 that top-flight ecologists such as Fraser Darling and Kenneth Mellanby lent their support to the organic philosophy. It is also interesting to note that no less a figure than Professor J.D. Bernal contributed to a series of Soil Association lectures on organic husbandry given in London during the winter of 1956–57, and, as we saw in Chapter 2, John Davy, Science Correspondent of *The Observer,* was prominent in the Soil Association. Taverne's arrogance in dismissing such a roll-call of scientists is all the more remarkable given his own lack of scientific qualifications.

So, let us turn to an 'evidence-based' look at some of the other scientists and at some of the scientific activities which played a part in the organic movement between the mid-1940s and the mid-1990s. By doing so, we can assess the validity of Taverne's claim that the organic movement does not believe in scientific method.

Science in Mother Earth

The first point to note is that during its first six or seven years, the Soil Association's journal *Mother Earth* paid considerable attention to the organic movement's attitude to science. One immediately anticipates the objection that the very name *Mother Earth* indicates a sentimental, mystical bias, and certainly there were those on the Association's Council who were unhappy about it on these grounds. Ten years after the first issue there was much debate about whether the title was 'a distinct handicap' when trying to arouse interest in the Association among business people and scientists. One anonymous member of the editorial board put the case for getting rid of the title; another put the case for retaining it, arguing that the word 'Nature' had long been associated with poetical imagery of babbling brooks and primrose banks, but was nevertheless the title of a highly respected scientific journal. A mandate of members proved inconclusive, so the title was kept. In one of the earliest issues of *Mother Earth,* Jorian Jenks summarized a lecture given by Professor J.A. Scott Watson, Chief Education and Advisory Officer to the Minister of Agriculture, approving the Professor's call for 'further research, of a truly fundamental kind, in agricultural problems', particularly by those who had knowledge of farm practice. The Soil Association maintained that existing knowledge of agriculture was limited but that the dogmatic attitudes of certain agricultural scientists stood in the way of its development.

In an editorial the following year — 1948 — Jenks expressed a certain sympathy with the cynical view that all agricultural science had achieved so far was 'to discover abstract explanations for practices which farmers and gardeners worked out for themselves long ago'. But perhaps the greater problem was that agricultural science proceeded from the wrong set of assumptions, in regarding natural processes as mechanical operations, its recent development being influenced by the intensive expansion of manufacturing industry. There had been plenty of investigation into the chemistry of soil fertility, but next-to-nothing on the biological effects of mechanization; though a study of biology is no less 'scientific' than a study of chemistry. What, after all, is science? Dr W.G. Macdonald addressed this question in a lecture given at a Chase Open Day in 1949, applying the adjective 'mystical' to the belief in the supposed infallibility of what happens in the 'glittering apparatus' of the laboratory. The truly scientific approach to cultivation was to observe and follow the laws of Nature, whose experience of plant production greatly outweighed that of those who manufactured National Growmore fertilizers. *Mother Earth* quoted with apparent approval the agriculturalist and broadcaster Ralph Wightman, whose attitude to science in farming was somewhat less provocative than Macdonald's. Wightman agreed that science did not ('yet') know all the answers, but warned against sneering at the advances which had been made. Nevertheless, adherents of 'the NPK mentality' had underestimated the importance of soil bacteria and humus.[3]

The supposed organicist objection to science was in fact an objection to the dogmatism of certain scientists and was, therefore, entirely pro-science. As Jenks pointed out in his Editorial Notes for the Autumn 1950 issue of *Mother Earth:* 'The moment that knowledge crystallizes into set formulae and techniques ... it begins to lose some of that vital dynamism which is one of science's most cherished characteristics.' Within the world of science there was an important battle between the dynamic and the statistical interpretation of natural processes: a battle which was evident at the annual meetings of the British Association for the Advancement of Science, whose proceedings *Mother Earth* monitored. The important point was that some scientists represented the general outlook of the Soil Association in their awareness of the harm which chemicals might inflict on insects, and their insistence that much more biological experimentation was essential. Jenks pointed out that official agricultural scientists admitted to complete ignorance of the field of soil ecology and of its relationship to plant and animal health. He was asking for more scientific research, not rejecting it. The history of science is, after all, a tale of discarded theories, and of evidence and insights which took many years to be appreciated. In his 1950 Sanderson-Wells Lecture 'Food and Folly', Sir Cedric Stanton Hicks told his audience that only 25 years earlier a distinguished medical colleague had denied the existence of vitamins; in the development of nutritional theory, medical policies had 'been laid down with dogmatic confidence at each stage'.[4]

The Soil Association's attitude to science, as evidenced in the writings referred to above, seems reasonable and balanced; but it was far from satisfying an agricultural student at New College, Oxford, David Henriques. Henriques — whose father Col. R.D.Q. Henriques, was a soldier, novelist, Gloucestershire farmer and Soil Association Council member — deplored the Association's attitude to scientific evidence and to orthodox agricultural scientists. While praising Eve Balfour for her caution in referring to certain data as 'indications' he condemned other supporters of the organic movement for their tendency to consider those data not merely as firm evidence, but even as clear proof. Henriques felt that *Mother Earth*'s tone displayed a clear element of bitterness towards the very people it should have been seeking to persuade, while his experience as an agricultural student had revealed that in fact a good deal of attention was paid to soil ecology. He concluded that some of his fellow students regarded *Mother Earth* as 'a scurrilous and irresponsible journal' (though it seems debatable whether this judgement reflects more on the journal or on the mind-set of the students in question). In reply, C.D. Wilson, a former industrial chemist, wrote a lucid outline of the Soil Association's position, arguing that the Association had every right to criticise an established hypothesis, patiently collecting facts which might point in another direction; just as defenders of the established hypothesis had a right to criticize its potential successor.[5]

Although there were religious, ethical and economic arguments for organic husbandry, Julian Herbert was convinced that the case for it would 'ultimately stand or fall by its scientific justification'; it would be an error to establish 'permanent "unknowables", shrouded in magic or mysticism'. But Herbert perceived something of a reaction against science, on account of its destructive power and commercial prostitution. In order to regain its status, science must cease to be 'the exclusive prerogative of specialised institutions' and return to close observation of natural phenomena: this meant studying things in their interconnectedness, rather than in isolation. Research workers should operate together, and the organic movement should encourage the establishment of research into hitherto neglected areas.[6]

As we can see, implicit in the debate about science is the question of how it should be defined. Not long before his death, Scott Williamson attempted to answer the question in a review of Anthony Standen's book *Science is a Sacred Cow*. It is interesting to note that, just as people of Lord Taverne's outlook condemn the organic movement as a 'religion', so Scott Williamson condemned the elevation of science into a religion at seats of learning and described Soil Association members as 'merely non-conforming protestants against this sacrilegious procedure'. Scott Williamson's idiosyncratic style, with its generous scatterings of capital letters, hints at the obscurity which would mark his posthumously published work *Science, Synthesis and Sanity* (1965); but his review-article makes an important distinction between the practice of science and the use to which technologists and commercial interests put its results. The issue of what counts as science was far from academic in Scott Williamson's own life, proving crucial in the closure of the Pioneer Health Centre. For Scott Williamson, the Centre was a scientific experiment, but the surgeon Sir Ernest Rock Carling had made the damning judgement: 'This is not science as I know it!' Following Scott Williamson's article, philosophical discussion of science disappeared from *Mother Earth* for many years; but the journal continued to carry many scientific articles, to the extent that it could be regarded as primarily a scientific journal. This was certainly how two Australian members regarded it when they wrote in 1964 to complain about the supposedly unwarranted intrusion into its pages of Rolf Gardiner's religious philosophy of agriculture. For those who care more about substance than labels, there is no question that *Mother Earth* dealt seriously with many scientific issues. To keep the evidence for this statement within reasonable bounds, we shall list some topics from the journal's first five years, and from the first half of the 1960s.[7]

Between 1946 and 1951, one can find in the pages of *Mother Earth* items and articles relating to ammonia, antibiotics, bacteria, calcium, carbon dioxide, composting, experiments agricultural and nutritional, insecticides, isotopes, lime (application of), magnesium, minerals, municipal composting, mycorrhiza, nitrogen, pests, potash, proteins, rotations in farming, soil structure and deficiencies, sulphuric acid, trace elements, viruses and vitamins. Between 1962 and 1965: the Advisory Council on Scientific Policy, algae, blister blight, cancer, coddling moths, composting, entomology, grassland management, hydroponics, irrigation, Lysenko's genetic theory, nitrogen, organo-phosphorous sprays, pollination, the rumen, soil fungi, superphosphates, warble fly and water pollution. These are not topics one would generally expect to find in 'mystical' publications, and perusal of the book reviews provides plenty of additional evidence that *Mother Earth* took an interest in developments in the world of scientific investigation. Above all, it needs to be remembered that the Soil Association grew, to a considerable extent, out of Eve Balfour's conviction that the nutritional thesis about the relationship between soil, crops and health which Howard and McCarrison were advancing in the 1930s required scientific investigation. The Haughley Experiment, which she began to prepare shortly before the outbreak of the Second World War, was the practical expression of her belief.

The Haughley Experiment

The Experiment, which looms so large in the story of the Soil Association's first 25 years, must be dealt with relatively briefly here, as but one aspect of the organic movement's relationship to science. The fullest account of it can be found in Eve Balfour's revised edition of *The Living Soil,* which contains as its second part 'The Story of the Haughley Experiment', a 180-page digest of the Experiment's post-war history. Lawrence Woodward's introduction to the Soil Association's 'Organic Classics' re-print of *The Living Soil*'s first edition contains a valuable assessment of the Experiment's purpose, principles and significance. *Mother Earth* kept Soil Association members up to date with the Experiment's progress and with the financial problems which it generated and which persistently threatened the Association's survival; the October 1956 issue (in fact headed *Journal of the Soil Association*) was dedicated to the Experiment. And in 1962 the Association published a substantial booklet summarizing the Experiment's progress since Eve Balfour first conceived of it.

The purpose of the Experiment was to investigate the 'ecological interplay' between the health of soil, plant, animal and man; it was 'a search for conditions attendant upon biological wholeness'. Considering nutrition as a cycle, it aimed to study this cycle as a whole through successive generations, on farm scale, under three different land-use systems. The first, a 'Stockless' section, carrying no animals and receiving no manure, would be fertilized by its crop residues and chemicals. The other two sections were stock-bearing. One, the 'Mixed', would receive the crop residues, and the animal manure produced on it, plus applications of chemical fertilizers; the other, the 'Organic' section, would receive only its crop residues and animal manure. This was a form of comparative research which did not exist in the world of agricultural science, and Balfour believed it was urgently required; the problem was how to fund it. One of the reasons why Howard did not join the Soil Association was his shrewd recognition that the cost of running the Haughley Experiment would be too great, given its highly ambitious nature. As Balfour explained in retrospect, 'it involved study of the interrelationships of soil, crops, and grazing animals in a fully rotational farm system': not merely expensive in itself, but unconvincing to the orthodox agricultural mind-set on account of its rejection of 'the conventional small randomized plot method' with its concomitant slavery of statistics. Attempts to eliminate variables — an essential aspect of the statistical approach — 'would destroy the "whole" [the Experiment] intended to study'. Furthermore, the Experiment intended to go beyond purely agricultural considerations: Balfour described it as 'an ecological search for quality in food [whose] objective was to enlarge our knowledge of the nature of health itself'. Since orthodox agricultural research was concerned almost entirely with increasing quantity, the Ministry of Agriculture was happy to dismiss Haughley as a matter for the Medical Research Council, which in turn regarded it as matter for the Ministry of Agriculture. The concept of agro-medical research was simply too novel for the conventional agricultural or medical scientist to take on board; but scientific advances are often made by pioneers working outside the bounds of officialdom and Balfour, strongly supported by Scott Williamson and Pearse in particular, persisted with the project.[8]

The Organic section, of which Friend Sykes and Deryck Duffy were co-directors, was the key to the Experiment. Its aims were twofold: to act as a 'control' against which to observe the effects of introducing chemical fertilizers on the Mixed section and of depriving the Stockless section of animal residues; and to see how far nature and man working together could establish in the agricultural food chain a biological balance similar to that found in wilderness areas. This necessitated the creation of a system which was as near as possible to a closed cycle, in order to study an ecological whole rather than a fragmented situation. Balfour explained the rationale behind this approach in the revised edition of *The Living Soil;* of particular interest is the orthodox agricultural assumption that such an agricultural cycle must result in a dwindling of productivity. Unfortunately, the Experiment was abandoned before any clear conclusion either way could be drawn.[9]

The Haughley Experiment was problematic from the outset. It was a huge drain on Soil Association funds, and many members were doubtful not merely about its financial implications but about whether the Association should be involved in research; on the other hand, there were those who thought that the Experiment was the main justification for its existence. The Editorial Notes to the special Haughley edition of the journal in October 1956 argued that it was wrong simply to assume the validity of organic principles and be content with demonstration farms. If the Soil Association wished to persuade agricultural scientists of its views, it must be willing to submit important issues to controlled experimentation. (The obstacle was, of course, that the Haughley Experiment by definition would not meet the criteria of scientists who had no interest in an ecological approach: the problem lay perhaps as much in the realm of the philosophy of science as in scientific practice.)

Dr Reginald Milton and Professor Lindsay Robb

Preparatory work for the Experiment was complete by 1952, and early that year Dr Reginald Milton was placed in charge of the sampling and analysis. A highly capable scientist, Milton had links with the organic movement which went back to the late 1920s, when he had worked with Scott Williamson on the biochemistry of the thyroid gland. During the war he was Biochemist in Charge of the Medical Research Council's Department of Industrial Medicine; after it, he returned to his role as a consultant biochemist. He had spent a short time on the Soil Association's advisory panel of experts for the Haughley Experiment before being invited to participate in its research work, and was familiar with the criticism that it was not susceptible to statistical treatment. Against this, Milton argued that the statistical approach to wider biological experiments was of limited value, since in any experiment involving a biological organism the variables could never be assessed — let alone controlled. Replicated plot experiments in agricultural research (Milton would surely have had in mind the organic movement's bête noire Rothamsted Experimental Station, with its famous Broadbalk field) often produced results irrelevant to conditions beyond the plot. Given the complex interplay of factors involved in the Experiment at Haughley, its limits could not be controlled; the Experiment was a pioneer venture which might produce data 'that could never have been obtained by plot or controlled agricultural experiment'.[10]

If Milton's views might be dismissed on the grounds that he was not an agricultural scientist, it was harder to make the same objection to Professor R. Lindsay Robb, who enthusiastically supported the work at Haughley. Robb was one of the world's top agricultural scientists and in the mid-1950s he was working for the Food and Agriculture Organization of the United Nations in Central America. Brought up on his family's dairy farm in Scotland, Robb, like Sir Albert Howard, had first-hand knowledge of agriculture, During his career he was appointed head of agriculture at London University's Wye College, then joined the agro-chemical corporation ICI, for which he toured Australia, New Zealand and South Africa. As ICI's chief grassland advisor, Robb fully understood the arguments for chemical farming, which makes his later support for the organic movement all the more notable. During the Second World War, he was director of agriculture for the British military in North Africa and used farming as a means of rehabilitation for sick soldiers. They were removed 'from an environment of lifeless barren sand into one of living soil, teeming with life in a setting of beauty and peace'. He later wrote of this period: 'Our agricultural approach to the problem of restoring the soldiers to health was entirely ecological. The problem had to be seen in its entirety as an inter-relationship with health as a whole, and man as a whole within a particular environment.' It was this ecological approach to problems of health and resources which attracted Robb to the Soil Association in the early 1950s: in other words, it was the Association's scientific philosophy which appealed to this widely experienced agricultural scientist. Robb believed that *problems* rather than *subjects* should be studied. He did not deny the value of the 'small plot' approach in agricultural research, but believed there was also a need for what he termed 'ecological exhibits' which would make it possible 'to explore and study what goes on within the pattern of Nature as we adapt and modify that pattern to our needs'. Robb also shared the central organicist interest in whether methods of cultivation affected food quality, concluding that the Haughley Experiment deserved support and envisaging that it might become a pilot experiment for larger-scale investigations.[11]

Robb's interest in Haughley was in fact shared by at least some scientists who did not support the organic movement. The journal *Nature* noted the Experiment's uniqueness, saying that whether or not it proved successful, it could 'scarcely be shrugged aside as of no importance'; and the organicists' long-term adversary D.P. Hopkins, whose book *Chemicals, Humus, and the Soil* was re-issued in 1957, wrote in the fertilizer trade press that year in support of expanding the Haughley research. There was evidently a feeling abroad in the world of agricultural science that the biological aspects of soil fertility required considerably more research.[12]

Evidence from Haughley

In June 1962, the Association published a survey of the Haughley Experiment's work since 1938, providing a summary of its farming operations since 1952 and drawing some tentative conclusions. The caution which Milton displayed belongs to a world remote from the sorts of claims which less scientifically minded supporters of the organic movement have sometimes made, and it would no doubt have disappointed those of a more enthusiastic disposition; but this was — so the Soil Association hoped — only the beginning of a long-term project, whose results might become much more significant with the passing of time.

We can note here that by the early 1960s there was evidence as follows: that the different forms of management produced differences in soil structure; that on the Stockless section the top soil turned more rapidly to dust; that water-holding capacity was greater where humus content was higher; that in a fertile soil there was a symbiosis between soil organisms and the growing plant; and that seed saved from the sections treated with chemical fertilizers tended to deteriorate more than seed from the Organic section. One major fact which was emerging at this time was that 'in both health records and quantity of milk, the Organic herd appear[ed] to be superior to the Mixed herd'. Given that differences between the sections did exist, the researchers were anxious to proceed to further questions: why did these differences exist?; what did they portend?; and would it now be possible to 'investigate nutritional quality in food in relation to the soil on which it [was] grown'? This last question was of course the one which the organic movement had been hoping to answer since the 1930s, and the 1962 booklet outlined a number of projects it hoped to initiate in order to do so. The first of these recalls the work on the diet of rats which McCarrison had undertaken in India: 'feeding the produce of the three sections to small laboratory animals of quick breeding turnover for at least 10 generations'. This idea subsequently took shape as what was informally termed 'the mousehouse', under the direction of Trevor McSheehy.[13]

Before long, however, the Experiment was again hit by financial difficulties: they concerned the security of the farm and threatened this time to be terminal; but benefactors appeared in the form of the property developer JACK PYE and his wife Mary, whose generosity — Pye gave the Soil Association £60,000 to buy the farm — was celebrated in an understandably fulsome editorial tribute to 'Our New Friends' in the July 1967 issue of *Mother Earth*. Jack Pye's money made a huge difference: according to Michael Allaby he pressed money on the Soil Association faster than it could be spent, funding the film 'The Secret Highway' and supporting the Conservation Corps. Some alleged that he was interested in getting hold of land which could later be turned over to building sites for his property company; but Siegfried Rudel, who knew Pye well, did not doubt the genuineness of his commitment to the organic movement and its research into health. Pye had been an early member of the Soil Association and, hearing years later from naturopath friends that it was in difficulties, decided to visit Haughley. Evidently he liked what he saw there.[14]

Douglas Campbell

Heartened by the receipt of the Pye Trust's money, the Haughley researchers could anticipate expansion of their work, under the direction of the agricultural scientist Douglas Campbell. Campbell had been appointed the Research Director at Haughley in April 1965, following many years overseas with the Colonial Office, including ten in southern Africa. In 1959 he was appointed Director of Agriculture for Basutoland (now Lesotho), a post which included responsibility for a 600-acre research farm. Also that year, he joined the Soil Association, *Mother Earth* attributing this decision to his 'ecological outlook'; but many years later his widow Peggy — herself a very able farmer, who during the war had run a dairy farm — expressed uncertainty about the degree of her late husband's commitment to the organic movement. The Haughley Experiment interested him because, as a scientist, he would have wanted to consider questions of soil fertility from every possible angle; but he felt that some of those involved in running the Experiment (Mrs Campbell was, regrettably, unable to remember who) had little interest in the efficient running of the non-organic sections of the farm. Despite the great improvements in the farm buildings which the Pyes' money facilitated, Campbell became impatient with what he perceived as — in his widow's words — a 'dreamy' approach to the Experiment, and was happy to return to agricultural development work abroad. The Experiment was in fact under pressure from within the Soil Association Council to concentrate on applied research for the benefit of the Association's farmer members and act as a demonstration farm. Those holding this view wanted to abandon the 'closed cycle' concept and pointed to indications that the Organic section was 'running down'. But as Eve Balfour pointed out in her account of the Haughley Experiment, the 'running down' of the Organic section in itself provided valuable data for further study and analysis.[15]

The Pye Research Centre

In the end, though, the voices of pragmatism prevailed over those who wanted to continue studying the closed system, and the Haughley Experiment, as it had been conceived, was abandoned. A further financial crisis in 1970 revealed that the Soil Association was in danger of bankruptcy, and the Pye Charitable Trust saved it from this fate by taking over the farms, now re-named the Pye Research Centre (PRC). It assumed financial responsibility for the Farms in October 1971. Although the PRC was a separate undertaking from the Soil Association, and was under separate management, it inevitably maintained close links with it. Its stated research aims were 'fully compatible with the ideals and beliefs of the Soil Association'. One of these aims was 'to identify and define the correlation between the way in which food is grown and processed and the health of consumers, both animal and human'. The other was to 'study those land treatments and farming systems that will preserve indefinitely the capacity of the soil to produce crops of the desired nutritional quality and that will maintain a healthy and aesthetically pleasing environment'. Colin Fisher, the PRC's General Director, stated unequivocally that the Centre would have 'no bias toward any existing farming system'. It was impartial and rejected 'any suggestion that the truth is known and requires no more than verification'. The Centre looked to collaborate with government, university and industrial research stations. After five years, Fisher was able to report that the Centre had developed a number of projects. The three different sections continued to be monitored; the entomologist Dr Robert Kowalski had joined the staff as an agricultural ecologist to study the effects of wild plants on weed control; Dr Mark Cowan was studying soil fungi, and the laboratories were expanding, in order particularly to facilitate bio-chemical analysis. The Centre had extended its parameters in order to examine various factors affecting the composition, morphology and yield of the crops, but regarded as most important of all the study of the chemical composition and biological value of the harvest crops to assess their nutritional status.[16]

Once the Elm Farm Research Centre (EFRC), of which more below, had been founded in 1980, Jack Pye became interested in its work and the research at Haughley was wound up in 1982, its scientific equipment being transferred to Elm Farm. This apparatus — beyond the financial resources of the Progressive Farming Trust which supported Elm Farm — made possible the development there of a soil testing service, based on continental techniques, offered to practising and potential organic farmers. A plaque in memory of Pye, who died in 1984, was unveiled in the courtyard of Elm Farm in the winter of 1985–86. Owing to the threat of bad weather, Eve Balfour was unable to attend the ceremony.

****

Lawrence Woodward has identified four factors which worked against the Haughley Experiment: lack of money; organizational troubles; poor management, and the major, methodological problem 'of studying ... whole systems ... without destroying the functional relationships that make up the whole'. In Woodward's view, Balfour's hypothesis of 'a health continuum' between soil, plant, animal and man remains unproven, though not without contemporary evidence in its favour. Balfour herself was always careful not to claim too much for the Experiment, which 'had only just begun before it ended'. She regarded it as over once the closed cycle on the Organic section had been abandoned. Nevertheless, she felt that the Experiment had demonstrated a number of worthwhile findings, among the most important of which were that levels of available minerals in the soil fluctuate according to the seasons, and the self-supporting nature of biological fertility. The various discoveries raised further questions which required investigation. The Haughley Experiment's greatest achievement was, it seemed, the circular one of having demonstrated the need for a 'fully ecological' experiment to investigate the wholeness of the soil-plant-animal-man food-chain.[17]

Dr Norman Burman

The termination of the Haughley Experiment disappointed many Soil Association members, among them Dr Norman Burman. Burman was not one of those who felt that the Experiment was inherently flawed on account of the range of variables involved, but he had no doubt that it needed to be long-term if its findings were to have any significance. A look at Burman's commitment to the Soil Association offers another example of how a very capable scientist responded to the organic philosophy, and of a now forgotten dedication to the organic cause over many years.

Born in 1915, Burman entered employment as a hospital laboratory technician and was required to remain there during the war. He then went to London's Metropolitan Water Board (MWB) as a bacteriologist, later becoming Senior Bacteriologist and then Scientific Assistant to the Director. When the MWB became the Thames Water Authority, Burman was appointed Manager of Scientific Services. He had gained his B.Sc. through evening classes and researched for a doctorate as an external student of London University, obtaining his Ph.D. — its subject was the survival of *E. coli* — in 1954. Burman's predisposition towards the organic approach stemmed from a love of gardening: he read in gardening magazines about the benefits of composting and so was led to Sir Albert Howard's *An Agricultural Testament* and Balfour's *The Living Soil*. He attended the Soil Association's inaugural meeting, but when I interviewed him in June 2006 could not recall anything about it. He went to lots of meetings in those days, he told me, and didn't know that this one would turn out to be something special.[18]

As a microbiologist, Burman was interested in the 'living soil' approach to cultivation, instinctively feeling that the productivity of all the soil's living creatures was fundamental to the health of plants and animals dependent on the soil. But he appreciated that there was no scientific proof of the superiority of organic methods, and no properly conducted research into them: hence his interest in the Haughley Experiment. Although Howard had said that the way forward was to write the organic case on the land, this would not, in Burman's view, satisfy scientists.

Burman joined the Soil Association's Middlesex Group and came to know Dr J.W. Scharff, who undertook consultancy work for the MWB on the risks posed by mosquitoes. Lecturing to horticultural and allotment societies became Burman's speciality, and he attended open days at the nurseries of J.L.H. Chase. During the 1950s and '60s he was also a regular contributor to *Mother Earth,* writing chiefly on the science of composting and reviewing a range of books on biology and ecology. Dr G.E. Breen wrote warmly of Burman's doctoral research, and Burman became at various times a member of the Association's Advisory Panel, its Editorial board, its Council and its Standards Committee. But Burman's MWB colleagues were sceptical of his support for the organic cause; they regarded the movement as unscientific and he had to tread carefully in order to avoid damage to his professional reputation. In fact, the worst he suffered was a lot of ribbing. Burman has never accepted the view that the organic approach is unscientific, and regrets that there has been so little research undertaken into the effects of organic methods. Too much of the organic case has relied, in his opinion, on hearsay evidence rather than recognized scientific procedures. The Soil Association, which in its early days was largely concerned with gathering evidence for the benefits of organic methods and seeking out relevant research, turned increasingly to campaigning. Burman stopped working for the Association in the early 1970s, and was not involved in the generational struggles which began to manifest themselves later that decade.[19]

Municipal Composting

One other aspect of the organic movement's scientific activities with which Burman concerned himself was municipal composting. Given that the case for the value of composting was accepted, how were sufficient quantities of compost to be produced? It was impossible to follow the example of Howard in India, where agriculture was the primary occupation and there was a vast pool of cheap labour; so the organic movement in Britain, with Howard's full support, looked to encourage the composting of urban wastes and the use of sewage.

The use of wastes to create fertility is integral to the organic approach, and advocacy of it can be found in the movement's earliest canonical texts, most notably G.V. Poore's *Rural Hygiene* (first published in 1893), F. H. King's *Farmers of Forty Centuries* (first published in 1911) and *The Waste Products of Agriculture: Their Utilization as Humus* (1931) by Albert Howard and Y.D. Wad. Just as, in Jungian psychology, the darkness of the unconscious mind can be turned into valuable psychic energy, so in the organic philosophy all that appears worthless and unpleasant, or is hidden in the darkness of the soil, can be transformed into fertility. During the post-war years the organic movement worked hard to promote the treatment of sewage and the composting of urban wastes, and monitored schemes, of which there were several, overseas. In Britain, J.C. Wylie, County Engineer for Dumfries, was a leading figure in the science of municipal composting, a Soil Association member who presented a paper on 'Composting Domestic Wastes' to the 1951 conference of the Institute of Public Cleansing and whose state-of-the-art treatment plant at Kirkconnel was opened in May 1953. Wylie went on to write two major studies of the subject for Faber and Faber. The first, *Fertility from Town Wastes* (1955), on which he was assisted by Lawrence Hills and which he dedicated to Louise Howard, featured a foreword by Sir Cedric Stanton Hicks. The second was *The Wastes of Civilization* (1959), an historical study of man's wastefulness and a manifesto for better use of the title subject. (This is perhaps a convenient place to note, albeit briefly, that the organic movement was concerned not just with the soil, but with the pollution of water, both by agro-chemicals and by wastes which should have been used to enhance soil fertility. Hence the support which the Soil Association received in the 1960s and early '70s from the noted angling writer Bernard Venables.) Other Scottish ventures in municipal composting included the scheme run by Cdr Robert Stuart at Gifford in East Lothian, who, with the approval of the County Sanitary Authority, manufactured composts from the sludge of settlement tanks and waste vegetation; and the scheme run by the City of Edinburgh. In England, there were notable schemes run by J.S. Townley at Nantwich in Cheshire; J.L. Davies, the Borough Engineer of Leatherhead, Surrey, and by John L. Beckett, City Engineer for Leicester. Such posts no doubt sound unglamorous to twenty-first-century readers, who are more used to organic publicity campaigns issuing from Kensington restaurants. But our concern here is with science and we should therefore be aware that, for instance, Beckett was a man at the top of his profession, in the late 1950s President of the Institution of Municipal Engineers.[20]

Municipal composting was not just a concern of the organic movement. In the early 1950s, the government's Ministry of Material kept an Advisory Panel on Waste Materials, to which in 1952 the Soil Association and the Albert Howard Foundation of Organic Husbandry jointly submitted a memorandum on the value of wastes in agriculture and horticulture, urging their systematic use and listing examples of already existing successful schemes. Two years later, the Zuckerman Report on the use of towns' wastes in agriculture appeared; but, as so often, the organicists were disappointed by officialdom's approach to such matters, considering the Report's tone generally negative despite its willingness to examine the progress of a newly installed plant in Jersey and its admission that composted town wastes could be of special value to horticulture. In the journal *The Surveyor*, a 'reasoned criticism' was advanced by two experienced sanitary engineers, one of whom, L.P. Brunt, was a Soil Association Council member and worked for a firm called Compost Engineers Ltd, for which Norman Burman undertook consultancy work. Also in 1954, the Soil Association held its first conference on municipal composting, chaired by Louise Howard and attended by various local councillors and sanitary experts. The conference's approach to the problem was, the Association believed, much more progressive than that of the Zuckerman Report, and those who attended, while not underestimating the practical difficulties involved, agreed that municipal composting provided a hygienic and economic alternative to other forms of waste disposal, whose product could be sold at satisfactory prices. A.G. Davies, surveying the situation in 1961, pointed out that since in England refuse disposal had never been regarded as a profitable undertaking, composting could not be condemned on the grounds of doubts about the sales potential of compost. Other countries were running successful schemes, and if wastes and sludge could be disposed of 'by processing to the advantage of agriculture, what more could one ask?' Davies, who worked for Woking Council in Surrey, was Honorary Secretary to the Joint Working Party on Municipal Composting. This body, chaired by organic veteran Lord Douglas of Barloch, consisted of representatives from the Soil Association, the Association of Public Health Inspectors, the Institute of Public Cleansing and the Institute of Sewage Purification, while the Society of Medical Officers of Health appointed an official Observer of its deliberations.[21]

So the organic movement continued, throughout the 1960s, to take an interest in the virtuous circle of municipal composting and the treatment of sewage, processes which could reduce water pollution, perform the alchemy of turning waste materials into the gold of soil fertility, and earn money for those councils which established the necessary plant. A new problem had arisen, though: by 1967, there was increasing anxiety about the toxic metal content of refuse and sewage sludge, and the Soil Association undertook trials at Haughley in order to assess how much of the toxic metals vegetables would absorb. The field officer for this composting project was C.D. Wilson, the Association's assistant treasurer, who 'mingle[d] the mysteries of high finance with a vast knowledge of municipal composting' and on several occasions taught composting at Findhorn.[22]

The research of Dr Ken Gray and Dr A.J. Biddlestone

From the mid-1960s onwards, much work on scientific composting was undertaken by Dr Ken Gray of Birmingham University, in conjunction with A.J. ('Joe', and later Professor) Biddlestone. Gray inherited from his father his love of gardening and his appreciation of the need for compost; as a young married man, he bought a secondhand copy of H.J. Massingham's *This Plot of Earth* (1944), which introduced him to the writings of the organic movement. His professional career seemed, though, to be heading in quite a different direction: he trained as a chemical engineer at the huge Esso oil refinery on The Solent before, in 1963, being appointed to the Department of Chemical Engineering at Birmingham. However, he was able to take up a project there on the composting of organic wastes and his reading on the topic soon led him to the Soil Association's work. He wrote asking what research had been undertaken on toxic metals and detritus like glass, and was invited to initiate some at Haughley with a grant from the Pye Trust. Thus began a long relationship with the Soil Association, though it was some while before Gray actually joined; as a scientist, he preferred to maintain neutrality. He had contacts with ADAS and some senior staff at Rothamsted, and got on well with both camps, trying to act as a mediator between them. Gray came to know many of the organic movement's leading figures, including Schumacher, Sam Mayall and Lawrence Hills, and he read widely in the early literature: Howard, Sykes, Stapledon and R.H. Elliot. His wife Cynthia was a dietician, and they read McCarrison, Picton, Wrench and Doris Grant. In the winter of 1969–70, the Agricultural Research Council awarded Gray a grant of £7,592 — an odd sum, but very substantial in those days — for an investigation into farm waste disposal. It was around this time that Joe Biddlestone joined Gray in his research work.[23]

Biddlestone had taken a degree in chemical engineering at Birmingham in the late 1950s and stayed on to do a doctorate before joining ICI and then transferring to Dunlop, who needed a small team of engineers to help build a polymer plant; Biddlestone ran the plant for about nine months. In 1965, he was invited back to Birmingham University to take up a lectureship in his old department, working with a mass spectrometer until it upped and went to Edinburgh when Biddlestone's senior partner became a professor at Herriot-Watt. Biddlestone knew Ken Gray and was interested in the idea of using wastes productively: not through any family background in farming, but partly through his wartime childhood experiences and partly because he was interested in systems and processes, biological as well as chemical. The department at Birmingham favoured this sort of research and was enrolling an increasing number of environmentally aware post-graduate students, so Biddlestone was able to team up with Gray: the start of a long productive academic partnership and of a friendship still strong in 2009. Unlike Gray, Biddlestone never joined the Soil Association. His training in orthodox science and industry, though, make his interest in the Association's composting research all the more significant; he remains convinced that it was a worthwhile area in which to experiment.[24]

Frequent contributors to the Soil Association's journal from 1970 onwards, Gray and Biddlestone provide evidence of the Association's continued interest in scientific research and are also among the comparatively rare instances in organic history of people who could mix comfortably in both the organic and the orthodox camps; their work appeared in the journals of the Royal Agricultural Society of England and the Royal Horticultural Society. Following the end of the experiments at Haughley, they carried out further work on poor soils in Birmingham, and were involved in the design, construction and operation of various large-scale municipal composting plants both in the UK — at Leicester in particular — and overseas, in Teheran, Libya and Hong Kong. In the 1970s they built a composting unit for the biodynamic farmer David Clement at Broome Farm, Clent, in Worcestershire, which won first prize at the Royal Agricultural Show at Stoneleigh; and Gray worked with David Stickland of Organic Farmers and Growers on an estate which had converted to organic methods. Gray and Biddlestone were also active in Stickland's project the International Institute of Biological Husbandry (IIBH), of which more below. Gray became something of a national figure in the mid-1970s, featuring in a *Sunday Times* article (11.5.1975) as the faintly sinister-sounding 'Doctor "Compost" Gray', a crucial figure in the fight against world starvation. It provided good publicity for the Soil Association, whose booklet *Garden Compost* Gray and Biddlestone had just updated. Another project on which they worked, during the 1980s, was for the agricultural machinery company ARM Ltd of Rugeley in Staffordshire, improving the design of elevator and spreading machines. This led on to some interesting work with reed beds as water purifiers, which was to prove commercially successful. Ken and Cynthia Gray spent many wintry hours 'harvesting' seeds from the banks of local canals and producing reeds in the greenhouse of their large garden.

Gray and Biddlestone also had links with India, where Howard

Dalzell was teaching dairy farming and organic cultivation to peasant farmers. With him, they wrote the booklet *Composting in Tropical Agriculture* for the IIBH; later, they expanded it into the United Nations Food and Agriculture Organization's Soils Bulletin No. 56, *Soil Management: Compost Production and Use in Tropical and Subtropical Environments*. The request to write the book came in the wake of the Ethiopian famine of the mid-1980s, its object being to promote the use of locally available organic wastes both as a source of plant nutrients and as a means of increasing humus content in order to combat soil erosion. The book was translated into French and Spanish, and went out to more than 150 Ministries of Agriculture around the world. As Gray and Biddlestone wrote: 'The Soil Association and the stalwarts of the organic husbandry movement can take quiet satisfaction from the fact that the United Nations FAO came to this country for their manual on composting.'[25]

Dr V.I. Stewart and the Bryngwyn Project

In Wales, the organic movement was responsible for an excursion into applied science, the Bryngwyn Project, which was under the direction of Dr Victor Stewart of Aberystwyth. Stewart became a soil scientist largely by chance, following war service which, he maintained, had consisted largely of playing rugby: a fellow soldier suggested soil science to him as a worthwhile subject of study. Stewart took a degree in agricultural chemistry at Bangor University, specializing in soil under the tuition of the noted authority G.W. Robinson. He then contracted TB, spending three years in hospital before resuming his career at Aberdeen, where he gained his doctorate for ecological research into the connection between soil development and types of tree: a fundamental organicist concern, which would have much interested Howard and St. Barbe Baker. He shared the organic movement's view of the importance of fungi, which linked trees and soil into one huge feeding-system, an ecological community. In 1956 he was appointed to a lectureship at Aberystwyth: the first there in soil science rather than soil chemistry, and therefore one which enabled him to pursue his ecological interests. With his research students, he studied the re-planting that was being undertaken in the Dovey forest, and discussed the problems of farming on hillsides. He was also interested in drainage, and later made a reputation for himself as a consultant on the quality of sports pitches, his work including the writing of reports on Test Match wickets for the MCC. For twenty years he studied another central organic topic: the role of earthworms, and exactly why they were so important to the soil. He found that exhausted soil could be restored by restoring earthworms, and developed an organic system to encourage them back. This work led him to the Soil Association: he had learned much from Dinah Williams, whom he came to know soon after arriving at Aberystwyth, and he was impressed by Eve Balfour when she spoke to the Agriculture department there — though his colleagues gave her a rough time. Through this contact, Stewart was invited by the Soil Association to give lectures on soil topics, and he spoke at the conference weeks at Ewell Technical College, which Dr Anthony Deavin organized. He found Balfour an excellent person to discuss with, and not at all dogmatic. He could not, though, share her religious perspective.[26]

Stewart is another figure who links the different generations. Nic Lampkin was one of his keenest students, and Stewart also thought highly of Peter Segger and other members of the West Wales group of the Soil Association, despite their 'brash' approach. The period of the late 1970s and early 1980s was one of generational tension and, notwithstanding his closeness to Eve Balfour, Stewart tended to side with the newcomers, who lived or died by their businesses. He spoke up for them, in fact nominating Segger for the Soil Association Council.

The Bryngwyn Experiment stemmed from Stewart's discussions with Eve Balfour, from whom he learned about the Keyline System developed by P.A. Yeomans. In August 1977 the Soil Association entered into a five-year agreement with the National Coal Board (NCB), through Schumacher's contacts, to carry out a research project on a 314-acre farm near Llanelli in South Wales. 250 acres had been restored after open-cast mining; the other 64 were still derelict. The Experiment's purposes were to devise techniques of accelerating the rehabilitation process and to monitor the effects of organic farming on the soil. In Eve Balfour's view, speaking as President to the Soil Association AGM in October 1978, this was 'just about the most important project the Soil Association ha[d] ever undertaken and, if successful, [would] do more to advance the cause of organic farming than anything else'. This was, perhaps, the sort of thing that Howard had in mind when he spoke of 'writing his answer on the land'. There were high hopes that Bryngwyn would provide 'visible evidence, backed by scientific observation, that organic farming methods [could] restore a filled-in open-cast site to grade one farming land much more quickly than orthodox methods that have so far been tried, and failed'. The farm's manager was David Entwistle, while the Soil Association's farm consultants were Dinah Williams and her son-in-law Gareth Rowlands. The Soil Association liaised with the NCB, and, given the vast areas of land which the Board had found difficulty in restoring by conventional methods, it seemed possible that there would be scope for young organic farmers to set about restoring those areas.[27]

In the end, the Soil Association's direct involvement at Bryngwyn lasted seven years, until in July 1984 responsibility for the farming operation was transferred to the university at Aberystwyth. J. Scullion, a doctoral student there who was involved in the project from very early on, offered various preliminary conclusions on its data in the June 1985 issue of the *Soil Association Quarterly Review*. Evidence suggested that poultry manure was more valuable for soil management than artificial fertilizers were, but that there was little to indicate that artificials had an adverse effect. It also appeared that earthworms indeed played a crucial role in soil rehabilitation, though further study of this hypothesis was required. The management team had not been rigid in its application of organic principles, recognizing that 'economic and practical considerations' demanded a certain flexibility; but the researchers would undertake fertilizer trials which might well be relevant to the process of converting from orthodox to organic methods. The farm, which was eventually reinstated in 1995, later went on to win recognition as a showpiece of environmental enrichment, under the care of the Tancock family.[28]

Dr David Hodges

Over on the eastern side of Britain, at Wye College near Ashford in Kent, another scientist was combining his academic career with an active commitment to the organic movement. This was Dr R.D. (David) Hodges, a zoologist gifted in anatomical work, who had undertaken his doctoral research at a specialist London hospital, St. Paul's, and worked at the Harwell Atomic Energy Establishment during his 'cushy' national service. There he tested the effects of radiation on goats: 'a ghastly business' which helped turn him into a pacifist and, subsequently, a member of the Society of Friends. In 1962 he took up a lecturing post at Wye College in the department of poultry research, which led to him writing his *magnum opus* The Histology of the Fowl (1974), a project which took several years. Hodges also worked on intensive poultry research; he did not entirely regret this, as it opened to his eyes to what was happening in the world of industrial agriculture. At around the same time, he and his wife Ursula read *Silent Spring*. Hodges had no background in farming, but these two experiences triggered an interest in the changes which intensification was bringing about and their detrimental effects on the animal kingdom and the environment. Hodges' response was as much intuitive as intellectual: he was convinced both that the industrial approach was morally wrong and that in practical terms it would cause more problems than it would solve. Although he did not join the Soil Association until 1973, he was sympathetic to its work from the mid-1960s onwards and for being so was regarded by his colleagues as eccentric. But 'there was, at that time ... something called academic freedom', and he was able, so long as he undertook his main work conscientiously, to 'wander into strange areas'. Hodges certainly took an unfashionable view of agriculture as it was developing in the 1960s, considering that agricultural chemistry and, especially, agricultural economics were distorting the nature of farming. Together these disciplines treated a complex biological system as if it were a sort of factory floor from which products were turned out conveyer-belt fashion, with no thought given to external costs.[29]

Hodges came to know Cdr Noel Findlay, who since 1949 had farmed organically at Hastingleigh, near Wye. Findlay persuaded Hodges to lend him support in a televised debate about organic farming. The programme, in the 'Farm Progress' series, showed a film of Findlay's farm, which was followed by a debate between Hodges and an agricultural officer from ADAS. Hodges was also in demand with the Soil Association, being persuaded to stand for its Council, to which he was elected in 1976 and on which he served until 1993. Following the inter-generational upheavals of the early 1980s, Hodges became a member of the *Soil Association Quarterly Review*'s editorial board and regularly contributed a column of technical abstracts, summarizing relevant scientific research, as well as writing book reviews and articles, these latter being chiefly about soil. With Charles Arden-Clarke, he produced a detailed review for the Association called *Soil Erosion in Britain*.[30]

This work for the Soil Association did not exhaust Hodges' energies — or at least, not until the 1990s. From 1982 until 1994 he was also a member of the HDRA Council, and from 1975 until 1987 was prominent in the activities of the International Institute of Biological Husbandry (IIBH), working as a colleague of David Stickland, whose commercial concern Organic Farmers and Growers we described in Chapter 5.

The International Institute of Biological Husbandry (IIBH) and Biological Agriculture and Horticulture

Stickland began to develop the idea of the IIBH in 1974. As Hodges recalls, at first it seemed that Stickland was looking to form a Research Committee for the Soil Association, and that September he called a meeting (held at the School of Economic Science in London), attended by six scientists. (Three others who had been invited were unable to attend.) The minutes described the event as 'the first meeting of a group of University Scientists and other specialists for the purpose of setting up a Research and Advisory Committee for biological husbandry'. These developments led to much bad feeling within the Soil Association, as it seemed clear that Stickland was in fact looking to establish his own organization independently of it; but Hodges and other Association activists, including Anthony Deavin, Hugh Coates and George McRobie considered the initiative worthwhile and joined the Board of Management when the Institute was formally established late in 1975. Since the Soil Association has always tended to resent any organic initiatives independent of itself, Hodges and Coates, who were both IIBH Board members and Soil Association Council members, had some difficulty in persuading Schumacher and the Soil Association Vice-President Sam Mayall that there was room for both organizations; but eventually a truce was established.[31]

Hodges brought his colleague at Wye Dr A.M. (Tony) Scofield

on to the IIBH Board and with him undertook most of the projects directly related to the scientific development of the Institute. Among the most important of these were, firstly, the Annual Lectures, on the theme of scientific approaches to organic agriculture, which Hodges and Deavin proposed as a means of raising the Institute's profile. The Inaugural Lecture, given in March 1977 by Dr Tilo Ulbricht of the Agricultural Research Council (ARC), was on 'Western Thought, Farming Systems and the World Food Problem', and argued that western thought had been dominated for three hundred years by a particular kind of approach: the analytical, which assumed, unlike previous civilizations, that the whole can be understood by analysing its parts. Sir Emrys Jones, the recently retired Principal of the Royal Agricultural College, Cirencester, chaired the second lecture, in which Dr G.R. Potts, Director of Research at the Game Conservancy, traced the ecological effects of chemical farming techniques. But attendance at both these events was disappointing, and plans for further lectures were put on hold.

More successful, was a conference at Wye in 1980 on 'An Agriculture for the Future'. Hodges particularly cherished remarks made in public, later confirmed in a letter, by Dr. G.W. Cooke of Rothamsted, who had previously been a senior member of staff at the ARC and apparently — as one would expect — a firm opponent of the organic movement. Cooke perhaps underwent some sort of Road-to-Damascus experience at the conference, Hodges speculated, 'realising that the organic movement was not completely cranky'. In his letter to Hodges, Cooke said he felt that many factors were leading people originally of different views to be converging in their approach to common problems, and that he had been glad to meet so many people who were concerned about the future of agriculture. The conference proceedings were edited by Bernard Stonehouse, one of the Institute's Board members, and published the following year as *Biological Husbandry: A Scientific Approach to Organic Farming*. A second international conference was held at Wye in 1984, on the subject 'The Role of Micro-organisms in a Sustainable Agriculture'. Organized by Hodges' colleague at Wye, Joe Lopez-Real, it emulated the success of the first and demonstrated that agricultural scientists from many countries considered organic farming a subject worthy of study.[32]

Like the Soil Association, the IIBH undertook a farming project in

East Suffolk which proved financially impossible to maintain; though it was short-lived in comparison with the Haughley Experiment. The intention was to convert Melton Lodge Farm — already semi-organic — to a permanent organic system and develop a farm and market garden which could demonstrate organic methods and become a centre for teaching and meetings. Stickland in particular put much hard work into it, but the family who owned the farm did not fully co-operate; the financial burden this placed on the Institute eventually proved too much to bear. Among other factors leading to the Institute's demise was disagreement among its Board's members as to the nature of organic systems. Hodges has described himself as an 'organic traditionalist', believing that any introduction of chemical fertilizers or pesticides into the complex web of a biological system risked upsetting its balance. David Stickland was more of a pragmatist, seeing no harm in using some short-lived chemicals, particularly during the period of transition from a conventional to an organic system. There were some who came to believe with hindsight that one of Stickland's motives for establishing the IIBH was possibly to explore more marginal practices that might have been useful to Organic Farmers and Growers: to learn, for instance, how rapidly a chemical would break down in the soil. The 'organic traditionalists' felt that this was not what the Institute had been set up to do, and that its scientific basis risked being compromised by commercial considerations. Stickland contested this view, saying that the aim of the IIBH was to try to provide a scientific basis for the practical aspects of organic farming and to explain the reasons for them. He felt that there was no proven scientific basis for organic methods. In fact, he disliked the very term 'organic', feeling that it was far too vague, and preferred the term which the French use, biological agriculture.

David Hodges' account of the Institute's demise contains a number of interesting details: for instance, there was an attempt (which came to nothing) to team up with Emerson College, the Biodynamic Agricultural Association's training centre at Forest Row in East Sussex, in order to create a base for the teaching of biological husbandry courses. The Institute's eventual closure in 1987 left some bad feeling between Hodges and Stickland. Hodges, Scofield and Julian Wade felt that, after ten years, the IIBH had made little progress along the path which had originally been mapped out; but with hindsight Hodges considers that the initiative was ahead of its time and, despite some support from David Astor, was unable to attract the necessary financial resources. He also feels that Stickland was at times too sanguine about the Institute's prospects, and that this ill-founded optimism created disillusion among his colleagues. And so, another organicist initiative ended in a rather confused anti-climax.[33]

A more enduring project was the scientific journal *Biological Agriculture and Horticulture (BAH)*, which was an indirect offshoot of the IIBH. When the 1980 conference was being advertised, Hodges was approached by AB Academic Publishers, who suggested that a new journal on organic farming systems should be established. Stickland and the Board members welcomed this idea and the journal was sub-titled 'The Official Journal of the International Institute of Biological Husbandry', though in fact the Institute had no financial or editorial control; the chief benefit of the connection was publicity value. Hodges was appointed editor, with Scofield and Lopez-Real as his deputies; but Hodges found that 'in reality it was a one-man band' and an enormous demand on his time and energies. They put together an advisory board of 23 members from around the world. Hodges wanted to make an official connection with IFOAM, but differences of approach made this impossible. Nevertheless, Hardy Vogtmann and some other senior IFOAM members sat on the board; other members included Ken Gray, Lawrence Hills, the Steinerian Dr H.H. Koepf, Dr. Willie Lockeretz, George McRobie, Victor Stewart, and Professor Colin Spedding of Reading University, who was interested in agricultural systems and took a detached view of the case for organic farming. The journal declared as its aim, 'to act as the central focus for the wide range of studies into alternative systems of husbandry [and to] publish ... work of a sound scientific or economic nature ... related to the many factors contributing to the development and application of biological husbandry in agriculture, horticulture, forestry etc. in both temperate and tropical conditions'. The areas covered included soil and environmental management, biological control methods, energy utilization studies, and the development of appropriate agricultural technologies. Hodges opened the first issue by presenting the case for a more biological approach to cultivation and inviting high-quality submissions from members of the scientific community. The journal also occasionally published pieces on the context of biological agriculture: Margaret Merrill contributed an article on 'Eco-Agriculture: A Review of Its History and Philosophy', and Tony Scofield wrote on the origins of the term 'organic farming'. The familiar organicist topics of soil ecology and erosion, the use of herbs in pastures, the potential of composting sewage sludge, and ecological approaches to farming, all found a place in the pages of *BAH*. Hodges' achievement in establishing the journal and keeping it afloat for thirteen years was remarkable; the effort involved being not just mental and physical, but emotional, as Nic Lampkin recognized. 'I am all too well aware,' he wrote to Hodges after hearing that the editorship would be changing hands, 'what a lonely path it can be pursuing the cause of a biological/organic/ecological approach to agriculture, and how little progress there sometimes appears to be, particularly in UK universities.' From Canada, Dr Stuart Hill praised Hodges for ensuring that *BAH* had 'published most of the really outstanding scientific papers on organic/ecological/sustainable agriculture'.[34]

Lawrence Woodward and Elm Farm Research Centre

In the meantime, a base for scientific research into organic farming had been established in the Berkshire countryside and was set to outlast the IIBH; this was the Elm Farm Research Centre (EFRC) at Hamstead Marshall, above the Kennet valley between Newbury and Hungerford. Its co-ordinator was Lawrence Woodward.

Woodward's path to prominence in the organic movement was an unusual one. From a South Yorkshire mining background, he originally regarded farming as an unattractive and unacceptably right-wing way of life; though he did register as significant the fact that his grandfather grew vegetables and kept chickens. He was fortunate to attend a high-quality state school, and it was this school which changed the direction of his life, by entering into an exchange with the school at Dartington Hall. Woodward spent a year at Dartington and while there met David Astor's daughter Alice, whom he married in 1972. Through his father-in-law, whose concern about environmental issues his friend E.F. Schumacher had aroused, Woodward became aware of the problems which faced conventional agriculture in a world of finite resources, where oil would become scarce, water supplies were being depleted and the soil was increasingly degraded. Astor also foresaw the potentially dire social and political consequences of a failure to protect primary resources, and the extreme authoritarian measures which were likely to ensue. In 1975, Astor and Woodward met Schumacher to discuss 'the need to develop "preliminary examples" ... of technologies and approaches that could bring about a society where production and consumption were more appropriate to a world of finite and diminishing resources.' The first subject they considered was organic farming. Woodward took a basic course in agriculture and in 1976 began farming organically on part of the Springhead estate at Fontmell Magna in Dorset, which had once belonged to Rolf Gardiner. This was arranged through Schumacher, who was Chairman of the Springhead Trust. Given Woodward's eagerness to downplay connections between his generation and its predecessors, there is a certain irony about him beginning his farming career at the Gardiners' estate; and there is further irony in the fact that he was helped by David Stickland, with whom he would later fall out terminally over issues connected with organic standards. (There is also a fine irony in David Astor supporting the organic movement, when his father had written so vigorously in favour of orthodox methods of agriculture.)[35]

Although Schumacher was its President, Woodward did not find the Soil Association particularly helpful, and sought help abroad. He went to Switzerland and met Dr Hartmut Vogtmann, as a result of which he came to realize the importance of biological systems and cycles; Vogtmann subsequently joined the Council of the EFRC. David Astor's brother Jacob, who was chairman of the Agricultural Research Council, made it clear that no official body was likely to undertake scientific research into organic farming — such was the intellectual curiosity of officialdom — so Woodward, with his father-in-law's support, established the Progressive Farming Trust and bought Elm Farm (previously a conventionally run dairy farm of 232 acres), registering it as a charity in 1980 and holding meetings that year with various research establishments, thanks to the Astor connection. David Astor's home at Sutton Courtenay in Oxfordshire provided a space in which overseas researchers into organic agriculture were able to meet orthodox British agricultural scientists. Elm Farm had as its enterprises dairying, beef, sheep and cereals, and was run 'as a total biological system, with an emphasis on developing techniques to meet the economic constraints of British agriculture'.[36] The Centre was keen to learn from mainland Europe, where such techniques were generally more advanced than in Britain.

The EFRC published its first report in April 1981, based on a colloquium held the previous November. These discussions, which followed preparatory visits to farmers, growers and scientists, were on the topic *The Research Needs of Biological Agriculture in Great Britain*. Woodward identified these needs as existing in the areas of soil analysis; comparative and investigative research into the products which organic farmers were using (Chilean nitrate and foliar feeds, for instance); and systems of 'integrated farming' (a basically biological approach given a boost by the use of certain chemicals). But whether farmers were organic or 'integrated', they needed to be familiar with the characteristics and limits of the biological approach. Problems of nitrogen supply were paramount for organic farmers, but although much knowledge already existed about legumes and the use of slurry, it needed to be applied in 'the context of a working biological farm with economic constraints', or else it would be of little practical value. Other organic techniques, such as the operation of herbal leys, were little understood. 'We need,' Woodward said, 'both an academic, theoretical investigation of the plant/herb/nutrient link and a development of the practical efficient use of this knowledge as an everyday part of our farming techniques and systems.' The same was true of rotations. As for the organic movement's central claim that organic produce enhanced the health of those animals and humans who consume it, Woodward agreed that it was 'hard to dispute the opinion that if it were scientifically shown that organic produce was better for people, then the organic movement would achieve tremendous momentum'; but, so far, the theory had 'little proven scientific basis'. Most of the techniques of organic farming were forms of a conventional wisdom which had not changed or developed and 'ha[d] never been rationally examined and tested.' This wisdom needed to undergo evaluation and 'then to be practically translated to the farm in such a way that it [made] sense agriculturally and economically.' Other contributors to the discussion included Michael Rust on weed and pest control, Peter Segger on the lack of advisory facilities, and Angela Bates on 'Nutrition, Health and Fertility'; as well as Hartmut Vogtmann and other European scientists. The Report concluded that Britain was lagging behind Europe and

North America in its research into biological agriculture, and that Elm Farm would play an important role in remedying the situation. Fundamental beliefs about organic practice would be modified or discarded altogether if found wanting. Woodward saw the EFRC as a bridge between mainstream agriculture and the organic movement: not just through the Astor connection, but also through the organic arable farmer Barry Wookey, who had contacts in the Conservative government and persuaded ADAS to monitor what was being done at Elm Farm. But the EFRC also supported the radicalism of the Organic Growers Association, and helped initiate British Organic Farmers. (Wookey, too, supported the younger generation of farmers and growers.)[37]

Elm Farm is in a sense an offshoot of the Haughley Experiment: Woodward came to know Jack Pye, who asked the EFRC if it might be able to take over the Experiment. This was not possible, but, as we saw earlier, Pye passed on the equipment at Haughley to Elm Farm (and also the title of the company, Haughley Research Farms). During its first few years, the EFRC's activities rapidly expanded, so that by 1986 the *Soil Association Quarterly Review* was celebrating its 'key role in many of the organic movement's research and advisory activities'. The previous year, it had set up the Organic Advisory Service (OAS), the first such service in the UK, headed by Mark Measures; this was a response to the increasing demand for organic food and the patent inability of the Ministry of Agriculture to help any farmer wishing to convert from the chemical approach. The OAS co-ordinated the work of advisory teams of experienced producers and provided technical information and a computer programme for preparing conversion plans (thereby helping dispel any criticism of organic farming as Luddite in tendency); it also offered information on markets and supplies. It operated on a commercial basis and its advertising was uncompromisingly commercial in tone, informing its potential clients: 'We are here to help you profit from the expanding demand for organic produce'.[38]

As the Soil Association has enjoyed a higher profile than Elm Farm, it is important to emphasize the EFRC's achievements. During the first fifteen years of its existence, in addition to establishing the OAS, it developed a soil analysis service; advised the Prince of Wales on converting his Home Farm; was involved in the establishment of the British Organic Standards Committee and the deliberations of UKROFS; worked with the Ministry of Agriculture on a nitrate monitoring research project and on investigating the use of leguminous green manures for cereals; undertook farm-scale monitoring of biodiversity and homoeopathy; participated in an EU project on weed control and nutrients; with North Wiltshire District Council set up a pilot scheme for composting household waste; and ran an eleven-year research trial on stockless cultivation. Woodward and Vogtmann addressed themselves to the looming issue of genetic engineering, drafting the organic movement's position in a form accepted by the Soil Association, UKROFS and IFOAM. Sir Colin Spedding congratulated the EFRC on its 'dedicated and independent research'.[39]

Over the years, Woodward gave much thought to the science of organic farming. He believed that the orthodox scientist, 'used to separating and analysing fragments' found difficulty in understanding biological agriculture. Awareness of its holistic nature was essential: an understanding that 'everything affects everything else'. This was in fact the primary recommendation of the US Department of Agriculture's Study Team Report into organic farming in 1980; new methodologies were required in order to investigate highly complex systems involving 'unknown or poorly understood chemical and microbiological interactions'. More than two decades later, as he reflected on the history of the organic movement and his own work at Elm Farm, Woodward felt that these methodologies had still to be perfected, and that the delay in this process was the consequence of 'the concern of those much maligned "organic" researchers that such methods should pass muster with respect to attributes of scientific rigour such as reproducibility, transparency, and accessibility to appropriate statistical treatment'. Indeed, the organic movement had always displayed a commitment to science and an insistence that its ideas should be explored and would 'eventually be explained through appropriate scientific analysis'. Woodward believed that much of the organic pioneers' research seemed inadequate in retrospect: experiments poorly designed or badly reported, and suffering from the inherent technical limitations of their time.

But the chief problem was intellectual: the need to communicate the ecological concept of holism, without which the idea of a farm as '"a living entity" operating self-regulating cycles within a relatively closed system' cannot be grasped. While the Haughley Experiment had (as we saw above) produced certain indications which might have been valuable to the organic cause, the analysis and reporting 'were not robust enough to satisfy modern investigators'; but this was not the case with more recent research, some of which appeared to support the organic hypothesis that methods of cultivation affect food quality.[40]

Dr Anthony Deavin

As we saw at the start of this chapter, Lord Taverne has claimed that Rudolf Steiner was the original inspiration for the Soil Association. As we approach the chapter's end, we can consider the complex figure of Dr Anthony Deavin, a scientist whom Steiner's ideas influenced considerably and who, in later life, moved into the field of alternative healing.

Deavin's scientific background was impeccable: first-class honours in Chemistry at Queen Mary College, leading to a doctorate at King's College, London; research at Heidelberg and a lectureship in biochemistry at St Thomas's Hospital Medical School. In tandem with this career, Deavin developed more esoteric interests. He had been taught Economics at the School of Economic Science by a land reform enthusiast, and through him was introduced to the ideas of Henry George and the Natural Law tradition. When he first came across *Mother Earth* he recognized the same tradition in it: the need to work in harmony with nature. Although the natural law philosophy is an attitude to life, not a scientific theory, Deavin was struck by the idea that the Haughley farms could be used as a resource, and that he might be able to provide a bridge between the natural law tradition and Reginald Milton's measurements. His holistic philosophy was already in place before he joined the Soil Association, which he did in 1969, the same year that he was appointed to the staff of the North-East Surrey College of Technology at Ewell, as Research Director in the Department of Biological Sciences. His work during his eleven years in that post concentrated on research into soil fertility, following the concepts of Dr Hans P. Rusch on the relationship between soil fertility and health; some of this work was undertaken in conjunction with Rothamsted Experimental Station and was described by him in scientific papers published in Germany. Deavin also published several articles in the *Soil Association Quarterly Review* in the mid-1970s, on soil fertility, nitrogen fixation, plant growth and associated matters.[41]

Like David Hodges, Deavin was a scientist who combined a conventional career with dedication to the organic movement. After his departure from Ewell, Mary Langman wrote to Bryn Lewis, the Soil Association's General Secretary, recording her view that the Association owed Deavin a great debt for the courses which he had organized at Ewell during the 1970s and the lectures he had given. He was an Association Council member, an architect of organic standards with Hugh Coates, scientific advisor to the HDRA from 1973 to 1979, and also a member of the Biodynamic Agricultural Association. He undertook a good deal of biodynamic training, particularly in the making of sprays, at the Steinerian Camphill Village Trust in North Yorkshire. Open to new ideas as any scientist should be, he found that contact with biodynamic ideas widened his perspective, and he undertook experiments in chromatography of the kind which had been pioneered by the Koliskos, exhibiting chromatograms of humus extracts and plant juices in an exhibition on soil fertility for the Scientific Section of the Chelsea Flower Show in 1972.[42]

Deavin's career from the mid-1980s would scarcely gain the approval of Lord Taverne and his colleagues in Sense About Science. He lectured in Biochemistry and Histology at the School of Herbal Medicine, completed four years' study of herbal medicine in 1991, went on to train in Polarity Therapy, a form of alternative medicine, and subsequently practised spiritual healing. According to Mary Heron, who worked at Haughley in the early 1960s, Eve Balfour maintained that there were no materialists in the Soil Association: a proposition which might sound paradoxical but which Deavin had no difficulty in understanding. For him, the early Soil Association in particular was a forerunner of contemporary spiritual movements, working to provide not just physical, but spiritual nourishment. For this reason, he had no problem in relating to the younger generation of enquirers who attended the Ewell courses and whose appearance so offended Brigadier Vickers; Deavin understood their deeper motives. Nevertheless, his spiritual concerns did not blunt his scientific criticism of the Haughley Experiment; he was sceptical about its value and followed Sir Albert Howard in considering that a well-run demonstration farm, with healthy animals and a healthy balance-sheet, would have better served the organic cause. The amount of work required to satisfy the demands of orthodox science was way beyond the Association's resources. Deavin sympathized with Reginald Milton, whose task he considered impossible: the difficulty of producing statistically valid results was 'horrendous'. The results from Deavin's own trial plots at Ewell, although these plots were on a far smaller scale than Haughley, had proved dauntingly difficult to interpret. He concluded that you could not establish the truth of a philosophy through scientific measurement.[43]

Michael Allaby's case for science

This chapter began with Lord Taverne's rant against the organic movement. We can bring it to a close by looking at a much more thoughtful and well-informed piece of writing, by someone who had known the organic movement from the inside: this is Michael Allaby's book *Facing the Future: The Case for Science*, which appeared right at the end of the period we are studying, in 1995. Like Deavin, Allaby, who spent several years at Haughley, was sceptical as to the value of the Experiment and of the reams of data which Reginald Milton supplied. Unlike Deavin, he was not a trained scientist; also unlike Deavin, he did not believe that the scientific case for the organic approach was ultimately less important than the spiritual philosophy. By 1995, Allaby was in full reaction against his erstwhile colleague Edward Goldsmith; tired, as Kenneth Mellanby had become twenty years earlier, of *The Ecologist*'s dislike of scientific progress and its hankering after the stability of tribal societies. Allaby aimed much of his book against environmentalism in general and its associated 'New Age' attitudes, but the chapter on reductionism and holism is particularly relevant to, and critical of, the organic movement's philosophy of science. Allaby criticized the use of the term 'reductionism' as an all-purpose insult, defended the use of the reductionist approach in science, and claimed that its supposed antidote, holism, 'reduce[s] substantially the amount of understanding that can be extracted from ... research.' Such holistic modelling of eco-systems as computers can now create 'depends on an extreme reductionism', which Allaby defined as a detailed knowledge of every element in a system. (Whether this is what agricultural scientists in the organic movement actually mean by 'reductionism' is of course open to question.)[44]

But where the organic movement can agree with Allaby is in his insistence that science is not about certainty: indeed, the absence of certainty, he says, 'is central to the scientific philosophy'.[45] We should bear in mind that the Haughley Experiment was intended as an investigation of a theory which, if validated, would be of great importance for national and individual well-being. Very few scientists of the agricultural and medical establishment evinced any interest in supporting this work, thereby, one might think, failing to demonstrate the intellectual curiosity which should be a genuine scientist's hallmark. The organic movement was thus hampered by lack of funding and was not able to undertake the sort of research it ideally would have done. Lack of funds is not the same thing as a rejection of science. A refusal to support such research, though, betokens precisely the sense of certainty that Allaby condemns as being against the provisionalist spirit of open-minded enquiry.

8. The Politics of the Organic Movement: An Overview

Organic politics: a sensitive issue

To mention politics in connection with the organic movement is to risk the disapproval — or even, in some cases, the wrath — of its senior figures. Following two Radio Four programmes on organic history in September 2000, with which the present author was involved, Dr Walter Yellowlees wrote to the then editor of *Living Earth* Catharine Stott to complain of irrelevant and repeated 'political waffle', suspecting (quite erroneously) the existence of a 'secret agenda to discredit organic farming'. Nearly fifty years earlier, in her preface to *Prophecy of Famine* (1953) by H.J. Massingham and Edward Hyams, Massingham's widow Penelope had written that her recently deceased husband considered the land 'a matter of such vital importance … "as to be above and below politics"'. Mary Langman was insistent that the Soil Association was founded as a completely non-political body; Newman Turner wrote in the mid-1950s of *The Farmer*'s non-political nature, and *New Farmer and Grower,* in the mid-1980s described the organic approach as politically neutral.[1]

Writing towards the end of the twenty-first century's opening decade, one has to recognize that the organic movement has, during the past eighty years, attracted support from every colour of the political spectrum: Greens, inevitably; Tories, both High and free-market; senior Labour politicians; LibDem activists, CND supporters, UKIP members, Fascists and BNP members, members of the Socialist Alliance, advocates of 'regional socialism' and anarchists; there has even been at least one hard-line Communist lurking, rather improbably, in the Cotswolds. Since the organic movement is capable of appealing to such a wide range of outlooks, does this not in fact establish its non-political nature? There need be nothing strange about this situation — so the argument goes — if the essence of the organic movement is identified as a concern for the environmentally sensitive production of healthy food: this is an aim with which someone of any political commitment or none might sympathize.

But, even if one defines the organic movement in this limited way, it cannot be considered non-political. Among the various dictionary definitions of 'politics' are 'the policy-formulating aspects of government'; 'the civil functions of government'; 'the complex or aggregate of relationships of people in society, especially those relationships involving authority or power', and 'the art and science of government'. To bring about a change in national agricultural policy for the sake of the nation's health is clearly a political objective. Its achievement would require government willingness to encourage research into, and the economically viable practice of, organic husbandry; a shift in national health policy to encourage greater emphasis on nutritional preventive medicine or to create Peckham-style health centres; and greater emphasis on the importance of farming for the nation's security. In his letter referred to above, Walter Yellowlees drew attention to the decline of the countryside, brought about by the 'money power of a wealthy chemical industry [and] of an equally wealthy pharmaceutical industry, both often acting covertly through craven politicians' — thereby conceding that the problems facing the organic movement cannot be considered in a political vacuum.[2]

But the organic movement's aims have never been simply about farming and health: or, rather, because its thought is holistic, it has always recognized that these issues are inseparable from wider social and economic, and therefore political, questions. In recent years, I have heard Lawrence Woodward say that the purpose of organic farming is to change the world, and been told by one of the movement's most prominent figures that it is 'revolutionary'. If this is so — and the propositions are certainly arguable — does this mean that the politics of the organic movement should be placed somewhere on the traditional Left, and that the radical Right-wingers and Tories and moderate liberals who have supported the movement are mistaken and inconsistent? At the 2006 Soil Association conference in London, the media pundit Claire Fox provoked a growl of audience disapproval when she gave her opinion that the organic movement seemed to be returning to its natural home on the Right. Historically speaking, there is much truth in her view, though to draw attention to that truth

8. THE POLITICS OF THE ORGANIC MOVEMENT: AN OVERVIEW

can attract, in this writer's experience, strongly disapproving responses from both the older generation of organicists and their successors in the Seventies Generation. But how helpful are the terms 'Left' and 'Right'? Jorian Jenks, the Soil Association's first editorial secretary was a leading member of the British Union of Fascists and therefore 'right-wing' in conventional parlance; but he opposed free trade, believed finance capitalism was destructive of the national interest and the environment, and deplored the power of Big Business: positions which, once upon a time, were held by the Left. The generation of organic activists who in the 1970s were influenced by John Seymour's self-sufficiency ideas and the commune movement, distinguished themselves from the older generation, whom they regarded as essentially reactionary; but a Marxist might well argue that the ideal of self-sufficiency amounts to a bourgeois retreat from the class struggle. Whether rightly or wrongly is another matter, but the point is that such a movement cannot be unequivocally deemed 'left-wing'.

The whole issue is complex to the point of impenetrability, and this chapter will make no attempt to impose any abstract framework on the inchoate variety of political positions adopted by organicists during the half-century following the war; though at the end there will be one suggestion as to a persistent thread which can be perceived. The chapter's chief purpose is to demonstrate just how confused the organic movement's political history really is and what views some of its leading figures actually held, as opposed to what views they ought to have held. It should provide material for anyone sufficiently ambitious to want to scale the greater heights of organic political theory.

Right and Left in the 1940s and '50s

Anyone familiar with my earlier book *The Origins of the Organic Movement*, or with the work of Richard Griffiths, David Matless, Richard Moore-Colyer and Dan Stone, will know that Claire Fox's jibe at the 2006 Soil Association conference carried an uncomfortable reminder that the organic movement's roots went consistently deeper in the 'radical Right' of politics than in liberalism or the Left, to use the conventional terminology. Where the Soil Association was concerned, Mary Langman was correct to say that it was non-political in the sense that membership was open to supporters of any political persuasion; but there is no question that the radical-Right element was strongly represented in its higher echelons for some years after the war. Many in the Association today would prefer to forget about the presence of the Mosleyite Jorian Jenks as Editorial Secretary from 1946 until his death in 1963, and those who remember him stress his (undoubted) gifts and hard work as editor of *Mother Earth* and his tweedy, country-dancing persona. His politics are referred to as something separate, yet one wonders, given the holistic philosophy which the organicists articulated, whether this could have been so. His 1950 book *From the Ground Up* suggests it was not. Jenks also edited *Rural Economy*, journal of the pro-organic Rural Reconstruction Association/Economic Reform Club and Institute, and so was a key figure in the post-war organic movement. Then there are the Soil Association Council members, which in the Association's first five years included such noted 'fellow-travellers of the Right' as the President, Lord Teviot; the Earl of Portsmouth; Lord Sempill, and Rolf Gardiner. A list of contributors to the Development Fund, dated 31 October 1950, includes the names of the leading pre-war fascist Captain R. Gordon-Canning; Colonel Hardwick Holderness, a member of Viscount Lymington's far-Right English Array in the late 1930s; and the Marchioness Dowager of Londonderry, who, with her husband, had been a close friend of Ribbentrop. Although it is interesting to come across these names, they are only three from a long list, so it is best not to read too much into them. As far as Eve Balfour's politics are concerned, we should similarly avoid reading too much into the fact that she was the niece of a Conservative Prime Minister; few of us hold the same views as our uncles. Erin Gill, who is researching Lady Eve's life, considers her to have been sympathetic towards moderate forms of socialism, and Chapters IX and X of the first edition of *The Living Soil* contain passages reminiscent of the ideas of R.H. Tawney. We also know that she was close to the post-war Labour Chancellor Sir Stafford Cripps, a friendship which dated back to the Tithe War of the 1930s, and that he was a Soil Association member. Another prominent Labour politician, F.C.R. Douglas (later Lord Douglas of Barloch) was instrumental in establishing the Association's constitution.[3]

The Soil Association, then, was non-political in that it crossed party lines, but no-one reading the later chapters of *The Living Soil* could fail to see the book as a manifesto for post-war reconstruction.

Indeed, about the time that it was published, in the autumn of 1943, Lord Portsmouth and some fellow peers, Glentanar among them, debated the issues it raised in the House of Lords: one could hardly get more political. Organic husbandry was also a main feature of the policy of the British People's Party (BPP), the Duke of Bedford's fringe national-socialist group. The post-war BPP is of interest to historians of the organic movement, as Dion Byngham, under the pseudonym 'Miles Yarrow' wrote a regular column for its newspaper *Peoples Post* entitled 'Countryside Causerie', which promoted organic ideas and literature; Byngham also worked with Edgar Saxon on *Health and Life* and Dr Siegfried Marian on *Soil Magazine*. As for the Duke himself, he was a leading figure in the monetary reform movement, which was intimately linked with the organic movement in the 1930s and '40s. The *New English Weekly,* founded to promote the doctrines of Social Credit, became from the late 1930s onwards the leading 'forum for organic husbandry'. Although the emphasis on Social Credit dwindled during the 1940s, the paper remained politically committed, looking (quite unrealistically) for a Britain in which the imbalance of urban and rural would be redressed and the state be run according to Christian social principles. It regarded Soviet Communism and environmental destruction as the two great 'diseases' which the post-war world faced. Its editor Philip Mairet belonged both to the 'Chandos Group' of Christian Sociologists who in effect ran the journal, and to the Kinship in Husbandry, whose members Rolf Gardiner and H.J. Massingham, among others, frequently wrote for it. The Kinship was centrally important in developing the organic movement in the 1940s, attempting during the war years to formulate a ruralist vision of post-war British society: de-centralized and as far as possible self-sufficient in food. Such a position could not be identified on the conventional Left-Right spectrum, some organicists argued, since both poles were themselves located within the over-arching ideology of industrialism. Edgar Saxon had proclaimed during the war that the organic movement's political path should be 'Not Right nor Left, but Straight' — whatever that means — but he unambiguously rejected the State Socialism of the post-war Labour government, with power concentrated in the hands of what he considered to be a bureaucratic slave-order. Anticipating the rise of Green politics in the 1980s, Saxon said that Britain was required to choose, not between Red and Blue, but between Red and Green. The latter was more important: the redness of blood could not exist without the primordial greenness of the leaf to feed it. 'Merlin or Marx: which is it to be?' he asked.[4]

Although the dominant tone of the organic movement under Labour was opposition to State socialism, the organicists also disliked finance capitalism and the power of Big Business. We saw in Chapter 1 how the case for a large-scale, industrial, efficient agriculture was made by the Fabian socialist F.W. Bateson; P. Lamartine Yates was another Fabian advocate of that approach, to which, of course, the organic husbandry school was implacably opposed. Perhaps one can even trace the origins of this antagonism to the split in the Labour movement between Guild Socialism and Fabianism, early in the twentieth century. Massingham, Mairet and Gardiner had all at one time been Guild Socialists, but Massingham demonstrated that he put the soil above political differences when he responded enthusiastically to a *New Statesman* article by the Jewish socialist Edward Hyams and ended up co-authoring *Prophecy of Famine*. This collaboration did not imply that the two men saw eye-to-eye on everything, but they were united in their dislike of 'the unconscionable middleman and the commercial violator of the soil'. Mrs Massingham emphasized that her late husband would never have agreed to Hyams' proposal that the land should be nationalized, but the book made it clear that Massingham was critical of 'the industrial mind', whether revealed by Left or by Right. Although Penelope Massingham described him as 'essentially non-political', it is evident that he felt strongly about politics and the dangers which the power of the State posed.[5]

In Chapter 4 we saw how state health provision aroused the organicists' opposition, particularly as it found no place for experiments such as the Pioneer Health Centre or Kenneth Barlow's Coventry family health club. Yet when the book on *The Peckham Experiment* by Innes Pearse and Lucy Crocker appeared in 1943, the journal *Official Architect* described the Centre as an example of 'Real Socialism', and George Scott Williamson in *Physician, Heal Thyself* (1945) put forward the view that socialism was inevitable. The Socialism that he wanted to see, though, was not State Fabianism; rather, it was 'Liberal Socialism as opposed to Monopoly Socialism', producing an equilibrium between the group and the individual: socializing the means of production for the needs of the person, in the way that the home ('the archetype of

Socialism') provides 'the only means to meet the needs of individuality, that is to say, Liberty'. So it seems that some form of socialism might have been acceptable to the organic movement in the 1940s: whether Kenneth Barlow's 'Regional Socialism', the medievalist Guild Socialism still lingering on among the *NEW*'s editorial board, Scott Williamson's 'Liberal Socialism' or even the national-socialism of the BPP; but Soviet Communism and the Labour government's State Socialism were objects of hostility for journalists such as Massingham in his column in *The Field* or Ronald Duncan in his role as the *Evening Standard*'s 'Farmer Jan'. (Duncan later ended up in the company of various proto-Thatcherite hard-line right-wingers.) Since the Labour Party drew most of its support from urban voters, it chiefly served the interests of urban-industrial society; but even when it produced a major piece of legislation to protect farming interests — the 1947 Agriculture Act — the organic school was only partly impressed. Not only was there a strong degree of centralized control, but the capitalist profit-motive — which in those far-off days the organicists, Sir Albert Howard in particular, abhorred — was enabled to flourish through the drive to industrialization in agriculture and the promotion of pharmaceutical solutions in the National Health Service.[6]

As it took shape in the post-war period, then, the organic movement was, plainly, political; and, in so far as many of its leading figures had scant sympathy with the 1945–51 Labour government, could in fact be considered negatively party-political. There were certain exceptions: Douglas, Cripps and Hyams most notably. *Mother Earth* was scrupulously non-committal about the defeat of Labour in October 1951, merely commenting that the general Election campaign had revealed a general agreement on the need for increased food production, and that urban voters were uninterested in the relationship of people to the land. The Age of Affluence was now dawning, and the organic movement's more overtly political associations faded. Rolf Gardiner was an almost permanent fixture on the Soil Association Council during the 1950s and '60s, but Portsmouth and Sempill had disappeared from it by the middle of the former decade. When the Rural Reconstruction Association's Research Committee produced its report *Feeding the Fifty Million* in 1955, Committee members included a sprinkling of the old far Right in the persons of Jorian Jenks, P. C. Loftus, Robert Saunders and D.R. Stuckey, but the report was presented as 'free from any party bias' and was in any case somewhat half-hearted in its support for organic farming. Britain was settling down into the comforts of increased consumerism and the ideological compromise of 'Butskellism'; this was not a period favourable to the sorts of groups which had flourished during the turbulence of the previous two decades. That is not to say that the movement was uninterested in political issues: agricultural policy must always be one of its prime concerns, and Jenks kept a watchful eye on developments under the Conservative government, noting in 1957 that, as a result of its restoration of open-market trading in home-grown and imported agricultural produce, it had 'a strong incentive to curb rather than stimulate home production': the opposite policy to that which the organic movement has always favoured. It may also be somewhat ironic that wartime State Socialism had ensured a healthier population than Britain has known since, and that the Conservative government appeared to be more interested in promoting commercial profit than in maintaining nutritional standards. (Margaret Thatcher's government was quick to do away with such standards for school meals at the start of the 1980s.) On the face of it, the organic movement should welcome government control of nutrition for the sake of the greater good.[7]

The 1960s: a leftwards shift?

Perusal of membership lists for the 1950s suggests — one must put it no more strongly, for fear of stereotyping — that Soil Association members were likely to be conservative or Conservative in outlook, given the prominence of figures from the military, the aristocracy and agriculture. The Association certainly did not look like a revolutionary organization, and in the 1960s this supposed conservative aura became a source of irritation to two of its members, who felt sufficiently strongly about the matter to express their unease in *Mother Earth*. From Australia, C. Halik and Dr Clive Sandy wrote to complain of two tendencies evident from time to time in the journal, both of which they considered 'quite impermissible in view of the stated aims of the Soil Association' and potentially 'most inimical to its growth'. These were the tendencies 'to align the Soil Association with conservative politics, and the Christian religion'. (We shall consider the second tendency in the following chapter.) In particular, they had disliked a reference to the late President Kennedy's 'courage' during the 1962 Cuban missile crisis, a term they would have replaced by 'irresponsibility'. They contended that politics and religion should be kept out of *Mother Earth*, which was 'a basically scientific journal'.[8]

The letter appeared at an interesting time, as in 1964 the editorial department passed into the hands of Robert Waller and Michael Allaby, neither of them a conservative, where it remained for nearly a decade. Waller was a liberal rather than a socialist, and had scant regard for the aristocrats, county types and military figures that he was required to deal with; he found the political views of people like Rolf Gardiner and the Earl of Portsmouth rebarbative. Allaby was a CND supporter who later joined the Labour Party; he has recalled how in his early days with the Soil Association the office received considerable amounts of mail from far-Right groups. Waller, along with Eve Balfour, Sam Mayall, Donald Wilson, the Association's President Lord Bradford, its Treasurer Edward Clive and its Secretary Constance Miller, signed the official response to the two Australians' letter. Waller had not in fact edited the January 1964 issue of the journal, to which they had objected, but neither of the two emergency editors for that issue was conservative in politics. Balfour and her fellow-writers pointed out that the Soil Association boasted 'an extraordinary richness of variety' in its membership, covering the entire range of political views, Jew and Arab, negro and white extremist, Orangeman and Sinn Fein supporter. Members sometimes resigned through personal dislike of other members but never, apparently, 'because they objected to the politics, religion or race of other members'. Political differences were superficial in comparison with what members had in common: primarily, the letter suggested, a reverence for life, 'rebellion against the tendency of authority to ignore ecology' and a desire for information as to how ecology worked. This is in itself an interesting choice of attitudes. It incorporates a vague sense of revolutionary dissatisfaction but leaves open the question of which political approach might best protect ecological systems. But as Waller and Allaby began to promote a broader environmentalist agenda, it became essential, as one American member pointed out, 'to explore the social, economic and political implications of our ideas ... Otherwise the implementation of these ideas will be stymied by the social, economic and political barriers of existing society upon which we impinge'. By appealing to people of all political shades, the Soil Association hoped to broaden its membership (although that remained stuck for years around four to five thousand); but it may have limited its effectiveness as well, making it able only to deal with certain specific issues rather than offering — as its philosophy implied that it should — a holistic vision of an ecologically sound society. How quickly a movement can lose the sense of its own history: little more than twenty years earlier, this was exactly what the Kinship in Husbandry, the ERCI and the *NEW* had been trying to do.[9]

Waller's first editorial essay, in the April 1964 edition of *Mother Earth*, sketched out the problems facing the Soil Association as it sought to communicate the social implications of its ecological perspective. But Waller was aware of the dangers lurking in the desire 'for a leader, like General de Gaulle, who will embody some attempt at seeing the nation as a whole, transcending the chaos of political ideologies'. This was 'a dangerous solution, substituting cultural dominance for harmonious integration.'[10] On the other hand, the see-saw of conventional politics had become wearisome, and those who rebelled against the establishment tended to share its materialist outlook. Although the Association could attract the occasional Marxist, such as Tony Stephenson — an intriguing figure who combined dedication to left-wing politics with entrepreneurial commitment to the commercial and manurial benefits of seaweed-based fertilizers — the Association was never likely to engage with political issues from the perspective of dialectical materialism. Nevertheless, the capitalist profit-motive, with its consequent reluctance to consider the harm its products might cause, was certainly an enemy. It might be challenged through conventional parliamentary methods, as was the case with the Farm and Garden Chemicals Act of 1967: this piece of legislation, guided through the Commons by the Labour MP Mrs Joyce Butler, was the result of pressure applied by Lawrence Hills and the HDRA — an effective single-issue campaign. On the face of it, an organization dedicated to gardening and researching the virtues of comfrey should be as non-political as one can get; but pesticides and the pollution caused by nuclear fall-out were issues which affected gardeners, and action against them necessitated political engagement.

As the new environmental movement began to emerge in the 1960s, it became clear that a new approach to politics was required: one which would reject the materialism exemplified in both capitalism and socialism. A 'third way', perhaps — steering a course distinct from American free enterprise and the State Socialism of the Iron Curtain countries.

Waller and Allaby set about trying to make the Soil Association, with its strong but little-known record of ecological concern, a leading environmental organization. As we saw in Chapter 6, they discovered that E.F. Schumacher, who was writing impressive pieces for David Astor's *Observer,* was a long-term Association member, and proceeded to involve him more actively in its work; he became its President from 1971 until his sudden death in 1977. His close friend George McRobie, a former colleague at the National Coal Board, has described Schumacher as a socialist who believed that his hand would drop off if ever he cast a vote for the Conservatives. Schumacher distrusted Edward Goldsmith, partly on account of his wealth and perhaps — given his own personal history as a refugee from Nazism — on account of his politics: according to Michael Allaby, Goldsmith at one time put out feelers to the National Front, wondering whether the far-Right party might be a vehicle for organicist ideas. Given the organic movement's early history, of course, Goldsmith was not wrong to wonder, and today's British National Party continues what is sometimes called the 'Brown/Green alliance'.[11]

There is another indicator of political attitudes within the Soil Association around this period which helps confuse the picture — if such confusion were needed. Tapes have survived from Cdr Robert Stuart's Soil Association Weeks at Chirnside in the late 1960s and early '70s, and, from listening to many hours of recordings of these occasions, one forms the impression of an implicitly pro-Conservative tone, with the occasional jibes at Mr Wilson's 'socialist' government, a rather precious concern about the nation's moral decline and the suggestion that an organic diet for the workers might ensure that they took fewer days off sick.

A call for 'eco-politics'

Meanwhile, in the pages of *Span,* edited by Michael Allaby, there was discussion of the need for a new 'bio-politics', which the *Torrey Canyon* disaster had made evident: national planning must henceforth pay heed to the possible biological consequences of technological advances.

Some areas of science policy required a bio-political perspective: the expense of research into pesticides and fertilizers, for instance, might with advantage be reduced by adoption of agricultural systems relying on minimal use of chemicals. *Span* identified the recently established *Resurgence* magazine as 'devoted to what we might now call "bio-politics"'. The pages of *Span*, though, do not present a consistent political approach. An editorial note prefacing an article on herbicides and de-foliants stated bluntly that the Soil Association had no political views and that *Span* did not favour — or, more accurately and equivocally, did not intend to appear to favour — one side or the other. The Conservatives' victory in the June 1970 general election led John Davy to speculate on what this might mean for the cause of conservation. He appeared reasonably well-disposed towards the outgoing Labour government's initiatives, but noted that environmental issues had played little part in the campaigning; this was not really surprising, as 'Both parties take for granted ... that a central aim of any government must be to promote economic growth. The nation is seen as a vast business enterprise, with the Cabinet as the Board of management'. (*Private Eye*'s 'Heathco' column proceeded to mock this tendency over the next three and a half years.) So, something different was needed: rather than 'bio-politics', this alternative was now called 'eco-politics', and the Soil Association's conference in October 1970 considered whether it might become a mass movement, and, if so, whether an eco-political party could be formed to direct its energies.[12]

The essence of eco-politics as discussed at Attingham Park that autumn was stabilization: of both economic and population growth, whose effects were socially and ecologically destructive. Mary French, author of *The Worm in the Wheat* (1969), described the decay of the Cornish rural society of her childhood; Dr Hugh Nicol, using what the philosopher John Macmurray termed 'the organic analogy' (the application — illegitimate and dangerous, in Macmurray's view — of biological concepts to human society), said that the concept of the stationary state was 'truly scientific'; while Edward Goldsmith 'delighted the audience ... with his essentially light-hearted description of impending catastrophe'. (A catastrophe that would no doubt have been easier to bear for anyone of Goldsmith's wealth and social connections.) Goldsmith argued that primitive societies are more stable than industrial societies, called for a halt to economic growth and traced the origins of the Roman Empire's collapse to the disappearance of small-scale farmers. The opening speaker Michael Allaby expressed faith in the younger generation of 'eco-activists' who were more environmentally and politically aware than their elders; but such optimism was dismissed as 'sheer apathy' by H.W. Sowden, a correspondent in *Span* who placed the blame for the environmental crisis squarely on capitalism and urged the creation of 'an uncompromising political party like the SPGB [Socialist Party of Great Britain]', which would 'abolish the present system and plan our environment for the health and happiness of everybody.' The historian Tony Judt described the SPGB as 'a happy marriage of doctrinal purity and political irrelevance'; one can hazard a reasonable guess as to what the older generation of gentleman farmers and nutritionists in the Soil Association made of such proposals. And yet, one could argue that the implications of their own views on food and health were themselves far-reaching. When another correspondent, Horace Jarvis, condemned the corruption of the capitalist food industry, he was only re-affirming the sorts of points which Eve Balfour had made nearly thirty years earlier. 'Our priority,' Jarvis concluded, 'should be the health of the people from the soil and not for the profits of the few.'[13]

For the recently appointed General Secretary of the Soil Association, Brigadier A.W. (Bill) Vickers, who had held a high-level post in military security, the Association's involvement in 'eco-politics' and its attempts to attract a younger generation of hippie-ish environmental activists were distinctly undesirable. Vickers purported to know that Waller and Allaby were Communists whose names were on a government list of dangerous subversives. But Waller and Allaby had both left the Association by 1973, *Span* was closed down, and the scope of the re-formatted journal, under its new editor David Stickland, contracted, with explicit eco-politics now excluded. Ironically, this change occurred largely because of Schumacher, who thought that the Association 'should be true to its name and concentrate on agriculture and only that.'[14]

Such is the confused, complex, incoherent nature of organicist politics that it is bound to produce such anomalies. Another of them is that the magazine *Seed*, designed to reflect and appeal to the values of the 'counter-culture', of which Vickers was so wary, expressly rejected, in the early 1970s, the conventional political Left. It considered itself on the Left in an entirely different sense from communism, re-defining as left-of-the-spectrum anyone close to Nature or to what Lyall Watson called 'Supernature'; on the Right were all 'conditioned, nose-to-the-ground realists' and particularly technologists and military commanders. This was in effect to remove the conventional concepts of Left and Right from politics and re-locate them in culture and psychology. Under such a system of classification, Communism and Fascism would find themselves together on the Right. *Seed* was concerned with cultural politics rather than party politics, though John Fletcher contributed a historically contextualized analysis of what he saw as Britain's drift, fuelled by the demands of economic growth, towards the nightmare of a centralized communist state: the sort of argument which might have found its way into the *New English Weekly* in the 1940s. And there was a poignant *cri de coeur* from the Truscott family, who longed for an Organic Living Party to represent the views of those who wanted nothing more ambitious than 'to lead a simple, healthy life on some land where the air is clean and full of the sounds of nature, not diesel fumes and juggernauts'. This was in the mid-1970s, at a time when mainstream political events, triggered by the Arab-Israeli War of autumn 1973 and consequent oil crisis, seemed to be unwittingly assisting the organic cause by drawing to national attention (as Schumacher had been trying to do since the 1960s) the danger of the UK's dependence on fossil fuels. The struggle between conventional Left and Right — unions vs. Conservative government — was in a sense irrelevant to the organic movement. Lawrence Hills, in the HDRA newsletter, declared a plague on both their houses: only a programme of ecologically aware policies could offer a solution to Britain's underlying vulnerability. Schumacher's friend and patron David Astor was afraid that if the industrial system broke down as a result of energy shortages, the ensuing social unrest would necessitate some sort of extreme authoritarian government to deal with it. For the sake of democratic stability, it was imperative to encourage a form of food production which would reduce dependence on finite resources.[15]

The politics of the Seventies Generation

Astor's son-in-law Lawrence Woodward was privy to his thoughts, as we saw in Chapter 7. Following the publication of *The Origins of the Organic Movement,* Woodward launched an off-the-cuff verbal assault against the present writer at the Soil Association conference of January 2002, saying that the figures who had featured in the book had little relevance to the organic movement of the late-twentieth and early twenty-first century. Woodward's comments were followed in due course by a more considered response from David Frost and Carolyn Wacher. Ten years earlier, Tracey Clunies-Ross had attacked my anthology *The Organic Tradition* along similar lines. All four critics seemed uneasy about connecting the post-1970s organic movement in any way with the various crypto-fascists and paternalist squires who had been involved in the early days. Instead, it was argued, the contemporary movement was the creation of a group of radical, left-leaning anti-urbanites who rebelled against Eve Balfour and her aristocratic cronies, established the more vigorous and practical British Organic Farmers and Organic Growers' Association, and staged a coup within the Soil Association which re-vitalized it and made it more relevant to the late-twentieth century. Some of the leading figures — Peter Segger and Nic Lampkin, for instance — were active in CND and at the beginning of the 1980s were clashing with the Association's President Lord O'Hagan, a Conservative MEP. Angela Bates, then a Soil Association Vice-President, interpreted Segger's behaviour as 'class war', while not denying Eve Balfour's tendency to turn for aid to fellow aristocrats. Lawrence Woodward disagrees, pointing out that the younger generation happily worked with and respected the Tory farmer Barry Wookey and the aristocratic Mary Langman.[16]

As ever, one finds a complex history. Certain figures in the Seventies Generation were from left-wing backgrounds: Frost was an academic, reading *New Left Review,* and Woodward came from a South Yorkshire mining community. Various factors — the oil crisis, the American 'counter-culture', John Seymour's books on self-sufficiency — led this generation to regard organic cultivation as a rational, far-sighted response to an unsustainable, oil-based, consumerist society. Patrick Holden was influenced by Charles Reich's *The Greening of America,* moved to West Wales and set up a commune on a bleak hill farm near Tregaron, to experiment with an alternative form of social living. (One should point out that there is nothing remotely new about such an idea; the late-nineteenth century saw plenty of such experiments.) Frost, as we saw in Chapter 3, also moved to West Wales, in order to seek a more viable way of life than that which consumer capitalism offered. In what sense can such commitments be regarded as 'left-wing'? There are serious conceptual and definitional problems involved here, to say nothing of the irony that once this generation of organicists had realized they could not survive without selling their products, they found themselves having to try to establish organic produce in the commercial mainstream. With certain exceptions, such as Lawrence Woodward and Iain Tolhurst, this generation was solidly middle-class and, despite their aversion to the aristocratic network of which Eve Balfour was a part, some of its members proved more than willing to mix with another sort of aristocracy in the period following that which this book covers. Indeed, before the 1970s were over, the self-sufficiency movement was being criticized as élitist, while opposition to industrialism and technological development was, in some quarters, labelled 'eco-fascist'. Michael Allaby and Peter Bunyard of *The Ecologist* wrestled with these problems in a series of lectures for Exeter University's Extra-Mural Department in 1978, which they expanded into *The Politics of Self-Sufficiency* (1980). The book is a detailed, wide-ranging and closely argued piece of social and historical analysis, and does not evade the awkward questions posed by environmentalist politics. Allaby concluded by challenging the environmental groups to cease being factions and adopt a coherent set of political policies, clearly defining their objective.[17]

The Ecology/Green Party emerges

Such a development was already under way in the form of the Ecology Party, though of course such a name was unlikely to mean much to the public at large and it was not until it became the Green Party in the mid-1980s and began to formulate a comprehensive programme, put forward most notably by the brilliantly articulate Jonathon Porritt, that it started to make any significant impact. (In 1979 the Ecology Party aroused the enthusiasm of organic veteran Adrian Bell, who recorded that at last there was a political party for which he could vote.) *Vole* magazine, edited by the anarchist Richard Boston, showed interest in the emergence of environmentalist political groups, though it was not necessarily favourable. Martin Stott, writing just before the 1978–79 'winter of discontent', argued that a clean break from conventional parties carried disadvantages, not least the fact that the environmental movement was philosophically heterogeneous, ranging from 'ill-disguised eco-fascist to the revolutionary utopian'. (Some would argue that that is not actually a great distance.) Stott thought it wiser either to work through existing sub-groups, like the Liberal Ecology Group or the Socialist Environment and Resources Association; or, for those who rejected the establishment game, to work through force of example in farms and communes. Stott's views provoked a variety of responses.[18]

This recognition that people of different political affiliations could share environmental concerns was given formal expression the same autumn by the formation of the Green Alliance, whose aim was to 'ensure that the political priorities of the UK are determined within an ecological perspective', by seeking to influence decision-making in government, finance and industry, and to ensure 'far deeper awareness of our need to live in harmony with the environment than industrial nations have yet allowed'. These ambitious aims were to be achieved from a non-ideological standpoint representing common sense and the wishes of 'the person in the street'. Such a claim might have tactical merit, but from a theoretical perspective the notion that government decisions should be subject to ecological considerations can hardly be regarded as non-ideological. This is the Soil Association's original dilemma re-surfacing in a different guise: on the one hand, the appeal to potential supporters of various political persuasions through the claim to be 'non-political'; on the other, the stubborn reality that organicist or environmentalist values must imply changes brought about by political means and may well imply a different conception of politics. In this matter, Jonathon Porritt was more uncompromising, drawing a distinction between environmental policies, to which all the major political parties could pay lip-service when times were affluent, and ecological, which rejected their growth-oriented assumptions. The Ecology Party supported the organic movement's long-held position on the need for a much greater degree of agricultural self-sufficiency, and wanted to see thousands more people employed in rural regeneration.[19]

Growing Concerns and the organic movement

The Ecology/Green Party produced a quarterly newsletter on agricultural matters, *Growing Concerns,* edited by Betty Whitwell and strongly reminiscent of *Rural Economy* in both its stapled, duplicated format — impossibly amateurish to our contemporary, digitalized eyes — and its wealth of material. It offered remarkably detailed coverage of all developments relating to environmental and agricultural policy and of all meetings and societies likely to appeal to anyone seeking an alternative to agri-business. Topics covered included vegetarianism, wholefood, national self-sufficiency, factory farming and animal welfare, forestry, genetic engineering, urban food production, low-input farming systems, organic food standards, BSE and pesticide residues; book reviews were thorough and letters were long and well-argued. Like the HDRA newsletter and *New Farmer and Grower*, *Growing Concerns* was a labour of devotion. To describe it as the organic movement's political house-journal would be inaccurate, but it was undoubtedly a close ally; only a list will serve to demonstrate this. It gave news of government funding for organic agriculture through Manpower Services Commission (MSC) schemes, referring to the work of Tony Wigens in Lincolnshire, which was making use of this opportunity; it advertised the value of WWOOF-ing, and it summarized Peter Segger's presentation of the Soil Association's *Charter for Agriculture* at the Autumn 1984 party conference. Bill Starling provided minutes of the January 1985 OGA/BOF conference at Cirencester, and argued for genuine wholefood co-ops as against corporate health food stores. Robert Waller provided minutes of a conference on 'Food Production and Our Rural Environment', and wrote pieces on countryside and resources issues. The newsletter noted the visits to England of Sego Jackson and Bill Mollison to talk about Permaculture. Francis Blake, the Soil Association's Symbol Co-ordinator, wrote in to defend the Association's stance on spray-drift, and Nic Lampkin's address on 'Organic Economics' to the Royal Agricultural Society of England/ Agricultural Development and Advisory Service conference on organic farming in March 1986 was summarized. Eve Balfour's death early in 1990 was noted and regretted by Helen Woodley, who described her as 'a marvellous person', and perhaps the only person to have mentioned the Keyline System on British radio. Woodley later wrote a piece on the Green Party's approach to water policy, recommending the integration of Keyline and Permaculture into its agricultural policies and appealing to the example of Sir Albert Howard, 'who we've largely forgotten'. She also gave an account of Arthur Hollins' low-input foggage system.[20]

Organic responses to Green Party wooing

The Green Party's firm commitment to organic and sustainable cultivation was not universally reciprocated. Certainly, there were members of the organic movement who supported the Green Party, but there persisted the belief that, strategically, it was unwise for the movement to identify itself with any particular organization. *New Farmer and Grower,* in its first issue, in the summer of 1983, stated that the organic approach was 'politically neutral, and therefore acceptable to all shades of opinion'; though it admitted that the effects of conventional agriculture were creating the raw material for a 'green vote'. Eighteen months later, Betty Whitwell wrote to the journal encouraging organic growers to support the Ecology Party. Patrick Holden, addressing the Green Party's annual conference in the winter of 1987–88 urged it to develop a full-scale organic policy for British agriculture. But one could not have concluded from this that the organic movement was about to identify itself with the Green Party, even though a specifically environmentalist party might have appeared to be just what the movement had been awaiting. Robert Hart still believed, as his hero H.J. Massingham had done, the organic philosophy to be beyond politics. A 'green' society in the full sense could 'never be achieved by merely legislative means. It can't be imposed from above. It must grow from the roots of our present society — organically.' (In Jorian Jenks' phrase of forty years earlier, 'from the ground up'.) Practical examples of 'self-sufficient, co-operative, democratic activity of all kinds' were essential as an alternative to the ruthless efficiency of the Greens' opponents. They had to precede electoral success, not wait upon it.[21]

Involvement in policy-making

A 'new realism'

Patrick Holden and other members of the Seventies Generation were meanwhile keeping the organic movement 'beyond politics' in a somewhat different sense, as they had begun to take it into the realms of policy-making, where advantage could be gained only through dealings with whichever political party had, or was likely to gain, power. They were required to become 'politicians' in the older meaning of the word, achieving their ends through shrewd diplomacy and persuasion. During the 1980s, this meant forging links with the Conservative government, and in fact some of the most powerful pro-organic writing of the decade issued from the pen of the Conservative MP, farmer and Euro-sceptic Sir Richard Body. In 1985, twenty years after the two Australians referred to above had expressed concern about the Soil Association's supposedly conservative stance, another member, Dr E.H. Eason, wrote to express his hope 'that the Soil Association ... [was] not sliding in the same direction as Oxfam', which was fast 'becoming a left-wing pressure group'. We do not know whether Eason would have considered it 'left-wing' to condemn the government for running down the Soil Survey; but David Hodges did so in no uncertain terms, writing of the Conservatives' 'obsessive fixation on the reduction of public expenditure as the panacea for all national ills; with eventual privatisation of what can't be cut — whether or not such changes are in the long term public interest'. Eason might well, though, have been discouraged by Colin Millen's letter expressing outright condemnation of profit-driven capitalism (yet Howard, too, had condemned it) and advocating 'radical changes to the fabric of society', in order to create, not a socialist state, but systems of mutual aid, free association, de-centralization and the re-establishment of community. Readers will again make the connection with certain organic pioneers.[22]

Millen's radical idealism was at odds with the Soil Association's policy, which, like the Labour Party, was adopting a 'new realism' as the Conservatives were returned to power in 1987 with another overwhelming majority. *Living Earth* outraged certain readers the following year by featuring Minister of State for Agriculture John Selwyn Gummer on its front cover, complete with super-imposed Soil Association symbol on his lapel. Ralph Maddern, who had been considering taking out life membership, instead resigned altogether from the Association, appalled by its naivety in falling for the Tory Party's bandwagon-jumping. The Tories' free-market policies were 'the chief threat to Britain's soil'. Dr Michael Smith condemned both Labour and Conservatives for letting down the ecology movement, and sought the 'middle ground' between them. *Living Earth*'s editor Geoffrey Cannon responded loftily: 'It's unlikely ... that abuse will help us gain our ends.' To condemn a politician simply because he was a Conservative was just silly, Cannon argued, as the case of Sir Richard Body proved. He offered a photograph of the celebrity Leslie Kenton on the cover of the current issue in the patronizing hope that the sight of her would 'soothe savage red breasts'. (It did not.)[23]

The politics of consumerism

The issue of *Living Earth* in which these letters, and Cannon's response, appeared (July-September 1988) carried an editorial of great significance for the Soil Association's future development, explicitly embracing what one might term (perhaps oxymoronically) the politics of consumerism. The new incarnation of the Soil Association journal as *Living Earth* had been launched that June at the Palace of Westminster, at a special meeting hosted by another Tory MP, Charles Irving. The idea of a 'living earth' was a campaign, too: one which encouraged the purchase of organic food as the means by which 'real' food could become more widely available and 'real countryside' be preserved or re-created. The editorial carried the sub-heading 'How we can use our power as consumers to transform our food and the environment'.[24] According to one's point of view, this development can be considered either as a pragmatic strategy for increasing the amount of organic food grown and sold, and thereby starting to reverse the damage caused by agri-business, or as the start of a process of compromise with the very system to which the organic movement had formerly expressed so much opposition. The movement — or certain influential parts of it — would work with whatever party claimed to favour its vision. If John Selwyn Gummer proved a disappointment — as, extraordinary to relate, he did — then perhaps the (as yet un-disgraced) Labour MP Ron Davies might be a useful ally should his party ever again form a government.

It is largely a matter of definition: in so far as the movement was liaising with political institutions and formulating organic standards for UKROFS, as was the case in the mid-to-late 1980s, then it was becoming increasingly political; and this was true even of a body such as Organic Farmers and Growers, whose founder David Stickland concentrated on marketing and had no interest in the organic movement's broader social aims. Nor was the political sphere confined to Britain: it inevitably expanded to include the European Economic

Community and with the signing of the GATT agreement in 1994 became world-wide. This agreement drastically reduced the amount of control that Britain could exercise over food standards and, in the view of the editorial for the May 1994 issue of *Living Earth & Food Magazine,* reduced the role of UK politicians in these matters to that of traffic wardens. If this were so, then 'there remain[ed] only one effective way to vote — with our purses'.[25] Shopping had now become political, and shoppers with a conscience would buy organic and fairly traded products (at least, for as long as the GATT bureaucrats would allow such products to be traded). The system of international trade, about which Jorian Jenks had written so critically at the start of the 1950s, was, it seemed, forcing the organic movement to conclude that its earlier critiques of the capitalist profit motive and consumerism were no longer relevant, and that its future lay in accepting the hegemony of market values and making them work for it. (Whether a parallel can be drawn with the New Labour project, which was taking shape at the same time, readers may decide for themselves.) From the mid-1990s onwards, the idea seemed to be that one could by-pass institutions, concentrating instead on increasing consumer demand for organic products; this would, so it was hoped, persuade farmers and market gardeners to convert to organic methods; and so a more ethical and environmentally friendly form of cultivation and animal-management would begin to push back the frontiers of agri-business. By no means all supporters of the organic movement agreed with such a strategy, and although the dissidents' voices did not prevail they offered an uncomfortable reminder that the movement was tying itself up with a system which was, from an ecological standpoint, indefensibly wasteful and therefore unsustainable.

The small-scale alternative

There may be an apt symmetry about the publication of John Seymour's novel *Retrieved from the Future* in 1996, just at the beginning of the rapid expansion in organic consumerism. Apparently written in response to the oil crisis of the mid-1970s, this futuristic work, while displaying few literary virtues, provides an unsettling reminder of our nation's fragility and imagines what might happen to Britain's civil and

8. THE POLITICS OF THE ORGANIC MOVEMENT: AN OVERVIEW

political structure if the oil supply were to be interrupted. Rather than taking free-market capitalism as the norm, Seymour places the events in historical perspective, recalling William Cobbett's condemnation of London as the 'great wen' on England's face, its political and financial institutions spreading their ugliness across the countryside. Perhaps surprisingly, Seymour evinces some sympathy for Marxism, envisaging a Marxist Minister of Industry in the post-reconstruction government and having one of his characters speak as follows: 'The Marxist interpretation ... and Marx would have been right ... would be that this war is a struggle between two interests — two cultures — economies — and ways of life ... Between the oil-users — the city-centred people — and the peasants. ... As long as the city people have oil — they will be unbeatable. ... If their oil dries up — they will wither away.' The new polity would be based on a peasantry ('an intellectually lively, spiritually alive peasantry') rather than a technocracy; and its roots would be regional, as they were before 'they were cut off by cosmopolitanism and industrialism'.[26]

This is, of course, in many respects the vision of organic pioneers like Massingham; and so we find, still present, half a century after Massingham wrote about Cobbett in *The Wisdom of the Fields* and celebrated the sustainable life of craftsmen and small farmers, perhaps the most clearly identifiable thread in the organic movement's confused political tapestry: the tendency towards the smaller-scale, the regional, the local, expressed by one of the movement's most influential personalities. It can be traced back to the anarchism of Kropotkin and Tolstoy, to Guild Socialism, to Distributism (which influenced Seymour's thought and in later life drew him to Roman Catholicism), to the Regional Socialism of Kenneth Barlow, to the estates and regional centres which Rolf Gardiner envisaged people like himself running, to the health centres described in Chapter 4, and to the communes movement and co-operatives of the 1970s and '80s. In our own time it is represented by the rise of farmers' markets, local economies and trading schemes, and Transition Towns. These various fringe movements and nostrums do not add up to a coherent political philosophy, not least because they would be at odds with each other over questions of emphasis and tactics, as fringe groups always are. But they have in common their resistance to Big Business, State Socialism and all forms of top-down management, believing instead in an 'emergent order' that derives from nature itself. This is why the anarchist Colin Ward was so impressed by the Pioneer Health Centre, where children were allowed to choose freely which activities they undertook and the result was a degree of harmony far from the chaos which disciplinarians would have predicted. More recently, Gregory Sams has expressed similar ideas.[27]

****

Now, we have seen that during the half-century which this book covers, members of the organic movement held a wide variety of political positions; that is a matter of empirical fact, however inconsistent with the organic philosophy being a Thatcherite — for instance — might appear. We have also seen that the movement's various bodies refused to throw in their lot with any particular party, not excluding the Greens; this was strategically a sound decision. But the movement has always been political in the sense of being concerned about government or EEC legislation on agriculture, food and the environment, and from the mid-1980s onwards became involved in the political activity of establishing organic food standards. And, given its holistic nature, the organic movement must, logically, imply a political philosophy: a theory of how Britain might best be arranged to ensure a fertile soil, a healthy population and a stable social structure. This was explicit in the writings of the organic pioneers, though the views that some of them embraced have provided ammunition for the movement's opponents. Some critics of environmentalism and New Age ideas consider that any holistic philosophy is implicitly fascistic, and the Seventies Generation was keen to separate itself from such ideas. On the other hand, a world in which the oil-dependent infrastructure is no longer viable might open up the prospect of a more democratic, locally based and convivial society, of the kind which 'Ernest Organic' (pseudonym of the North Wales farmer Patrick Noble) envisages — albeit idiosyncratically — in his book *Notes from the Old Blair and Bush*.[28] The politics of shopping will have little relevance to such circumstances.

9. Earth and Spirit

Christian influences on organicist thought

The term 'Orwellian' can be applied to the methods of the 'Big Brother' surveillance state, or to the obliteration of history carried out by the Ministry of Truth, in *Nineteen Eighty-Four*. It is in the latter sense that an 'Orwellian' moment occurred at the Soil Association conference at Cardiff in 2007, during a 'workshop' on the theme of Land and Spirit. Following various contributions — on Goethe, classical mythology, the intuitive capacity of chickens and the lifestyle of the Kalahari Bushmen — the 'ecological theologian' Dr Edward Echlin, backed by the present author, reminded the assembled company that many adherents to, or sympathizers with, the Christian spiritual tradition, had played major roles in the organic movement. As I argued in *The Origins of the Organic Movement,* there was a strongly Christian context to the early organic movement. We began the present book with the 'Encounter' organized by the Council for the Church and Countryside (CCC), and it is no coincidence that this conjunction between agricultural concerns and religious philosophy should have occurred. Some of the most thoughtful attempts to articulate a philosophy of 'Land and Spirit', and of humanity's relationship to Nature, emerged from the work of those active in the CCC. But the workshop at the Cardiff conference demonstrated no interest whatever in these organicist thinkers. There was not even any overt hostility or disputation: just complete blank lack of curiosity and a rejection of history. (At the previous year's conference in London, the 'culinary activist' Miche Fabre Lewin had spoken of her links with the Church, but these consisted of visiting a de-consecrated building in order to make 'sacred mayonnaise' with 130 eggs.)

Once more, we find ourselves in an extraordinarily complex area, and one containing plenty of ironies. While there are those among the Seventies Generation who have no interest in the spiritual dimension to the organic movement, the most frequently cited inspirers of that generation — Rachel Carson, E.F. Schumacher and John Seymour — were all, for considerable parts of their lives, active supporters of the Christian Church; in the case of the latter two in particular, this commitment was a product of their mature consideration. The Christian influence in the organic movement cannot therefore be dismissed as a feature of the bygone age in which the movement first took shape, though even if that were the case it would still merit attention as an aspect of the period covered by this book. Given the number of prominent figures in the organic movement during that period who were members of, or sympathizers with, the Christian tradition, some reference to that tradition's place in the movement is essential, particularly as writers such as Rory Spowers, in his environmentalist history *Rising Tides,* ignore it. The list of such figures includes A.G. Badenoch, Eve Balfour, Beata Bishop, Joanne Bower, Margaret Brady, Denis Burkitt, John Butler, Ralph Coward, Richard de la Mare, J.G.S. Donaldson, Laurence Easterbrook, Rolf Gardiner, Alan and Jackie Gear, Ken Gray, Ruth Harrison, Giles and Mary Heron, Lawrence Hills, David Hodges, Sir Albert Howard, Jorian Jenks, Riccardo Ling, Philip Mairet, H.J. Massingham, Sir Robert McCarrison, Philip Oyler, John Papworth, Lionel Picton, Jonathon Porritt, Lord Portsmouth, Prof. R. Lindsay Robb, W.E. Shewell-Cooper, Douglas Trotter, Hugh Trowell, Newman Turner, Kenneth Vickery, Robert Waller, Aubrey Westlake, C. Donald Wilson, and Dr Walter Yellowlees. If one includes the esoteric Christianity of Rudolf Steiner, then the Christian influence is even more far-reaching. One also finds admiration for Christ as healer expressed by Edgar Saxon and Juliette de Bairacli Levy. Of course, there have been many prominent figures in the organic movement who have not been sympathetic to the Christian tradition; but the above list, which is not complete, is nevertheless very impressive. It poses problems for those who, accepting the authority of Lynn White Jr, or paying disproportionate attention to Schumacher's essay on 'Buddhist Economics', take it as incontestable that the Judaeo-Christian tradition is incompatible with care for the environment. Rather than accept this assumption, it would show greater intellectual curiosity to look at the issue from a different perspective and investigate what it is in that tradition which has enabled it to produce or attract so many of the key figures in the organic movement. That is indeed a task for an 'ecological theologian'. As historians, we shall concern ourselves here first with a survey of the part which Christianity played in the organic movement and then look at the presence in the movement of other spiritual, and of esoteric, ideas.[1]

Mother Earth and Eve Balfour

We started this book with the always controversial figure of Rolf Gardiner, who organized the 'Encounter' between agri-culture and agri-industry. It was an article by Gardiner in *Mother Earth (ME)*, 'Towards a Sacramental Agriculture', which provoked two Australian members of the Soil Association, Dr C.E. Sandy and C. Halik into speaking their minds about a perceived tendency to align the Association with the Christian religion. Gardiner's 'attempt to link up sound farming practice with Christianity is not only absurd, but could be offensive to our (we believe) numerous non-Christian members'. While generously allowing Gardiner the right to hold his views, Sandy and Halik considered that *ME*, 'a basically scientific journal', should not contain articles about religion. As Chapter 7 has demonstrated, *ME* was indeed a strongly scientific journal; but it was by no means exclusively so. Given that the organic philosophy is holistic, its religious dimension might expect to find a place in the pages of what was in 1964 the most prominent organicist journal. *ME* had never made any secret of its religious inclinations: over the years, its front covers had featured quotations from Sir Thomas Browne, the books of *Genesis* and *Deuteronomy*, Alexander Pope, John Ruskin and Archbishop William Temple.[2]

There is nothing remarkable about making a connection between religion and agriculture: through most of history these two fundamental human activities have been intimately linked, and Gardiner contended that the separation between them opened the way to the exploitation of the soil which the organic movement was pledged to oppose. Whether or not this argument is valid, it was held by some of the most influential figures of the movement's early years. We have seen that from the word 'go' the organic movement was concerned to bring about social changes, and many of the early organic writings looked — albeit with greatly misguided optimism — to a post-war reconstruction of British society based on Christian social principles. I have provided evidence for this is *The Origins of the Organic Movement,* but it is worth re-emphasizing some of it. Eve Balfour, in her Postscript to *The Living Soil,* looked to the dethronement of '[t]he false idols of comfort and money' and their replacement by 'the Christian God of service'. Mary Langman told Allan Pepper that Eve Balfour had been 'brought up in the Church and in faith in a benign Deity whose purposes were working out', and that this faith would have played a part in her acceptance of ecological thinking. (It appears that later in her life, Balfour's beliefs grew somewhat less orthodox, showing signs of what might be termed 'New Age' thought; but her belief in a benign power at work in the world remained. Dr Anthony Deavin has recalled how she responded philosophically to criticism and problems by saying, 'It's all taken care of'.)[3]

The Council for the Church and Countryside

The *New English Weekly,* which was, as I have argued elsewhere, the leading forum for organicist ideas during the 1940s, was run by a predominantly Christian group of social figures; its editor Philip Mairet was a man of wide-ranging cultural, social and religious interests who in the 1940s provided some valuable, far-sighted essays on humanity's relationship to the environment. Mairet was prominent in the CCC, whose publications in the late 1940s included attempts at working out a 'theology of the countryside', as the Rev David Peck subtitled his booklet *Earth and Heaven* (1947). There is no need to repeat here the summary of Peck's argument to be found in *The Origins of the Organic Movement,* but the essay remains a learned and thoughtful piece of writing which any 'ecological theologian' might with value use as a starting-point for his or her own reflections; it is altogether a deeper and more considered piece of writing than Lynn White Jr's influential polemic. (We should remember, though, as the fact is usually ignored, that White was himself a churchman.) For Jorian Jenks, too, there was nothing 'absurd' about connecting organic cultivation with Christianity; organic husbandry was an agricultural form of humility, recognizing man's place as a member of the God-given natural order. Both Peck and Jenks made the point that human beings stand in an ambiguous relationship to nature, both being a part of it, as biological creatures, and distinct from it in their ability to study, manipulate, harm or improve it. Man's 'responsibilities for the study and observance of natural laws' was therefore all the greater: 'he expresses "dominion" only insofar as he is a just steward'. Jenks was pleased that — 'as yet' — the organic movement showed no signs of becoming 'a mere Nature-cult'; for 'anyone nurtured in the Christian tradition, Nature is so clearly a manifestation of the glory of God'. The CCC met opposition from within the Church of England because some felt it smacked of paganism at a period when the influence of Karl Barth — who rejected all possibility of natural theology — was very considerable. For an organicist such as Massingham, it was possible to combine the older sense of the divine spirit of the land with Christian social ethics, 'so that the two ... symbolized by the superimposition of churches upon pagan holy places, represented a perfect system of faith. This he evoked, against the "utter darkness and savagery" which he discerned in modern urbanized culture. He coined the slogan "Let the Church come back to earth".' Organic farming was one means by which the Church could do so. It is interesting to note that Jenks' article was followed on the next page by Richard St. Barbe Baker's environmentalist manifesto *The New Earth Charter*. Baker adopted the Baha'i faith in the 1920s, but was happy to preface the Charter with a piece of verse beginning: 'Through God's good grace, through strength of English oak,/ We have preserved our faith, our throne, our land ...' and to conclude by quoting Ruskin's belief that 'God has lent us the earth for our life. It is a great entail'.[4]

Christian influence was strong in other organicist publications. Newman Turner described himself as 'a person believing in God and advocating practice of His laws'. His approach to agriculture was directly derived from his Christian faith in natural law: 'It is patent that the basis of nutrition and health is biological, and those who seek to make it chemical have got to satisfy us that they know better than God'. North of the border, *Health and the Soil* was produced by the Scottish Soil and Health group, in which the former medical missionary A.G. Badenoch was one of the three leading figures. He became a Catholic and in the post-war years contributed to the Dominican journal *Blackfriars*. *Health and Life,* though less significant a journal than it had been in the 1930s and '40s, continued its commitment to the organic cause and became more openly Christian in the 1950s than it had previously been. Edgar Saxon had broken away from his Nonconformist upbringing but retained his admiration for the person of Jesus, seeing him as the fullest example of divine radiance and healing power. The influence of H.J. 'Dion' Byngham had ensured a more specifically pagan element in the journal. Saxon was succeeded as editor in the mid-1950s by James Gathergood, evidently a devout Christian. There was also *Star and Furrow*, the newsletter of the BAA, underpinned by the heterodox, esoteric Christianity of Steiner's Anthroposophy.[5]

The stewardship philosophy of Philip Mairet

The Australian humanists who took exception to Rolf Gardiner's Christian views appearing in *Mother Earth* were opposing what was actually a very marked feature of the organic movement. We are not concerned here with whether they were justified in their opposition to it; the point to note is just how strongly marked it was. As time went on, there would be more opposition to it, but it remained present. The same year that the humanists raised their objections saw the appearance of an essay by Philip Mairet which offered a Christian philosophy of the environment. Entitled *Bailiff for God's Estate on Earth*, it exists as a typescript with a printed cover indicating that it was circulated to those involved in 'Church and Countryside'. C.H. Sisson did not include it among Mairet's *Autobiographical and Other Papers*, and it seems unlikely that many people would have read it. Inherent merit and breadth of readership, though, are entirely distinct matters, and, in view of Mairet's erudition in the fields of both Eastern and Western religious philosophy and his importance to the organic movement, it is worth examining his approach to this controversial area.[6]

Mairet began by identifying the origin of the Judaeo-Christian tradition as the point at which the Hebrews repudiated the Nature gods worshipped by the nations of the Middle Eastern agrarian civilizations 'and acknowledged One invisible God as the sole Creator of all things, themselves included'. In this monotheistic system, Man was given 'dominion' (Psalm 8, v.6) over everything, in the earth in the water or in the air. But this authority was conditional, not absolute: a responsibility and an immense dignity. Mairet then jumps to the age of the Scientific Revolution, regrettably not looking at the centuries of medieval Christendom; though he does suggest that the increasingly humanistic tendencies in European culture can be traced back beyond the work of Copernicus, Galileo and Descartes. These tendencies accelerated once 'the works of scientific technics became so impressive'. (The use of the term 'technics' reminds us that Mairet was a good friend of the American polymath Lewis Mumford, one of the gurus of the 1960s 'counter-culture'.) But it was not merely Man's relationship to God which altered: 'something also happened to his conception of his natural environment'. There was a loss of a sense of kinship with it; natural things were thought of as mechanisms to manipulate or puzzles to solve. Enlightenment humanism looked for the meaning of all things in Man, 'making for a conception of Man in Nature radically different from what we find in the Judaeo-Christian Scriptures. The three terms: God, Man and Nature are reduced to two: Man and Nature'. Mairet's argument, then, is that it is the *abandonment* of the Judaeo-Christian tradition which has encouraged a ruthlessly exploitative attitude to the environment: 'ecological dominance' as he terms it, in which species depend upon Man's approval for their survival. Mairet describes the competition and precarious state of our 'globalized' economy, in which nations become increasingly dependent for their survival on supplies from beyond their own territories and into which it seemed certain the whole of humanity would be drawn. The technological revolution brought with it 'a portentous change in consciousness', dimming Man's awareness of God and distorting his relation to Nature. His chief focus of consciousness became his own works. Mairet prophetically saw that in time these works — consumer goods — would come to dominate the desires of countries beyond the West. What response could theologians offer to this seemingly unstoppable growth of the technological consciousness, with its bias against Nature?[7]

Ecology and 'the organic school of husbandry' were a corrective factor to this bias, for they viewed the part from the standpoint of the larger whole, seeking the truth not through a process of ever-more-minute analysis but viewing 'every organism in the context of the "functional" relations between them all'. Both organic husbandry and biology, Mairet believed, were tending towards a recovery of respect for Nature. In theology, the only influential thinker aware of such issues was the unorthodox Dr Albert Schweitzer, 'with his almost Buddhistic doctrine of "reverence for life"'. While some Christian thinkers had expressed concerns about atomic armaments and cybernetics, they were only vaguely aware of the 'cosmic impiety' of industrial civilization towards the natural creation. But the late Pope John XXIII's 1961 encyclical *Mater et Magistra* was a hopeful sign, giving as it did special recognition to the plight of agriculture in an industrial civilization and laying down social principles conducive to a stable rural community: as one would expect of an organicist, Mairet particularly approved of the document's emphasis on the importance of the family farm. But there remained an urgent need 'for a complete theological study and re-statement of the responsibilities of Man in Nature and towards Nature'. Such a re-discovery of principles was an essential prerequisite to the regaining of a sense of the sacramental in agriculture: Mairet referred specifically to the article by Gardiner with which this chapter opened. He was not so naïve as to think that the sense of the sacred experienced in the past could be recaptured; nor did he either wish to reverse the achievements of science or consider such a thing possible. (In this respect, his position was more optimistic than that later taken by *The Ecologist*.) But Man could free himself from slavery to his inventions only through love of God, and that love implied 'a depth of respect for his Creation'. An increasing disbelief in God had gone hand-in-hand with 'an increasing propensity to abuse and despoil Nature — including human nature'. Mairet evidently felt frustrated by the lack of a Christian philosophy of nature in the face of the challenge which technology posed: theologians needed 'to indicate something of the ways in which Man is expected to apply his powers to improve Nature both in himself and in the natural environment'. He saw ecology and organic husbandry as examples of what one might term practical theology, working 'against continual opposition from ... formidable anti-natural forces'. (I recall the aggressive secularist Claire Fox saying at the 2006 Soil Association conference that she was not concerned about the environment, only about people — as if people would be able to survive independently of a natural environment.)[8]

I wish to quote again here a passage which appears in *The Origins of the Organic Movement,* since it can be regarded as the mature expression of ecological faith by one of the few philosophers which the movement can claim.

> If there is to be any goal of collective salvation on earth
> for us to work for, it must include the perfecting of all the
> other life upon which Man's life depends and by which it is

enriched. ... And a people is seen to be on the road towards this goal if and when its powers are directed more and more to such things as the reclamation of wastes and deserts, the beautifying of landscapes, purifying of waters, perfecting of species of plants and animals. And, last but not least, when it employs not as few as possible but as many as possible of its people in this sanest, healthiest and happiest of all human occupations — that of improving, ordering and beautifying the surface of this planet, the Earth, which is the particular garden of God entrusted to our care. Here is infinite scope for all the analytical lore and technical mastery acquired since the scientific revolution. To what other end could they sensibly be employed?[9]

We see implied in this passage the organicist rejection of agricultural efficiency measured according to output per man, and the perennial concern for a populated, working countryside of rural crafts and industries.

Two Christian environmentalists: E.F. Schumacher and John Seymour

Many would see Mairet's approach as still too 'Western' and manipulative, and find in Eastern thought a more environmentally friendly philosophy. (Edward Goldsmith quotes Lao Tzu on leaving the world of nature entirely alone, since it would be impossible to improve it.) E.F. Schumacher's essay 'Buddhist Economics', first published in 1966, has been influential in this regard, but it is time that more balance was brought into discussion of Schumacher's religious philosophy. Rory Spowers, while paying considerable attention to this essay, simply ignores the fact that Schumacher committed himself to the Roman Catholic Church, to whose thought he had long been attracted, in 1971, during his first year as Soil Association President. To emphasize this point is not to deny in any way the importance of Buddhism in Schumacher's spiritual development; it is simply to offer a more accurate picture of the complexity of his thought. In this regard, it is also important to point out that Schumacher himself explicitly stated that his choice of Buddhism to demonstrate the economic implications of abandoning Western materialism for a spiritual tradition was 'purely incidental; the teachings of Christianity, Islam or Judaism could have been used just as well as those of any other of the great Eastern traditions'. Let us look at another of Schumacher's essays: *The Age of Plenty: A Christian View* (1974), first given as a lecture in Edinburgh in 1973.[10]

Although given less than a decade after Mairet wrote his essay, Schumacher's lecture was more cautious about scientific progress. In the intervening years, *The Ecologist's Blueprint for Survival* had appeared and the 'limits to growth' debate was by 1973 in full swing. The concerns of an environmentalist minority were being expressed much more widely, and industrial society was increasingly regarded as a threat to stability and even to survival. (By the time Schumacher's lecture was published, the 1973–74 oil crisis had magnified these anxieties.) In order to assess the condition of life in an industrial, consumerist society, some scale of moral values had to be applied. Schumacher regretted the lack of unanimity in Christian thought on economic life, and so drew on a 'Foundation' offered by St Ignatius of Loyola which demonstrated both 'an implacable logic and genial common sense'. It is interesting to see that Schumacher chose a piece of writing which displays one of the attitudes to which many environmentalists particularly object: St Ignatius takes it for granted that 'other things on the face of the earth were created for man's sake'. Schumacher noted that 'some of us' would have preferred to formulate this idea slightly differently, but he did not reject its essential meaning. From St Ignatius' 'Foundation' he concluded that Christians were called upon to '*strive*' to use the goods of this world '*just so far*' as they helped them attain salvation, and to withdraw from them '*just so far*' as they hindered (italics in original). Whether or not one agrees with St Ignatius' assumptions, this approach logically implies a very different attitude to the environment from a philosophy of unrestrained productivity.[11]

We saw in Chapter 1 how post-war agriculture had adopted a criterion of efficiency based on output per worker; as we know, the organicists challenged this criterion, also drawing attention to its effects on the rural population. Schumacher, too, felt that the concept of efficiency had become 'quite uncannily narrow and exclusive', relating only to quantity of material goods and monetary profit. The idea that a process might count as efficient because it made workers happy and fulfilled would be deemed sentimental nonsense: under industrialism, the spiritual condition of workers was disregarded, even if it was deemed to exist. Such systems of production moulded the society of which they were a part, and in order to escape slavery to them new types of organization and technology must be developed. Schumacher argued that, from a Christian standpoint, such new arrangements should be human-scale, simple and non-violent. Making a living was a means to an end, not the primary purpose of earthly existence, and the specialization which complexity necessitated rendered the attainment of any wisdom or higher understanding almost impossible; Schumacher quoted St Thomas Aquinas in support. Aquinas also reminded him that humans must never lose their 'sense of the marvellousness of the world around and inside [them]': an attitude which engendered non-violence. Schumacher dismissed as 'excessively superficial' the view that modern man's ruthlessness towards the environment stemmed from the Biblical teaching that Man was given 'dominion' over the creatures of the earth. The Christian view of industrial society rejected what Thorstein Veblen termed 'crackpot realism', which assumed 'that people really do not matter; that we are masters of nature which can be ravaged and mutilated with impunity; that some Divine Improvidence [sic] has endowed a finite world with infinite material resources; and that consumption is the be-all and end-all of human life on earth'. Finally, Schumacher distinguished between two responses to the predicament of technological society. On the one hand were the people of 'the forward stampede', with their slogan 'A break-through a day keeps the crisis at bay'; on the other, 'the homecomers: people striving to lead things back to their proper place and function ... They believe that the spiritual has dominion over the material ...' A great convergence was taking place, in which the 'language of spiritual wisdom [could] now be understood also as the language of practical sanity'.[12]

It seems to me that Schumacher's place in the history of the organic movement is not only as an influence on the younger, environmentally minded generation who joined it in the 1970s, but also in the stream of organicist thought represented by Mairet's generation. As a thinker, Schumacher came to have a good deal in common with Mairet. Both were influenced by Eastern philosophy and both had read the writings of Ananda Coomaraswamy. Both were familiar with the esoteric system of Gurdjieff and Ouspensky: this influence is evident in Schumacher's book *A Guide for the Perplexed* (1977). Both came to forms of Catholic Christianity: Anglican in Mairet's case, Roman in Schumacher's. Schumacher's posthumously published book *Good Work* (1979) addressed the same sorts of concern as Tom Heron, decades earlier, had done, and from a similar standpoint owing a good deal to medieval social thought. And there exists a strong family resemblance between Schumacher's Intermediate Technology and the small-scale machinery which L.T.C. Rolt — another organicist writer with sympathy for the medieval order — advocated in Massingham's 1947 symposium *The Small Farmer*. Schumacher was also familiar with the work of René Guénon, whom he described as 'one of the few significant metaphysicians of our time' and whose book *The Reign of Quantity and the Signs of the Times* Lord Northbourne had translated. It is not too difficult, in fact, to imagine Schumacher in the company of the Chandos Group and the *New English Weekly's* editorial board.[13]

We can now turn to another of the major influences on the Seventies Generation, John Seymour, whom we know to have been much influenced by Rolt. In 1988 Green Books re-printed Rolt's *High Horse Riderless* (1947) in its series of 'Organic Classics', and Seymour provided an introduction, clearly sympathizing with the book's final message 'of cheer and hopefulness ... Rolt accords to humankind a high and noble destiny: the Kingdom of God on Earth he claims: "is not a misty theological chimera, it is a practicable goal within the range of our conscious ability to attain"'. Such sentiments might come as a surprise to anyone whose knowledge of Seymour is confined to Benjamin Davis' obituary of him in *The Ecologist,* which makes no mention of his rootedness in the Christian tradition; but Seymour's biographer Paul Peacock explicitly states that he 'strongly believed in a personal God' and 'that we could get nothing right unless we managed to get our spiritual values right'. Like Schumacher, Seymour became attached to Roman Catholicism later in life, and Peacock believes that a major reason for this commitment was his discovery of Distributist ideas. If this is so, it provides further evidence for Seymour's closeness to the early organic movement, to which Distributism made a significant contribution. Like Mairet, he felt that God had given humanity the task of tending Earth's garden, and that the loss of Catholic traditions had brought in turn a loss of reverence for nature. If the substitution of a money economy for a subsistence one was occurring so rapidly in supposedly Catholic countries during the late-twentieth century, it was because they were no longer really Catholic: in England, Henry VIII's destruction of the monasteries 'ended a thousand years of slow organic development towards a fruitful and beautiful relationship between Mankind and the rest of Nature — a relationship which was possibly the most pleasing to the Life Force of any that had so far developed on this earth'. In such a passage we can see the spirit of Massingham living on. Seymour's active opposition to GM crops in Ireland later in his life was based on his Catholic beliefs. *'To mess around with God's Creation,'* by fundamentally altering the makeup of plants, was a *'mortal sin, and it's a mortal sin, if you understand that, not to prevent it'* (emphasis in the original). When in his late eighties, Seymour attended a one-day conference of the Christian Ecology Link at Ryton Gardens, organized by Alan and Jackie Gear: among the many people present were Ben de la Mare, chaplain of Collingwood College at Durham University and son of Faber's agriculture and horticulture editor Richard de la Mare; and the Green peer Tim Beaumont, who half a century earlier had invited Lord Portsmouth to address the Oxford University Plough Club, of which Beaumont was President.[14]

Mairet, Massingham, Rolt, Schumacher and Seymour were all thinkers who saw the organic movement from a historical and philosophical perspective: the same is true of Jorian Jenks, in *From the Ground Up* and in certain sections of *The Stuff Man's Made Of*. In the latter book he distinguishes between 'vertical' attitudes — flowing upwards from roots in villages, towns and churches towards Heaven — and 'horizontal' ones, determined to conquer other lands for the sake of political power and material wealth. Jenks was concerned to analyse the process by which Western civilization had adopted an exploitative attitude towards nature and to identify the intellectual underpinnings of this attitude. For all of them, the decline in influence of the Christian faith — the chief source in Western civilization of a sense of the sacred — was a crucial factor.[15]

Robert Waller's 'ecological humanism'

On his appointment as editor of the Soil Association journal, David Stickland at once began to lead it in a more practical and entrepreneurial direction, dismissing his predecessor Robert Waller's interest in philosophy as irrelevant. Such a narrowing of focus, though, was incompatible with the organic movement's traditionally holistic outlook, which sought to bring earth and spirit together. Waller had studied philosophy in the 1930s under Professor John Macmurray at University College, London and produced radio programmes on the subject for the Third Programme in the 1940s; he continued to read widely in it throughout his life. In 1973, the year that Stickland took over from him and that *Small is Beautiful* appeared, Waller published *Be Human or Die,* which he intended as a summation of his many years' study of cultural history and environmentalist thought. It made little impact and bears signs of the difficult personal circumstances in which it was written, but is nevertheless significant as one of a small number of serious attempts by leading organicists to create a comprehensive philosophy for the organic movement, and contains much that is still relevant. (The contrast between the editorials written by Jenks and Waller and the brief paragraphs in contemporary issues of *Living Earth* indicates all too starkly the extent to which the Soil Association has jettisoned its intellectual heritage.) *Be Human or Die* is perhaps over-ambitious for a book of modest length (260 pages), but any attempt to develop an organicist philosophy must of necessity range over all aspects of life. It is 'organic', too, in its attempt to do justice, and bring balance, to various features of human and natural life. Waller subtitled the book 'A Study in Ecological Humanism', indicating his debt to Stapledon's concept of 'human ecology' but not implying any kinship with the atheism prominent among today's media élite. Waller was a self-styled 'undogmatic Christian' who could never have followed Schumacher and Seymour into the Catholic Church but who like them 'rejected the materialism and empiricism of twentieth-century British society. In his view, the atrophying of a sense of reality beyond the confines or comprehension of the scientific method had devastating environmental and cultural consequences. He called ... for a new world view that recognised the limits of science and of human reason.' The recognition of 'limits' is a central preoccupation of organicist thought, as we shall see later on; and so is a mistrust of reductionism, the attempt 'to explain the higher dimensions of reality in terms of the lower'. Waller had imbibed this mistrust from Macmurray, who denied the adequacy of mechanistic concepts for understanding organic life. On the other hand — and here we see Waller's concern for balance and proportion — the wholesale rejection of reason, so evident in fascism or the drug culture, brought its own dangers. Waller believed that reason must consist, not in abstract speculation and utilitarian calculus, but in the capacity to think and act in terms of the Other, of that which is not oneself: God, nature and one's neighbour. This, for Waller, was 'human ecology': a right relation to the dimensions of spirit, nature and the personal life. Writing before the 1973–74 oil crisis intensified fears about the viability of industrial civilization, Waller perceived signs of hope in the increasing inability of that civilization to satisfy people's material, or their emotional and spiritual, needs. New ways of organizing society would be found, and people would 'seek a true international spirit that can be expressed in the old notion of the Fatherhood of God' (a prospect no doubt unacceptably patriarchal for the emerging generation of ecological feminists). Waller later wrote on the feasibility of a 'Green' Christianity with Anthony Etté in *The Ecologist Quarterly* and for a Green Books symposium: Massingham's influence was evident in the latter essay. Waller's impact was minor in comparison with that of Schumacher and Seymour; the point is, though, that someone who had thought deeply about environmentalist philosophy saw no necessary incompatibility between support for the organic cause and commitment to the Christian tradition. As the list given earlier indicates, he was far from alone in this. Cdr Robert Stuart even timetabled Sunday morning worship into the programme of his Soil Association Weeks, which implies that there must have been a fair number of guests who would have wished to attend.[16]

A brief note on Little Gidding and on Judaism

One link between the Church and the Soil Association should be particularly satisfying to the historically minded. At Little Gidding, in the remote Huntingdonshire countryside, a Christian community was established in 1977, and the following year John Robinson, who had spent eight years in agricultural extension work in India, joined it to take responsibility for its husbandry. Within a couple of years the community had become a Symbol-holder. Vivian Griffiths, later prominent in the Biodynamic Agricultural Association, spent some time there. That Little Gidding should have become a centre for organic husbandry is very apt, as T.S. Eliot's poem 'Little Gidding', one of his *Four Quartets,* first appeared as a special supplement in his friend Philip Mairet's journal the *New English Weekly*.[17]

There is little in organic writings on the more specifically Jewish element of the Judaeo-Christian tradition, but *The Ecologist* published an article by Nigel Pollard on 'Israelites and their Environment', arguing that the Israelites' attitude to nature was, on balance, ecologically sound, while Daniel Hillel's *Out of the Earth* provides a thought-provoking 'organic' analysis of the early chapters of the Book of Genesis. Evidently Juliette de Bairacli Levy and Col. Robert Henriques found no incompatibility between their Jewish origins and commitment to the organic movement.[18]

****

Even this very brief survey of the role which Christians have played in the organic movement may be a source of embarrassment to those who, for whatever reason, would prefer to ignore it. But that presence is there, and should not be airbrushed from history. Those who reject it are setting themselves at odds with a formidable array of the organic movement's most significant figures; nevertheless, there have been plenty of other significant figures for whom this Christian element in the organic movement is unnecessary or undesirable. To analyse why this element has been so strongly marked in the organic movement, when this has not been the case in the wider environmental movement, would be a worthwhile task, but cannot be undertaken here.

The esoteric Christianity of Rudolf Steiner

We are not quite finished with the Christian contribution to organicism, though, since the biodynamic movement derives from the esoteric Christianity of Rudolf Steiner's Anthroposophy. Gary Lachman describes Steiner's ideas about Christ as 'highly eccentric', but the fact remains that Steiner considered Christ to be of unique importance in spiritual history; Steiner's estrangement from the Theosophical movement arose on account of this belief. No doubt it is possible to practise biodynamic methods of cultivation without subscribing to or even knowing about Steiner's views on Christ; but this is not the issue.

Rather, we should note that the biodynamic movement was formed of people who subscribed to a form of Christian, rather than Eastern, esoteric philosophy. John Soper — who, as we saw in Chapter 2, was one of the most influential Steinerian agriculturalists — wrote in evangelistic terms in *Star and Furrow* of dedicating one's will to Christ, while Gertrude Mier saw in the threefoldness of the plant a reflection of the principle of the Holy Trinity. Lachman believes that Steiner's Christocentric beliefs can be put aside 'without losing what remains of importance in his work', but he admits that most Anthroposophists would probably disagree with him. Charles Waterman (nom de plume of Charles Davy) certainly did, making it clear in *The Three Spheres of Society* that Christianity held a 'central place' in Anthroposophy, while Wendy Cook, who studied with John Davy, has described the Steiner movement as 'essentially a new kind of Christian consciousness'.[19]

Most of the articles on biodynamics which appeared in the Soil Association journal during our period were on the methods rather than the philosophy, but Laurence Easterbrook, in an article to mark the centenary of Steiner's birth, was happy to refer to some of the occult knowledge which Steiner claimed to have been vouchsafed. As I mentioned in *The Origins of the Organic Movement,* Easterbrook must surely be the only Public Relations Officer for the Ministry of Agriculture to have written a pamphlet on reincarnation, and his autobiography was published by the Spiritualist Association of Great Britain. The autobiography 'included the text of a lecture he had given to the Association in which he sardonically revealed that in later life he had discarded the teachings of the Church of England in favour of the teachings of Christ'. As a young man, Easterbrook had struggled unsuccessfully with Steiner's *Outline of Occult Science* and given him no further thought until he met the biodynamic gardener Mrs Pease and then the farmer Dr Carl Mier. Their ideas changed his way of thinking about agriculture, but he did not join any Steinerian body, or consider Steiner 'as a repository of absolute wisdom.' He thought of him as 'a highly evolved being who volunteered for service on this earth to deliver us equally from the intellectual poverty of second-rate scientists and the dreary materialism of the established church. He came to demonstrate to us that religion and science are indivisible ...' By 'science', though, Easterbrook understood something far removed from reductionism and from anything that Lord Taverne (or, for that matter,

Sir Albert Howard) would consider genuine science. Steiner's was a 'far broader conception of science', dealing with 'all the realms of man's being and its close affinity with the invisible workings of the universe' and including study of non-material states of being. Easterbrook noted that 'Steiner's ideas on agriculture are very far from general acceptance', apparently implying that he found this surprising. But if he did find it so, that would have been because he believed that biodynamic methods were successful in practice, productive and aesthetically appealing.[20]

'New Age' spirituality

Findhorn

Moving on from esoteric Christianity to the more 'New Age' atmosphere at Findhorn, we see there, too, the belief that paying heed to the spiritual world could bring blessings in the biological world. The garden which Peter and Eileen Caddy created on a windswept wasteland of gorse and sand dunes produced remarkably large vegetables. Alexis Edwards, an astrologer at the Community, attributed this to 'simply ... listening to nature's voice and then giving her what she needs. Attunement to the devic and elemental kingdoms is simply a lost art which is now being revived ...' Edwards wrote an account of his Findhorn experiences — material needs provided by a miraculous 'divine supply' — for *Seed;* as we saw in Chapter 5, *Seed* wrote warmly, though perhaps rather flippantly, of the Findhorn Community's belief in nature spirits. Was such a belief — and might this also be true of Steiner's cosmology — a valuable heuristic concept: an experiment in what Vaihinger termed 'the philosophy of "as if"'? Edwards appeared to have taken the belief in nature's spirits literally, but what of Professor Lindsay Robb, who, like the leading biodynamic farmer George Corrin, showed much interest in what was happening at Findhorn? Did the remarkable success of Findhorn cultivation in fact provide evidence for the occult theory which was offered as the explanation of it? Certainly there was every reason for an agricultural scientist to investigate Findhorn's methods, which Edwards described as 'enlightened'. By 'enlightenment' he meant 'recognising and accepting a higher law at work within the earthy sphere': a law which would bless with abundance those who obeyed it.[21]

Readers will recognize immediately that we are once again in the realm of 'the natural order', albeit this time in a context different from that in which the organic pioneers developed their philosophy. Here indeed are some familiar organicist themes re-worked in a New Age guise: the 'muck and magic' for which the organic movement has so often been mocked (compost-based cultivation and nature spirits); the organic society (Findhorn as a seed, sending out many tender shoots); and the holistic vision of the world (its unity formed of many individual cells or 'seed patterns'), a world itself ultimately contained in the Oneness of God. (Those with a keen eye for the fascist strain in organicist thought will appreciate Edwards' image of 'unfold[ing] together the divine heritage inherent within the Oneness of our true group identity': truly, the 'organic society'.)[22]

The esoteric tradition

As we saw in Chapter 5, *Seed* contained a considerable amount of 'New Age' material and various contributions on Eastern philosophy and mysticism; items in the Soil Association journal on similar topics were extremely rare, though not unknown. A Miss Helen Macfarlane of Somerset informed readers that, inspired, by Findhorn literature, she had called on the nature spirits and 'divas' [*sic*] to protect her crops from wood pigeons and other pests, and that the results had been impressive. Her call for other Association members to carry out similar experiments met only with a frivolous response. J.R. Hughes of Sussex, who lacked Miss Macfarlane's faith in nature spirits, preferred to falsify the labels in his garden, so that, for instance, mice would think that peas were really potatoes, and leave them untouched. More substantial than this exchange, though, was a letter in the late 1970s from Maryel Gardyne which occupied three-quarters of a page. As a member of what was evidently a Christian group which discussed ecological issues, Gardyne had debated the Biblical idea of the Sabbatical Year, which was 'part of the cycle of Seven, on which ancient knowledge seems to have been based'. She supported the Research into Lost Knowledge Organisation, which was trying to draw attention to ancient scientific principles, particularly those based on numerology, and she was familiar with the esoteric works of Ouspensky and T.C. Lethbridge, and the new approach to physics recently outlined by Fritjof Capra.

Like Dr Anthony Deavin, who saw the Soil Association as a forerunner of contemporary spiritual movements, Maryel Gardyne believed that Association members, through their 'respect for natural order' were enabled to 'pick up one thread of the cosmic web' and draw closer to a 'deeper understanding of the whole of creation'. Through meditation, one could bring one's inner understanding into harmony with the laws of creation.[23]

Such esoteric ideas are of course a gift to those wishing to dismiss the organic movement on account of its association with 'mystery' or 'magic' or hippie philosophy, and can be an embarrassment or, at least, as in Carolyn Wacher's view, of little value to those in the movement who wish either to sell products in a secular market or make the case for organic methods scientifically. But esoteric thought is a perennial strand in organicist philosophy and, some would argue, essential to it. The case of Lord Northbourne is interesting in this respect. His 1940 book *Look to the Land* is one of the clearest early statements of the organic case, stating the nutritional, agricultural, economic and social dimensions of it and ending with a short passage which embraces the typically organicist theme of harmony and wholeness, and of man's arrogance in trying to 'conquer' nature. The present Lord Northbourne, his son, has described his father as a conventional, though thoughtful, church-goer at this time; as time went on, he grew increasingly interested in the 'perennial philosophy': the idea that within all the major religious traditions can be found the same core of spiritual truth. He was particularly attracted to the Sufi mystical tradition, though, perhaps inconsistently, he recommended spiritual seekers to keep to their own traditions and to encourage other seekers to keep to theirs. *Look to the Land* precedes the period which this book covers, but Northbourne wrote at greater length on the relationship between agriculture and religion in the 1960s and early '70s, and joined the Soil Association in 1969, publishing articles in its journal. Northbourne's objections to industrial agriculture were part of a broader critique of a civilization whose aim was 'to possess or command everything in its environment that has so far eluded its grasp' and whose cast of mind was 'separative'. 'Reality itself is departmentalized; it disintegrates, and man becomes ever more lonely and puzzled'. This alienation could be dispelled only by a re-discovery of 'the sacred centre' and a rejection of the modernistic belief in progress and perfectibility through technological control of the outer world. Such ideas are not very different from those of Schumacher, and they belong to the same period.[24]

9. EARTH AND SPIRIT

The Seventies Generation cannot, as a group, be identified with any particular spiritual or religious philosophy. John Humphrys drew a sharp distinction between the 'almost religious lines' along which the Soil Association had been run until the 'new breed of aggressive young farmers' appeared, and the new era of profit-turning which they inaugurated, referring to Peter Segger and Patrick Holden. But the strand of esotericism was still to be found: Segger was attracted to Steiner's ideas; Holden and Charles Wacher to Krishnamurti's. Holden has also been involved in Gurdjieff groups. The somewhat misleadingly named School of Economic Science, has also attracted members of the organic movement, among them John Butler and Anthony Deavin. The School was founded in the 1930s 'to develop thinking that would help bring about greater economic justice', but its main emphasis switched during the 1950s to philosophy. Its appeal to organicists would lie in its dedication 'to the quest for unity and balance in all aspects of life'. In the mid-1980s, one correspondent, Howard Case, complained to the *SAQR* that the organic movement's philosophical dimension was not as prominent as it had once been. He regretted that the *Review* was falling away into 'materialism albeit of the organic kind' and wanted it to resume its philosophical role rather than be merely 'the hand-maid of economics' — a role to which *New Farmer and Grower* was better suited. But in fact, the journal's previous issue had granted SIR GEORGE TREVELYAN a page on which he could outline his organicist spiritual philosophy: one consisting of a rediscovery of 'the Ancient Wisdom, which knew that the universe is Mind, a vast continuum of thought and life poured out from the Divine Source'. Like the veterinary scientist Reginald Hancock more than thirty years earlier, Trevelyan based his belief in the organic movement on theoretical physics, referring to James Lovelock's Gaia Theory and, by implication, to Capra's *The Turning Point* (1982) in support of his belief that contemporary physics equated with the wisdom of the mystics. Were such beliefs essential for those who supported the organic movement? Trevelyan claimed there was 'no need whatever' to hold those views, but nevertheless he felt that to do so immensely enhanced the significance of their actions.[25]

We find ourselves again in the presence of Eve Balfour's apparent paradox: 'There are no materialists in the Soil Association'. Trevelyan, who saw the Association as an organization 'at the forefront of a *redemptive* drive to rescue the planet from disaster brought about by human greed, fear and ignorance' (my emphasis) believed such work to be in tune with God's purposes, and to represent a transcending of the limits which rational materialism imposed on human consciousness.[26] His comment that his spiritual outlook was quite unnecessary for commitment to the Association's work is somewhat puzzling, as it is not clear whether he means that belief in the specific ideas of Capra or Lovelock, or in any spiritual philosophy, is unnecessary. One could have belonged to the Soil Association or the HDRA, or BOF/OGA or the McCarrison Society, or have supported Elm Farm without subscribing to any sort of spiritual philosophy. But given the weight which so many of the organic movement's most prominent figures have attached to 'the spiritual' — whatever that elusive term may mean — it is hard not to feel that those who lacked belief in a spiritual dimension would have been missing something of central importance to organicist thought. If the spiritual element *can* be discarded, though, the question arises as to where the essence of the organic philosophy resides. What possible answers can be found among the writings which the organic movement generated during half a century? Are there certain ideas to which all organicists can subscribe?

Bedrock philosophy: limits and wholeness

Two ideas stand out as possibilities, I would suggest, though both of them might be considered implicitly spiritual or religious; they indeed both appear often enough in the writings of spiritually minded organicist thinkers. These are the idea of limits and the philosophy of holism, both of which can in principle be accepted purely on grounds of pragmatism and of scientific observation and experiment. Perhaps one should add that they might be accepted 'negatively', as a reaction against the consequences — biological, physical and moral — of the opposing philosophy.

Recognition of finitude

The 1960s and '70s produced several books about 'limits': before *The Limits to Growth* (1972) there had been *The Limits of Man* (1967) by the scientist Hugh Nicol, and after it came Bishop John V. Taylor's *Enough is Enough* (1975), which Dorothy Paulin reviewed for the *SAQR,* and Ivan Illich's *Limits to Medicine* (1976). In organicist thought, the idea can be found at the beginning of our period, in Massingham's assertion that all 'genuine husbandry is "ecological"; it is conditioned by the nature of the land'; or in his introduction to Jenks' *From the Ground Up,* where he points out that the idea of ever-expanding industrial production is inseparable from a belief in unlimited natural resources. Jenks himself later argued that Western civilization had always been obsessed with the idea of expansion, the latest manifestation of this drive being space exploration. John Black, Professor of Natural Resources at Edinburgh University, saw ecological breakdown as the consequence of removing ethical limits of human dominion over nature, while Tristram Beresford applied the idea specifically to the techniques of industrial agriculture, with their arrogant drive to manipulate nature without social restraint. Above all, there was the ineluctable fact that the planet was finite, and that industrial civilization's profligate waste of its resources threatened to destroy Earth's life-support systems. Theodore Roszak, whose work influenced members of the Seventies Generation, wrote in his introduction to the 1995 edition of *The Making of a Counter Culture* (first published in the USA in 1969) that the oil crisis of the 1970s had made people aware of ecological restraints: '*The sky is not the limit; the earth is*' (emphasis in original). No longer could the Age of Affluence, secure in its ecological ignorance, go on promising a 'delusionary consumer's paradise'; the biospheric facts of life were reporting in. We can also add to the concern with limits the organicist conviction that human knowledge of the soil and of other natural systems is very incomplete and that we do not know how technological developments will affect them. Sir George Stapledon had posited a Law of Operative Ignorance, by which he meant that new discoveries automatically create new, unknown consequences and potential risks.[27]

Wholeness, holism and Smuts

That the idea of Wholeness is a perennial theme of organicist thought should be evident to those who have read this far. It is not quite the same as Holism, which has a more specialist meaning in the realm of the philosophy of science, but both concepts emphasize the necessity of regarding phenomena in relationship to each other rather than trying to understand them by analysing them into their component parts. Eve Balfour identified the philosophy of Wholeness with an ecological approach, but her view of Wholeness involved more than just the biological world: it included 'body, mind and spirit'. Laurence Easterbrook expressed the idea of Wholeness in more purely biological terms when he wrote: 'All life is one and ultimately we must share our environment with the rest of the animal and vegetable kingdoms'. Such a statement nevertheless carries distinctly mystical overtones, and we know Easterbrook to have believed in a spiritual dimension to life. Nevertheless, the holistic approach to agricultural science and to nutrition remains an option for those unhappy about involving the spiritual or religious in the case for organics. R.W. Widdowson, the Soil Association's Agricultural Advisor in the 1980s, put the case for a holistic agriculture, and Lawrence Woodward has suggested that the ideas of J.C. Smuts provide a valuable basis for a philosophy of organic agriculture, since they emphasize the need to recognize that organisms and living entities 'are not "entirely resolvable into parts"'; they do not 'exist apart from their surroundings (which are themselves complexes of wholes) but ... evolve and vary partly in response to the stimulus which comes from them'. More pragmatically than the metaphysician Smuts, Sir Colin Spedding — although not a supporter of the organic movement — has tried to encourage the study of agricultural systems, among which organic cultivation takes an important place.[28]

A belief in limits and a holistic approach might, then, be regarded as the irreducible core of organicist philosophy, acceptable purely on grounds of human self-interest, scientific observation and productive efficacy, without the need to introduce any reference to spiritual or religious ideas. Whether or not this is so, is a matter for organicist theoreticians to debate: our concern here is with how these concepts — limits and holism — were in fact regarded during the period of our study. The evidence suggests that belief in them was commonly associated with some form of religious or spiritual perspective. Respect for limits was thus an acceptance of creatureliness, or an avoidance of hubris, or a sign of reverence for nature's complexity, or the result of belief in a God-given natural order, or an expression of the virtue of humility: this last, given the central importance of humus in organic farming, being particularly appropriate. As for holism, Smuts offered two distinct senses of the term, one purely empirical and the other metaphysical. The latter expressed 'the view that the ultimate reality of the universe is neither matter nor spirit but wholes as defined in this book *[Holism and Evolution]*'. Holism was a theory of reality, a 'faith', as Smuts termed it, that — in language indebted to St Paul (*Romans* 8, v.22) — 'the groaning and travailing of the universe' would produce the 'realisation of the Good'. Smuts' book ends with an early expression of the organic movement's insistence that '[w]holeness, healing, holiness' spring 'from the same root in language as in experience' and are 'secure of attainment': a faith remarkably similar to that which Eve Balfour held.[29]

Aesthetic unity and mysticism

Another important element of organicist philosophy can be introduced here: the aesthetic. D.P. Hopkins perceptively pointed out that the organic movement 'tends to appeal to those whose leanings are somewhat artistic', and a holistic view of the world might be considered essentially aesthetic rather than spiritual. Terms like 'harmony', 'balance' and 'variety' frequently occur in organicist writings, and this is also the discourse of aesthetics, which speaks of variety of form and colour being harmoniously ordered into a satisfying, balanced whole. Much of the organicist critique of industrial agriculture is based on aesthetic valuation: the ugliness of agribusiness methods and of the landscapes and products they give rise to, contrasted with the beauty that mixed farming or permaculture can create, or the beauty of animals which are allowed to live freely and healthily, or the attractiveness of the foods produced by artisan specialists. The critique goes beyond the agricultural, to the wider social realm: Britain, the organicists believed, had become *unbalanced,* ignoring the importance of rural life. But they did not want a picturesque countryside of twee villages fit only for celebrities to retire to. Rather, rural life should contain a balance of farming, crafts and industries, and cities should be 'greened' through city farms, allotments and markets. Sir George Trevelyan's article, mentioned above, provides a classic instance of how the organic philosophy can result in an aesthetic vision of human life: in this case humanity is seen as 'a Unitary Being' with 'each of us ... a cell in the great Whole'. So the immense variety of humanity is harmonized into a cosmic art-work, just as the innumerable plants and creatures in the natural world contribute to the balance and harmony of an ecological system.[30]

If, as the philosopher John Macmurray maintained, mysticism should be regarded not as an aspect of religion but as an attempt to conceive the entire world as an object of aesthetic perception, then the jibe that the organic movement is about 'muck and mysticism' has some justice to it. What is Steiner's cosmology if not, at one level, an elaborate pattern in which all things find their place? Beyond the domain of esoteric philosophy, the organicist convergence of biology, holism and aesthetics can be traced back at least as far as Sir Patrick Geddes, and it is no doubt significant that Philip Mairet became a disciple of Geddes and designed charts and diagrams for him when he gave a course of lectures in London. This convergence can also be found in the writings of contemporary thinkers such as James Lovelock, whose Gaia Theory, while being a scientific hypothesis, also exhibits the mystical, aesthetic characteristics of a wholeness which is constituted by the harmonious balance of the Earth-organism's many and various parts. (This organic-aesthetic vision contains more than a hint of totalitarianism, reminiscent of inter-war ideas of the Organic State, with individuals as cells in the greater social organism. We should remember that Plato banned artists from his holistic Republic, since their visions might have conflicted with his own.)[31]

Edward Goldsmith and the natural order

The organic movement has produced few attempts at a serious philosophical study of its ideas, but one of them appeared almost at the end of our period: Edward Goldsmith's *The Way: An Ecological World-View* (1992; revised edition 1996). Goldsmith's former colleague Michael Allaby was scathing about this attempt to 'invent a new religion entirely from scratch ... purely for the purpose of social control', which he considered a thoroughly irreligious and eco-fascist project. It is in fact somewhat misleading to say that Goldsmith is working entirely from scratch: his book's title carries an unmistakable reference to the Tao of Lao Tzu, whom Goldsmith quotes, and to similar concepts in the religion of ancient Egypt (*Maat*), of Vedic

India (*R'ta*) and of Hinduism (*Dharma*). Allaby is right, though, that Goldsmith sees religion's true function as a means of ensuring social stability, with the gods of 'vernacular man' (a term Goldsmith uses to indicate self-organizing and self-governing cultures) reflecting the hierarchical structure of his society. But vernacular religion also included (Goldsmith quotes Robert T. Parsons) 'the relationships of people with the earth as a whole, with their own land and with the unseen world of constructive forces and beings in which they believe. *Religion brings them all into a consistent whole*' (emphasis in Goldsmith). A sense of one-ness is, Goldsmith believes, a 'fundamental principle of the vernacular world' and is closely related to aesthetic sensibility, which is 'an important means of apprehending and of understanding our relationship to the world around us', above all to the natural world.[32]

That which is beautiful is likely to be ecologically desirable, and Goldsmith follows Massingham and Massingham's hero Cobbett in finding the traditional small farm both beautiful and socially beneficial. And, like Massingham and his colleagues in the Kinship in Husbandry and the Council for the Church and Countryside (CCC), Goldsmith appeals to the idea of a natural order, whose processes result in the beauty of natural forms. 'The Way' consists in living in harmony with this natural order, and humanity's problems begin with departure from it, which 'threatens the order of the cosmos and must thereby give rise to the worst possible discontinuities'. The natural order is benign when followed but harsh when flouted; Goldsmith reminds us of the revenges which Mother Earth inflicts on Erysichthon and Orion in classical mythology. Technological civilization, which has deliberately set itself against the natural order, suffers instability, violence, ugliness and anomie as a result; while its members are the victims of physical and mental ill-health. For those who, like the Nobel Laureate James Watson, follow the anti-Way, any maladaptation of humanity to the new world of technology must be corrected: not by returning to a life in harmony with the natural order, but by altering humanity — presumably by genetic engineering — so that it can endure in a 'polluted and ecologically degraded world'.[33]

Goldsmith is dismissive of Christianity and other revealed religions: in contrast to Mairet, Schumacher and Seymour he considers them to have de-sanctified the natural world, leaving it open to exploitation and destruction. He takes a surprisingly Marxist view of religion's role in contemporary society, seeing its other-worldly emphasis as providing psychological compensation for the emptiness of an atomized culture. Nevertheless, at the end of our fifty years, we can see in Goldsmith's ideas certain themes to be found in the early organicist thought of the immediate post-war era, as expressed by members of the CCC at their 'Encounter' with agri-industry and in their publications. Belief in a natural order whose limits must be respected; the faith that such restraint will result in a harmonious and productive life; the rejection of all attempts to stand outside nature and manipulate it; and a sense of the need to reverse the direction which western society was taking, in order to return to The Way, or the Will of God. Goldsmith's call to strengthen the family and the community, and to create a localized and diversified economy is entirely in the tradition of the Kinship in Husbandry. For those who enjoy discovering continuities, it is particularly satisfying to see, as prefaces to Goldsmith's final chapter, quotations from Lewis Mumford and Richard St. Barbe Baker: Mumford, the friend of Philip Mairet and supporter of Kenneth Barlow's Coventry Family Health Club initiative; and Baker, who devoted sixty years of his life, from 1922 onwards, to the organic movement.[34]

The persistence of husbandry

In the end, it is probably best to sum up the organic movement's philosophy of earth and spirit in the simple word 'husbandry', whose meaning implies ideas of craftsmanship, reverence, responsibility, beauty, stability and fruitfulness. Agri-business has at times tried to identify the term with efficiency, as the 1947 Agriculture Act did, or with technical advances; but it denotes much more than this and remains central to organic thinking in the twenty-first century. Perhaps one should say that it is the essence of organicist thought, for it is a concept to which members of the organic movement have repeatedly appealed. Our period opens with Massingham's symposium *The Natural Order,* subtitled 'Essays in the Return to Husbandry'. In what would today appear to be an oxymoron, Massingham defines husbandry as 'loving management', or as treating nature in a family spirit and honouring the natural law. Such treatment necessitated personal and local responsibility, and its result was beauty, the ethical and aesthetic aspects of husbandry uniting to create a distinct culture. Massingham was at pains to emphasize that husbandry was intensely practical — its alternative being self-destructive — and also eternal: 'a means of recovering a certain order and mode of being which is timeless and universal'. In a later work, Massingham wrote that husbandry was a form of ecological awareness and suggested that its ethical basis probably explained 'why a peasant society is rarely other than a religious one': a remark that anticipates Goldsmith's emphasis on the devotion of 'vernacular man' to his gods. Jorian Jenks valued husbandry as a source of social stability, describing it as 'essentially conservative'. Indeed, he considered husbandry fundamental to civilization.[35]

Belief in the importance of husbandry by no means disappeared with the passing of the pioneers. For Joanne Bower and Anthony Etté good husbandry was a combination of 'Reverence and responsibility', of gratitude and skill, of balance between taking and giving, of awareness of the past and moral obligation to the future. Such qualities could be found in many pagan belief systems and survived to some extent in certain practices of the Christian Church. But the husbandman had given place to the technologist, and Sir Richard Body lamented that the term itself had disappeared from the vocabulary of agriculture. For Body, husbandry denoted a quality of care incompatible, for instance, with keeping large herds of cattle. It meant recognizing and working with natural cycles, and therefore resulted in sustainability and security. George McRobie saw non-violence as a prime feature of husbandry: for him, as for David Stickland, use of chemicals was the antithesis of good husbandry. And the Steinerian Charles Davy considered that the practice of husbandry involved aesthetic sensibility, calling for 'a certain intuitive, almost artistic feeling for what is appropriate and harmonious in the relations of human society to nature and the earth'.[36]

One other point needs making: husbandry also implies thrift, as in the idea of a husbanding of resources. Since 1945, agri-industry has had things almost entirely its own way in terms of policy, but this does not mean that it has finally defeated husbandry. Given the squandering of resources over which industrial farming has presided, it seems increasingly probable that 'the return to husbandry' which Massingham sought in vain at the end of the war will provide the only possible chance of a secure agricultural future. If this proves to be so, then the spiritual philosophy on account of which the organic movement has so often been mocked by its critics may turn out, in the last resort, to be practical wisdom.

Appendix A: Leading figures in the organic movement, 1945–1995

Names of people in CAPITALS indicate an entry in Appendix A. Names of groups, organizations, institutions or journals in CAPITALS indicate an entry in Appendix B.

An asterisk after a name indicates that a fuller account of this person's earlier involvement in the organic movement can be found in Appendix A of *The Origins of the Organic Movement*.

ALLABY, MICHAEL (b.1933) Editorial assistant at the SOIL ASSOCIATION from 1964 to 1972, producing *SPAN*. Journalist, author and left-wing environmental activist who joined the editorial board of *THE ECOLOGIST* and worked closely with EDWARD GOLDSMITH. Contributor to various pro-organic publications, including *RESURGENCE* and *VOLE*. Later became sceptical about the organic movement.

ASTOR, DAVID (1912–2001) Editor of *The Observer* who promoted environmentalism in its pages. Published many articles by his close friend E.F. SCHUMACHER and employed JOHN DAVY as Science Correspondent. His daughter Alice married LAWRENCE WOODWARD. Astor was a founding father and trustee of the ELM FARM RESEARCH CENTRE.

BADENOCH, A.G. (1896–1964) Doctor and medical missionary; authority on public health. Leading spirit in the SCOTTISH SOIL AND HEALTH GROUP, along with Dr Angus Campbell and CDR ROBERT L. STUART. Like many of the early organicists, an adherent of the economic reform theories of Social Credit.

BAKER, RICHARD ST. BARBE* (1889–1982) Founder of the MEN OF THE TREES movement in 1922 and founder member of the SOIL ASSOCIATION. Leading spirit in launch of the *New Earth Charter* in 1949. Devoted his life to the protection of the world's forests, and sought to resist further spread of the Sahara Desert. Helped to establish the FINDHORN COMMUNITY. Author of many books on trees and the environment. Influence on EDWARD GOLDSMITH among other leading organicists; still active in the late 1970s.

BALFOUR, EVE* (1898–90) With GEORGE SCOTT WILLIAMSON and FRIEND SYKES, one of the founders of the SOIL ASSOCIATION. She also set in motion the HAUGHLEY EXPERIMENT at her Suffolk farm. Travelled widely, forging links with pro-organic farmers and nutritionists overseas, particularly in the USA. Was sidelined in the Soil Association by the SEVENTIES GENERATION of younger organic activists in the early 1980s but remained actively involved in the organic movement until the late 1980s. An influence on SATISH KUMAR, who visited her in 1986.

BARLOW, DR KENNETH E.* (1906–2000) Friend and supporter of GEORGE SCOTT WILLIAMSON and INNES H. PEARSE, Barlow unsuccessfully attempted in the post-war years to establish the COVENTRY FAMILY HEALTH CLUB HOUSING SOCIETY as an extension of the community health principles embodied by the PIONEER HEALTH CENTRE. SOIL ASSOCIATION Council member for many years. Later a radiographer in the NHS, Barlow became Chairman of the McCARRISON SOCIETY and editor of the journal *NUTRITION AND HEALTH*. Contributed to *THE ECOLOGIST* and was still writing on health in the late 1980s.

BEDFORD, HASTINGS, DUKE OF* (1883–1953) His pre-war national-socialist BRITISH PEOPLE'S PARTY re-emerged in 1945, its membership including the pro-Nazi eugenicist ANTHONY LUDOVICI. Its newspaper *PEOPLES POST* took a pro-organic stance and featured DION BYNGHAM as a regular columnist.

BODY, SIR RICHARD (b.1927) Dairy farmer, Conservative MP and opponent of the Common Agricultural Policy. Author of influential books in the 1980s and '90s, critical of agri-business. Active in the SOIL ASSOCIATION and the FARM AND FOOD SOCIETY. Associated with DAVID and KATIE THEAR's publishing firm.

BOWER, JOANNE (1912–2006) Founder of the FARM AND FOOD SOCIETY and active opponent of factory farming, influenced by RUTH HARRISON's book *Animal Machines*. A good friend of RICHARD and ROSAMUND YOUNG.

BRADFORD, LORD (1911–1981) Formerly Viscount Newport. President of the SOIL ASSOCIATION from 1951 to 1970. Owner of extensive estates in Shropshire and Staffordshire; keen forester and organic farmer. Active in the Country Landowners' Association.

BRADY, MARGARET (1900–85) Nutritionist, journalist and author, advocate of vegetarianism and whole food, noted for displays of breadmaking. SOIL ASSOCIATION Council member for more than 25 years and active in the McCARRISON SOCIETY.

BREEN, DR G.E. (d. 1981) Epidemiologist, medical correspondent for *The Times,* physician to the Chelsea Arts Club, painter and sculptor. Breen edited the *Medical Press and Circular* and was a founder-member of the SOIL ASSOCIATION, serving on its Council and Editorial Board for a quarter of a century.

BROCKMAN, ALAN (b.1927) Biodynamic farmer in Kent, active in both the BIODYNAMIC AGRICULTURAL ASSOCIATION and the SOIL ASSOCIATION. Frequent contributor to *STAR AND FURROW,* and chairman of the Soil Association's Organic Standards Committee in the 1970s.

BROOKE, HILDA CHERRY (see **HILLS, CHERRY**)

BRUCE, MAYE E.* (1879–1964) A practitioner of RUDOLF STEINER's methods of cultivation, exponent of 'quick-return' composting and founder member of SOIL ASSOCIATION.

BURMAN, DR NORMAN P. (b.1915) Microbiologist and bacteriologist who held senior positions with the Metropolitan Water Board. Prominent member of SOIL ASSOCIATION's Middlesex Group; frequent contributor on scientific subjects to *MOTHER EARTH*. Chairman of Association's Executive Committee in 1960s, and member of Standards Committee. Took interest in municipal composting, and was a friend of organic market gardener J.L.H. CHASE.

BYNGHAM, H.D. 'DION'* (c.1893–1990) Wrote for Siegfried Marian's *Soil Magazine* in the post-war period and contributed pro-organic articles on agricultural topics to *PEOPLES POST.* Worked on *HEALTH AND LIFE* magazine, a close associate of EDGAR SAXON. Many years later, he contributed to *RESURGENCE.*

CANNON, GEOFFREY Journalist/author who wrote powerfully against the food industry. Active in the McCARRISON SOCIETY. Devisor of SOIL ASSOCIATION journal *LIVING EARTH,* and its editor in the late 1980s. Association Council member in early 1990s.

CHASE, J.L.H. (d.c.1992) Commercial market-gardener at Chertsey, Surrey and pioneer of cloche techniques. His gardens were visited by prominent members of the organic movement, including EVE BALFOUR, LAURENCE EASTERBROOK and HUGH SINCLAIR. Friend of JULIETTE DE BAIRACLI-LEVY, helping her research into herbalism through experiments carried out at his nurseries, and of DR NORMAN BURMAN. Knowledgeable about South American agriculture, he spoke at an IFOAM conference in 1984, and left a legacy to the SOIL ASSOCIATION, on whose Panel of Experts and Editorial Board he had served.

CLEAVE, SURGEON-CAPTAIN T.L. (d.1983) Director of medical research for the Royal Navy, a Fellow of the Royal College of Physicians and an authority on degenerative diseases. Joined the SOIL ASSOCIATION in 1969 and was active in the McCARRISON SOCIETY. His work on "saccharine disease" was much admired by DR WALTER YELLOWLEES.

CLEMENT, DAVID (1911–2007) Leading figure in the BIODYNAMIC movement, farming at Clent in Worcestershire for more than 40 years. A friend of EHRENFRIED PFEIFFER. President of the BIODYNAMIC AGRICULTURAL ASSOCIATION and a member of the steering committee of BRITISH ORGANIC FARMERS.

COATES, HUGH Son of SOIL ASSOCIATION member Lance Coates, Hugh Coates joined the Association in 1967. Land agent, farmer and miller, he was, with DR ANTHONY DEAVIN, one of the chief architects of organic standards. He chaired the Soil Association's Organic Marketing Committee and was on the committee of BRITISH ORGANIC FARMERS.

COPPARD, SUE Leading figure in the establishment of WORKING WEEKENDS ON ORGANIC FARMS, and associated with CRAIG and GREGORY SAMS through working on the production of *SEED* magazine.

COWARD, RALPH* (1902–90) Dorset organic farmer and founder member of the SOIL ASSOCIATION. In the 1930s he had belonged to Viscount Lymington's far-Right organization the English Array. Involved in the COUNCIL FOR THE CHURCH AND COUNTRYSIDE; member of the Soil Association Council and a leading spirit in its very active Wessex Group. Provided opportunities for members of WORKING

WEEKENDS ON ORGANIC FARMS (WWOOF). A television documentary on him, produced by KENNETH BARLOW's daughter Joanna, was broadcast in the 1980s.

DAVY, JOHN (1927–1984) Son of the journalist Charles Davy (who used the pseudonym Charles Waterman), John Davy was for many years Science Correspondent of *The Observer* and a protégé of DAVID ASTOR. A devotee of ANTHROPOSOPHY, he was also active in the SOIL ASSOCIATION, serving on its editorial board in the late 1960s/early '70s, and a good friend of MICHAEL ALLABY, contributing to the *SPAN* newsletter.

DE LA MARE, RICHARD* (1900–86) Edited the agriculture and horticulture lists at FABER AND FABER from the early 1930s to the early 1970s, being responsible for publication of many of the most important books in the history of the organic movement. SOIL ASSOCIATION founder member, and a long-term member of its Council.

DEAVIN, DR ANTHONY (b.1936) Highly capable scientist who ran the series of successful courses on Organic Husbandry at Ewell Technical College in the 1970s. MARY LANGMAN considered that the SOIL ASSOCIATION owed him a great debt for the work he put in, and ALAN and JACKIE GEAR were strongly influenced by him. Scientific advisor to the HENRY DOUBLEDAY RESEARCH ASSOCIATION from 1973 to 1979. With HUGH COATES was the architect of organic standards, and was a member of the Organic Marketing Committee. A member of the BIODYNAMIC AGRICULTURAL ASSOCIATION, owing to his scientific interest in biodynamic experiments.

DOUGLAS OF BARLOCH, LORD (1889–1980) Formerly Labour MP for Battersea, F.C.R. Douglas, and a frequent contributor to agricultural journals. A solicitor, who gave valuable legal advice during the establishment of the SOIL ASSOCIATION. A member of the Hopes Compost Club run by CDR R.L. STUART. Chaired the Joint Working Party on municipal composting.

EASTERBROOK, LAURENCE* (1893–1965) A top agricultural journalist; farmer; founder member of the SOIL ASSOCIATION. He served on various Association boards (Council, Panel of Experts or Editorial Board) from the Association's inception until his death. Easterbrook was sympathetic to the BIODYNAMIC methods of RUDOLF STEINER.

GARDINER, ROLF* (1902–71) SOIL ASSOCIATION Council

member almost uninterruptedly for a quarter of a century. Leading spirit in the COUNCIL FOR THE CHURCH AND COUNTRYSIDE and the KINSHIP IN HUSBANDRY. Organized a European Husbandry Meeting in 1950, which gathered at his Dorset estate and at London's Caxton Hall. Estate-owner in Africa and involved in European environmentalist initiatives. Worked for the Council for the Protection of Rural England.

GEAR, ALAN and JACKIE (both b.1949) Highly influential husband-and-wife team responsible for the greatest success story of organic horticulture, the HENRY DOUBLEDAY RESEARCH ASSOCIATION (now Garden Organic) headquarters at Ryton near Coventry. LAWRENCE and CHERRY HILLS were their mentors, and they were also influenced by DR ANTHONY DEAVIN and JOHN SEYMOUR. Key figures in the SEVENTIES GENERATION, they joined the HDRA in the winter of 1973–74, and supervised the move to Ryton in the mid-1980s. Alan became Chief Executive at Ryton in 1989, and Jackie Executive Director the following year. Alan was also involved in the establishing of organic standards and the work of UKROFS.

GOLDSMITH, EDWARD (1928–2009) Brother of multi-millionaire businessman, environmentalist and politician Sir James Goldsmith. Founder of *THE ECOLOGIST* and leading spirit in establishing the Ecology Party. Had close links with the SOIL ASSOCIATION, particularly around 1970 before he moved to Cornwall. RICHARD ST. BARBE BAKER was a strong influence on his environmental outlook.

GRANT, DORIS (c.1904–2003) Nutritionist, author and breadmaker: inventor of recipe for "the Grant loaf". Council member of the HENRY DOUBLEDAY RESEARCH ASSOCIATION; supporter of FRANK NEWMAN TURNER's journal *THE FARMER* and contributor to *HEALTH AND LIFE*. A friend of SIR ALBERT HOWARD and DR LIONEL PICTON, she influenced DR WALTER YELLOWLEES and many in a younger generation of organic activists. On the Committee of Management of the WHOLE FOOD SOCIETY.

GRAY, DR KEN R. (b.1931) Scientist at Birmingham University who, with Dr (later Professor) A.J. (Joe) Biddlestone, undertook long-term research into municipal composting and wrote articles on the issue for the SOIL ASSOCIATION. Spoke at various venues including the Ewell Technical College courses on organic husbandry. Worked with DAVID CLEMENT, with DAVID STICKLAND and with C. DONALD WILSON.

HARRISON, RUTH (1920–2000) Ruth Harrison's book *Animal Machines* (1964) might be regarded as the *Silent Spring* of factory farming; indeed, it contained a preface by Rachel Carson. Harrison was a member of the Brambell Committee on animal welfare and of the Farm Animal Welfare Council. In the 1970s she was active in the Council of Europe with work which led to the convention on the Protection of Animals kept for Farming Purposes. Member of the SOIL ASSOCIATION Council from 1966 to 1972.

HART, ROBERT A. DE J. (1913–2000) Exponent of forest farming and gardening, influenced by the pioneers of the early organic movement including SIR ALBERT HOWARD, H.J. MASSINGHAM and SIR GEORGE STAPLEDON. Farmed in Shropshire; prolific writer active in various branches of the organic movement. Interested in the KEYLINE SYSTEM of P. A. YEOMANS. An influence on the PERMACULTURE movement.

HARVEY, GRAHAM Journalist, hard-hitting author of books on agricultural topics, and agricultural story editor of Radio 4 soap opera *The Archers*. A member of the SEVENTIES GENERATION of organic activists, Harvey was editorial co-ordinator of the *SOIL ASSOCIATION QUARTERLY REVIEW* and contributed to *NEW FARMER AND GROWER*.

HERON, GILES (b.1928) and **MARY** (b.1922) Giles Heron, son of Tom Heron of the *NEW ENGLISH WEEKLY*'s editorial board, and history teacher at Abbotsholme School, married Mary Barran in 1963. Mary's father was a member of ROLF GARDINER's Springhead Ring and a friend of RICHARD ST. BARBE BAKER. She worked as an assistant to EVE BALFOUR in the early 1960s. The Herons set up a farm community in North Yorkshire in the 1970s and had close links with the STEINER community at Botton village.

HICKS, SIR CEDRIC STANTON (1892-1976) A forgotten giant of the organic movement: doctor, chemist, nutritionist, soldier, Food Consultant to the Australian Army from 1952 to 1973. Born in New Zealand, he was for more than 30 years Professor of Human Physiology at the University of Adelaide. While a Research Fellow at Cambridge in the 1920s he met GEORGE SCOTT WILLIAMSON and INNES PEARSE and was influenced by the work of SIR ROBERT McCARRISON. Founder-member of the SOIL ASSOCIATION, he was patron of its South Australia Group. Gave the ALBERT HOWARD Memorial Lecture in 1958.

Took a keen interest in the KEYLINE SYSTEM, in the HAUGHLEY EXPERIMENT and in experiments in municipal composting. Wrote on ecology and environmentalism.

HILLS, CHERRY (1896–1989) Hilda Cherry Brooke was an active member of the SOIL ASSOCIATION in South Africa who grew comfrey and corresponded with LAWRENCE HILLS. On a 1961 visit to England she met him and they married three years later. She was a self-taught nutritionist who cured her husband of coeliac disease, and his indispensable partner in running the HENRY DOUBLEDAY RESEARCH ASSOCIATION. A strong influence on ALAN and JACKIE GEAR, and also on CHARLOTTE MITCHELL.

HILLS, LAWRENCE D.* (1911–90) Horticulturalist who was a friend of RICHARD DE LA MARE and wrote books for FABER AND FABER, for which company he was also a reader. Founded the HENRY DOUBLEDAY RESEARCH ASSOCIATION in 1954. Gardening Correspondent of *The Observer* (1958–66) and of *Punch* (1966–70). Helped establish *THE ECOLOGIST,* on whose editorial board he served for many years. A prolific journalist, he contributed to the *SOIL ASSOCIATION QUARTERLY REVIEW* and *VOLE,* among many other publications.

HODGES, DR R. DAVID (b.1934) Animal physiologist who taught at Wye College of Agriculture; prominent in the organic movement from the 1970s to the 1990s. David Hodges was a SOIL ASSOCIATION Council member in the 1980s and '90s, and on the editorial board of its journal. He worked with DAVID STICKLAND in promoting the INTERNATIONAL INSTITUTE OF BIOLOGICAL HUSBANDRY and, with his colleague Dr A.M. Scofield, ran the journal *BIOLOGICAL AGRICULTURE AND HORTICULTURE.*

HOLDEN, PATRICK (b.1950) Holden took a course in BIODYNAMIC CULTIVATION at Emerson College in the early 1970s then moved to West Wales as a member of an agricultural commune. The commune was short-lived, but Holden remained on the farm and became, with NIC LAMPKIN, PETER SEGGER and other members of the SEVENTIES GENERATION, a leading figure in the West Wales Group of the SOIL ASSOCIATION. Also central to BRITISH ORGANIC FARMERS, of which he was Co-ordinator. Director of the Soil Association from 1995 until 2010. Much influenced by MARY LANGMAN and MRS DINAH WILLIAMS.

HOLLINS, ARTHUR (1915–2005) Shropshire dairy farmer; inventor; restaurateur; pioneer of the organic yogurt industry and of organic marketing. His autobiography, *The Farmer, the Plough and the Devil*, was published in 1984 with the support and encouragement of ROBERT WALLER.

HOWARD, SIR ALBERT* (1873–1947) Arguably the most important figure in the development of the British organic movement. An outstanding agricultural botanist who pioneered scientific composting techniques and whose work provided scientific evidence for the organic case. Influenced, among others, EVE BALFOUR, LAWRENCE D. HILLS, JULIETTE DE BAIRACLI-LEVY, F.C. KING, the EARL OF PORTSMOUTH, W.E. SHEWELL-COOPER, CDR R.L. STUART, FRIEND SYKES, FRANK NEWMAN TURNER and the work of the PIONEER HEALTH CENTRE. He took over his friend DR LIONEL PICTON's *News-Letter on Compost* and re-cast it as *SOIL AND HEALTH*. Following his death, his widow LOUISE E. HOWARD established the ALBERT HOWARD FOUNDATION OF ORGANIC HUSBANDRY.

HOWARD, LOUISE E.* (1880–1969) Second wife of SIR ALBERT HOWARD. After his death, she devoted herself to preserving his legacy and spreading his ideas, helping establish the ALBERT HOWARD FOUNDATION OF ORGANIC HUSBANDRY, producing the *Albert Howard News Sheet* until 1964, and writing an account of his achievements in India. A close friend of EVE BALFOUR. She supported the SCOTTISH SOIL AND HEALTH GROUP, who took over Sir Albert Howard's journal *SOIL AND HEALTH* and re-constituted it as *HEALTH AND THE SOIL*.

JENKS, JORIAN* (1899–1963) First editor of the SOIL ASSOCIATION journal *MOTHER EARTH*, from its inception in 1946 until his death in 1963. A Mosleyite who was still active in far-Right politics during the late 1940s. Jenks was also editor of the journal *RURAL ECONOMY* during the 1940s and '50s and helped draft *Feeding the Fifty Million* for the RURAL RECONSTRUCTION ASSOCIATION. He was also secretary to the COUNCIL FOR THE CHURCH AND COUNTRYSIDE.

KUMAR, SATISH (b.1936) Land reformer, peace campaigner and former Jain monk, Kumar settled in England in 1973 and took over the editorship of *RESURGENCE*. In 1986 he met EVE BALFOUR while on a pilgrimage round Britain, and became interested in the history of the organic movement. In 1987 he established the environmental publishers Green Books, and in 1991 founded the SCHUMACHER College in Devon.

LAMPKIN, DR NIC (b.1960) Influential member of the SEVENTIES GENERATION who established academic courses in organic agriculture at Aberystwyth University, where he had been a research student of DR V.I. STEWART, and who wrote a major textbook on the subject. Active in the West Wales Group of the SOIL ASSOCIATION and later a Council member; Development Director of the Centre for Organic Husbandry and Agroecology; involved in establishing of organic standards through UKROFS.

LANGMAN, MARY (1908–2004) Central figure in the organic movement's history, and perhaps the outstanding example of continuity between its early days and later development. Worked at the PIONEER HEALTH CENTRE before and after its wartime closure; founder member of the SOIL ASSOCIATION; farmed at Bromley, Kent in the post-war years; ran the Association's WHOLEFOOD shop in London with LILIAN SCHOFIELD; active in the establishment of IFOAM. A good friend of EVE BALFOUR, she nevertheless sympathized with the aims of the SEVENTIES GENERATION, and was a mentor to PATRICK HOLDEN and LAWRENCE WOODWARD.

LEVY, JULIETTE DE BAIRACLI (1912–2009) Trained in veterinary science, Levy began investigating alternative approaches to animal health and became an authority on the herbal lore of Gypsies, Bedouin Arabs and American Indians. SIR ALBERT HOWARD encouraged her studies, and she became a skilled practitioner of herbal treatment of animal illnesses. She worked closely with FRANK NEWMAN TURNER and J.L.H. CHASE, and through her study of New Forest Gypsies came to know DR AUBREY WESTLAKE and his daughter Jean.

MAIRET, PHILIP* (1886–1975) Writer, designer, translator and environmental philosopher who edited the pro-organic *NEW ENGLISH WEEKLY* from 1935 to 1949. Member of the KINSHIP IN HUSBANDRY and the COUNCIL FOR THE CHURCH AND COUNTRYSIDE. He was a friend of the polymath Lewis Mumford, whose views on technology influenced the 'counter-culture' of the 1960s.

MANSFIELD, DR PETER (b.1943) As a young GP, he was profoundly influenced by DR INNES PEARSE and the ideas of the PIONEER HEALTH CENTRE, setting up his own experiment in social medicine at his practice in Lincolnshire. He also admired the work of SIR ROBERT McCARRISON and of DR LIONEL PICTON. Author of various books on nutrition and health, and a frequent contributor to SOIL

ASSOCIATION journals in the 1980s.

MASSINGHAM, H.J.* (1888–1952) Prolific journalist and writer on agriculture, rural history and topography. Member of the KINSHIP IN HUSBANDRY, the COUNCIL FOR THE CHURCH AND COUNTRYSIDE and the Council of the SOIL ASSOCIATION. His work enjoyed a revival of interest in the late 1980s thanks to the 'Organic Classics' series published by SATISH KUMAR's Green Books.

MAYALL, GINNY (b.1959) Daughter of Richard Mayall and granddaughter of SAM MAYALL, Ginny Mayall was a central figure in the SEVENTIES GENERATION and the first organic farmer to win the *Daily Express* Young Countrywoman of the Year award (in 1990). She was particularly effective at promoting the Soil Association at the Royal Show in the 1980s. Chaired the editorial group in the early days of the *SOIL ASSOCIATION QUARTERLY REVIEW,* served on the SOIL ASSOCIATION Council, and chaired the Association's Symbol Certification Scheme. She and her father Richard were prominent in BRITISH ORGANIC FARMERS.

MAYALL, SAM (1900–1980) Organic farmer in Shropshire from the late 1940s, father of Richard Mayall and grandfather of GINNY MAYALL. Became an elder statesman of the SOIL ASSOCIATION, and was for many years a Vice-President of it. Close friend of EVE BALFOUR and also influenced members of the SEVENTIES GENERATION, including Julian Rose and RICHARD YOUNG. Spoke at Ewell courses on organic husbandry and during the 1970s featured in television programmes on organic farming.

McCARRISON, SIR ROBERT* (1878–1960) Major-General Sir Robert McCarrison was the major influence on the organic movement's philosophy of nutrition, and on the work of GEORGE SCOTT WILLIAMSON and INNES H. PEARSE at the PIONEER HEALTH CENTRE. In 1966 the McCARRISON SOCIETY was founded in order to promote his ideas on the relationship between diet and health.

McROBIE, GEORGE (b.1925) Worked in mining and as a journalist on *Coal* magazine; spent six years on the research staff of the thinktank Political and Economic Planning. In 1956 E.F. SCHUMACHER appointed him as his assistant at the National Coal Board. He and Schumacher established the Intermediate Technology Development Group in 1965. After Schumacher's death McRobie wrote *Small is Possible* (1981), continuing to promote his mentor's ideas. From 1987 to 1997, McRobie was President of the SOIL ASSOCIATION.

APPENDIX A: LEADING FIGURES IN THE ORGANIC MOVEMENT

MELLANBY, DR KENNETH (1908–1993) Distinguished scientist and naturalist who wrote *Pesticides and Pollution* for the Collins New Naturalist series and was the Director of Monks Wood Experimental Station from 1961 to 1974. He had also been Head of the Department of Entomology at Rothamsted. During the 1960s and '70s he was an active supporter of the SOIL ASSOCIATION, chairing its Research Advisory Committee. He saw scientific problems from an ecological perspective, but later distanced himself from the views of EDWARD GOLDSMITH and *THE ECOLOGIST*.

MILTON, DR REGINALD F. (1906–1986) In charge of the analytical work for the HAUGHLEY EXPERIMENT during the 1950s and '60s, and a member of the Advisory Panel of the SOIL ASSOCIATION throughout the 1950s. As a young biochemist he worked with DR GEORGE SCOTT WILLIAMSON and became interested in the concept of positive health. Before and after the war, an independent consultant; during the war he was in charge of the Medical Research Council's Department of Industrial Medicine. It is fair to say that Milton's work for the Soil Association was regarded with some scepticism by a number of people, including fellow scientists DR NORMAN BURMAN and DR VICTOR STEWART.

MITCHELL, CHARLOTTE (b.1953) Key figure in moving the SOIL ASSOCIATION towards a more consumer-oriented approach, with an emphasis on food marketing rather than farming. Worked for Real Foods wholefood business in Edinburgh, becoming Trading Director. Through friendship with CRAIG SAMS became involved in the Association's work and was appointed Chairman at the end of 1991. On the BBC's agricultural and rural affairs committee. Contributed to books on Green consumerism.

PEARSE, DR INNES H.* (1889–1978) Co-founder, with GEORGE SCOTT WILLIAMSON, of the PIONEER HEALTH CENTRE. In contact with SIR CEDRIC STANTON HICKS and SIR ROBERT McCARRISON from the 1920s, and a major influence on KENNETH BARLOW and EVE BALFOUR. A member of the SOIL ASSOCIATION Council from 1946 to 1972 and centrally active in the Pioneer Health Centre Ltd after the Centre itself had closed down, continuing to write on health in the 1960s and '70s. Her ideas and personality inspired DR PETER MANSFIELD.

PFEIFFER, EHRENFRIED* (1899–1964) Leading advocate of

BIODYNAMIC CULTIVATION following the death of RUDOLF STEINER. Based in the USA from 1933 onwards, he was visited there by EVE BALFOUR on one of her American tours. In 1950 he was a guest of honour at the European Husbandry Meeting, organized by the KINSHIP IN HUSBANDRY.

PICTON, DR LIONEL J.* (1874–1948) Picton was a founder member of the SOIL ASSOCIATION, a member of the COUNCIL FOR THE CHURCH AND COUNTRYSIDE and a close friend of SIR ALBERT HOWARD. He was the driving spirit in the publication of the 1939 *Medical Testament,* at whose launch both Howard and SIR ROBERT McCARRISON spoke, and which was revived by the Soil Association in 1957. His *News-Letter on Compost* became Howard's journal *SOIL AND HEALTH.* His legacy and example have been particularly admired by DR PETER MANSFIELD.

PORRITT, JONATHON (b.1950) Highly articulate advocate of environmental politics in the 1980s; prominent in the Ecology/Green Party; director of Friends of the Earth from 1984 to 1990. Argued for organic husbandry as an essential alternative to environmentally destructive industrial agriculture. Later a Patron of the SOIL ASSOCIATION.

PORTSMOUTH, EARL OF* (1898–1984) Gerard Vernon Wallop, formerly Viscount Lymington, was a founder member of the SOIL ASSOCIATION, a member of its Council from 1946 to 1950, and a member of the KINSHIP IN HUSBANDRY and the COUNCIL FOR THE CHURCH AND COUNTRYSIDE. A key figure in the coalescence of the organic movement in the 1930s and '40s, his 1965 autobiography *A Knot of Roots* throws light on the context within which the movement developed.

POWELL, L.B. Writer and agricultural journalist who linked the early organic movement to the 1970s. In the post-war period his Country Living enterprise promoted the work of writers such as H.J. MASSINGHAM; in the 1960s and 1970s he contributed articles to *SPAN* and to *THE ECOLOGIST.* A friend of ROLF GARDINER, he wrote a tribute to him in the SOIL ASSOCIATION journal in 1973.

PRICE, WESTON A. (d.c.1948) Eminent American dental scientist who studied physical anthropology, and whose masterpiece *Nutrition and Physical Degeneration,* first published in 1939, was a source of evidence for the SOIL ASSOCIATION's views on the relationship between diet and health. It was re-published in 1945 with a Supplement to which Dr

William A. Albrecht provided a foreword, and remains a classic text of the organic 'canon'. It profoundly influenced the organic farmer Michael Rust, who was active in both the Soil Association and ORGANIC FARMERS AND GROWERS.

PRINCE OF WALES (see **WALES**)

PYE, JACK (1905–1984) Wealthy property speculator whose interest in health issues led him to support the SOIL ASSOCIATION and the HAUGHLEY EXPERIMENT in the late 1960s. Also supported the HENRY DOUBLEDAY RESEARCH ASSOCIATION and was sympathetic towards BIODYNAMIC CULTIVATION. Following the Soil Association's relinquishing of the Haughley Experiment, the Pye Research Centre continued scientific work at Haughley. Later, the ELM FARM RESEARCH CENTRE benefited from Pye's support.

ROBB, PROF. R. LINDSAY (1885–1972) Agricultural scientist of international reputation, whose many posts included Head of Agriculture at Wye College in Kent and grassland advisor to Imperial Chemical Industries in Australia, New Zealand and Southern Africa. During the Second World War he was Director of Agriculture for British Forces in North Africa. During the 1950s he worked in Central America for the United Nations Food and Agriculture Organisation. After retiring in the early 1960s he worked for the SOIL ASSOCIATION as an advisor and lived at its Haughley headquarters. He was impressed by the food-growing achievements of the FINDHORN COMMUNITY. Following Robb's death, ROBERT WALLER devoted an entire issue of the Association's journal to his ideas and achievements.

SAMS, CRAIG (b.1944) Entrepreneur whose success with Green and Black's chocolate and ice cream in the 1990s facilitated his transition over 40 years from prominent figure in the London 'counter-culture' to millionaire media celebrity. With his brother GREGORY SAMS he was a key figure in developing a market for macrobiotic and whole foods in the late 1960s and 1970s. Had dealings with the WHOLEFOOD shop in London and with DAVID STICKLAND of ORGANIC FARMERS AND GROWERS, who supplied organically grown cereals. In the 1970s he was involved in the production of *SEED* magazine. Chairman of Whole Earth Foods, which he founded with his brother in 1970 as Harmony Foods. Met CHARLOTTE MITCHELL through his interest in marketing whole foods; in the early 1990s became Treasurer of the SOIL ASSOCIATION with Mitchell as Chairman.

SAMS, GREGORY (b.1948) Brother of CRAIG SAMS, working with him in the late 1960s/1970s to develop the market for macrobiotic and whole foods. Provided catering at large-scale rock festivals. Joined the SOIL ASSOCIATION in 1968 and had dealings with its WHOLEFOOD shop in London and with DAVID STICKLAND of ORGANIC FARMERS AND GROWERS, who supplied organically grown cereals. Inventor of successful brand the VegeBurger.

SAXON, EDGAR J.* (1877–1956) Saxon's journal *HEALTH AND LIFE* continued to support the organic cause in the post-war decades. Saxon was a SOIL ASSOCIATION member and his wife Winifred Savage was a leading figure in the Association's very active Middlesex Group.

SCHOFIELD, LILIAN (d. 1989) Pharmacist who worked for Boot's the chemists for 20 years but whose doubts about the drugs she dispensed brought about a personal crisis of illness. Influenced by reading SIR ALBERT HOWARD in 1945. Took job at the SOIL ASSOCIATION's WHOLEFOOD shop in 1961 and later became its Managing Director, working closely with MARY LANGMAN, and Director of the Wholefood Trust. Honorary Secretary of the McCARRISON SOCIETY. Was much respected by CRAIG and GREGORY SAMS, who dealt with the shop.

SCHUMACHER, E.F. (1911–1977) Ernst Friedrich ('Fritz') Schumacher was a brilliant economist who knew DAVID ASTOR from the early 1930s onwards and who was led towards the organic movement through an interest in farming (he was a farm labourer for a time during the Second World War) and compost gardening. He was Economic Adviser to the National Coal Board from 1950 to 1970, working with his close friend and disciple GEORGE McROBIE. He joined the SOIL ASSOCIATION in 1951 and the environmentalist ideas which later found fame through *Small is Beautiful* (1973) were developed in the 1950s and '60s. In 1971 he became Soil Association President and was a powerful influence on the SEVENTIES GENERATION of organic activists. Influenced by Buddhism and the esoteric philosophy of Gurdjieff and Ouspensky, but joined the Roman Catholic Church in 1971.

SEGGER, PETER (b.1945) Controversial leading figure in the SEVENTIES GENERATION of organic activists, instrumental in turning the organic movement's energies towards effective marketing of produce. A former businessman, Segger moved to West Wales in the mid-1970s to take up smallholding and became one of the chief begetters of the West Wales Group of the SOIL ASSOCIATION, along with PATRICK

HOLDEN. First of the Seventies Generation to be elected to the Association's Council, to whose discussions his contributions were seen by many as offensively confrontational. Nevertheless, he was also admired for his determination to bring new life to the organic movement, and had the support of older figures such as DR VICTOR STEWART and MARY LANGMAN. Was central to the establishment of the ORGANIC GROWERS ASSOCIATION and the journal *NEW FARMER AND GROWER,* and established the Organic Farm Foods business. Sympathetic to the methods of BIODYNAMIC CULTIVATION.

SEYMOUR, JOHN (1914–2004) The guru of 'self-sufficiency' who, like E.F. SCHUMACHER, found himself a cult figure to a younger generation comparatively late in life. Paul Peacock's biography of Seymour is essential reading to appreciate the breadth of Seymour's activities, and also pays tribute to the vital contribution of his wife Sally. Influenced by H.J. MASSINGHAM and L.T.C. Rolt, Seymour was a farmer and writer whose uncompromising opposition to agri-business attracted members of the 'counter-culture'. He in turn influenced the SEVENTIES GENERATION of organic activists, whom he knew when living in West Wales; he was also a strong influence on ALAN and JACKIE GEAR. He was still writing prolifically on environmental issues in the late 1980s. A long-term member of the editorial board of *RESURGENCE.* Like Schumacher, he was attracted to the Roman Catholic Church.

SHEWELL-COOPER, DR W.E. (1900–82) Distinguished horticulturalist and prolific writer whose skill at self-publicity also benefited the organic movement. While Superintendent of Swanley Horticultural College, Kent in the 1930s met SIR ALBERT HOWARD and was influenced by his writings. Ran Horticultural Advisory Bureau and Training Centre at Thaxted from 1950, moving in 1960 to Arkley Manor, Herts. Founded GOOD GARDENERS' ASSOCIATION to encourage organic methods. On Council of the SOIL ASSOCIATION from 1946–47 to 1962. He and fellow-organic gardener LAWRENCE HILLS were at loggerheads with each other. He contributed to *THE FARMER, SPAN, SEED* and *VOLE,* as well as to the Soil Association journal. His son Ramsay, another outstandingly able horticulturalist, also served on the Association's Council.

SINCLAIR, DR HUGH M. (1910–90) Noted academic nutritionist who was an authority on essential fatty acids; Fellow of Magdalen College, Oxford from 1937 to 1980. Student and friend of SIR ROBERT McCARRISON, whose papers he inherited. An active member of the SOIL ASSOCIATION's Oxford Group and of the McCARRISON

SOCIETY. In 1972 he set up the Association for the Study of Human Nutrition (later the International Institute of Human Nutrition).

SOPER, JOHN (1904–1998) Experienced agriculturalist for the British Colonial Agricultural Service in East Africa and Malaya, who was awarded the CBE and served as Honorary Secretary and Treasurer to the BIODYNAMIC AGRICULTURAL ASSOCIATION. From 1949 to 1958 he was Deputy Director, and then Director, of Agriculture in Tanganyika. Returned to UK and became active proponent of BIODYNAMIC HUSBANDRY, frequently contributing to *STAR AND FURROW,* which his wife Marjorie (1908–90) edited from 1969 until 1979. He developed and adapted the agriculture course of RUDOLF STEINER for a younger generation.

STAPLEDON, SIR R. GEORGE* (1882–1960) Outstanding agricultural botanist and grassland expert, respected both by orthodox agricultural scientists and by leading figures in the organic movement, particularly ROLF GARDINER. He did not join the SOIL ASSOCIATION, but sympathized with its aims. His philosophy of Human Ecology profoundly influenced the radio producer ROBERT WALLER, later editor of *MOTHER EARTH,* who wrote his biography *Prophet of the New Age* (1962). DINAH WILLIAMS, a mentor to some members of the SEVENTIES GENERATION, had been impressed by Stapledon in the 1930s, and his influence can still be seen today in the writings of GRAHAM HARVEY.

STEINER, RUDOLF* (1861–1925) Steiner was a scientist, mystic and educationalist whose methods of BIODYNAMIC CULTIVATION, outlined in lectures given the year before his death, have interested or influenced various figures in the mainstream British organic movement, including LAURENCE EASTERBROOK, PATRICK HOLDEN and PETER SEGGER.

STEWART, DR VICTOR I. (b.1924) Soil scientist and expert on sports pitches who taught DR NIC LAMPKIN at Aberystwyth, was active in the West Wales Group of the SOIL ASSOCIATION and a mentor to members of the SEVENTIES GENERATION. A close friend of EVE BALFOUR, he nevertheless encouraged the younger generation and nominated PETER SEGGER for the Soil Association Council. ALAN and JACKIE GEAR forged links with him. He was influenced by the KEYLINE SYSTEM in his work with the BRYNGWYN EXPERIMENT in South Wales, and spoke at a conference run by the INTERNATIONAL INSTITUTE OF BIOLOGICAL HUSBANDRY.

APPENDIX A: LEADING FIGURES IN THE ORGANIC MOVEMENT

STICKLAND, DAVID (1930–2010) Controversial figure in the history of the British organic movement since the 1970s. After more than twenty years of experience in farming and agricultural marketing, both in Britain and overseas, Stickland became editor of the SOIL ASSOCIATION journal in 1973 but left the post to establish his own marketing business ORGANIC FARMERS AND GROWERS. He became something of a bogeyman for the SEVENTIES GENERATION and for some older members of the Soil Association, who considered him to be compromising the standards of organic produce. This led to severe disagreements during the period when UKROFS was being established. Stickland was also instrumental in establishing the INTERNATIONAL INSTITUTE OF BIOLOGICAL HUSBANDRY, and he sold produce to the SAMS brothers when they ran the Ceres grain business. His admirers included Angela Bates and Michael and Deidre Rust, who felt that he had brought considerable energy and commitment to the organic movement.

STUART, CDR ROBERT L. (1900–80) Robert Stuart followed early service in the Royal Navy with time spent on a tea plantation in India, where he became aware of the problems of soil erosion. Later, he became a lawyer, and served again in the navy during the Second World War. From the 1940s, influenced by the work of SIR ALBERT HOWARD, he became an organic farmer and gardener in Scotland. He was centrally active in the SCOTTISH SOIL AND HEALTH GROUP, and formed the Hopes Compost Club, which attracted many leading members of the organic movement. He promoted composting of municipal wastes and offered his own composting service. During the 1960s and '70s he ran week-long SOIL ASSOCIATION conferences at his prize-winning country house hotel, which many prominent figures in the organic movement attended.

SYKES, FRIEND* (1888–1965) Farmer and racehorse breeder who, along with EVE BALFOUR and GEORGE SCOTT WILLIAMSON, was one of the three founders of the SOIL ASSOCIATION. His friend SIR ALBERT HOWARD considered Sykes' downland estate a better argument for humus farming than anything that the HAUGHLEY EXPERIMENT was likely to produce. His book *Humus and the Farmer* (1946) is commonly cited by farmers of a younger generation as an influence on them.

TEVIOT, LORD* (1874–1968) President of the SOIL ASSOCIATION from 1946 to 1950.

THEAR, KATIE and DAVID Husband-and-wife team who ran Broad Leys Publishing from their home near Saffron Walden, Essex, and produced the magazine *PRACTICAL SELF-SUFFICIENCY/ HOME FARM*, which promoted organic husbandry and the value of small farms. Friends of SIR RICHARD BODY, one of whose books they published.

TREVELYAN, SIR GEORGE (1906–96) Craftsman and teacher; Principal of Shropshire Adult College, Attingham Park from 1947 to 1971, where the SOIL ASSOCIATION held a number of successful conferences in the 1950s and '60s. A friend of Professor Edmond Szekely, who encouraged the herbal studies of JULIETTE DE BAIRACLI LEVY. Trevelyan founded of the Wrekin Trust in 1971, a body whose aim has been to promote non-sectarian spiritual education.

TURNER, F. NEWMAN* (1913–64) Farmer and journalist whose work and writings continue to be cited as influential and inspirational. He was active in a wide variety of organic initiatives: a SOIL ASSOCIATION Council member from 1952 until his death; Chairman and President of the HENRY DOUBLEDAY RESEARCH ASSOCIATION from its foundation in 1958 until his death; editor of his own journal *THE FARMER;* established the PRODUCER CONSUMER WHOLE FOOD SOCIETY LTD; leading figure in the ALBERT HOWARD FOUNDATION OF ORGANIC HUSBANDRY, and ran courses in husbandry at his Somerset farm. A herbalist, he worked closely with JULIETTE DE BAIRACLI LEVY and wrote on how to deal with foot-and-mouth disease.

VOGTMANN, DR HARTMUT (HARDY) (b.1942) A good friend of the British organic movement and frequent visitor to the UK, Vogtmann earned his doctorate through research into animal nutrition. Founder and first Director of the Research Institute of Biological Farming at Oberwil, Switzerland. The first person to become Professor of a university-level course in organic agriculture, being appointed to the Chair of Alternative Agriculture at Kassel, Germany, in 1980. A prime mover in IFOAM. A strong influence on LAWRENCE WOODWARD and Trustee of the ELM FARM RESEARCH CENTRE. Gave the first EVE BALFOUR Memorial Lecture.

WALES, HRH THE PRINCE OF (PRINCE CHARLES) (b.1948) The Prince of Wales came out in support of the organic movement early in 1983, expressing his approval of a conference on organic food production held at the Royal Agricultural College, Cirencester (of which he was President) and organized by BRITISH ORGANIC FARMERS, the ORGANIC GROWERS' ASSOCIATION and ELM FARM RESEARCH CENTRE. In 1985 he began to convert the Home Farm of his Highgrove Estate to organic, receiving advice from Elm Farm. In 1989 he visited the HENRY DOUBLEDAY RESEARCH ASSOCIATION headquarters at Ryton and became a Patron of the Association; he visited RICHARD and ROSAMUND YOUNG at Kite's Nest farm the same year. In a 1991 address to the Royal Agricultural Society of England, of which he was President, he rejected the misconception of organic farming as suitable only for ex-hippies. The Prince's support undoubetedly helped raise the organic movement's profile, but may also have played a part in creating a perception of it as something for the élite.

WALLER, ROBERT (1913–2005) Poet, journalist, radio producer and environmental philosopher. Influenced by the Human Ecology of SIR GEORGE STAPLEDON and by what he saw of the effects of agribusiness when he produced agricultural programmes for the BBC's West of England regional service. Largely as a result of writing Stapledon's biography, he was appointed editor of *MOTHER EARTH,* working for the SOIL ASSOCIATION from 1964 to 1972 and, with his assistant MICHAEL ALLABY, trying to position the Association in the vanguard of the growing environmental movement. Left the Association in 1972 following differences with senior staff, but returned as a columnist in *LIVING EARTH* in the late 1980s. Remained active in environmental causes through to the 1990s, and attempted a major work of ecological philosophy in *Be Human or Die* (1973).

WESTLAKE, DR AUBREY* (1893–1985) London doctor and Hampshire estate owner active in many environmentalist causes from the 1920s onwards. Member of the SOIL ASSOCIATION Council from 1953 to 1961 and a close friend of JULIETTE DE BAIRACLI LEVY. Wrote studies of health and the effects of environmental pollution in the 1960s. Interested in forms of alternative healing and particularly radionics; also in BIODYNAMIC CULTIVATION. An influence on Michael and Deidre Rust.

WILLIAMS, DINAH (1911–2009) Daughter of Abel Jones, Professor of Agriculture at Aberystwyth, and Elizabeth Lyon Jones, a lecturer in the Dairy Department there who remained a farmer after her husband's early death. Mrs Williams knew SIR GEORGE STAPLEDON and agricultural scientist Sir John Russell in the 1930s, and was cured of health problems by Dr Cyril Pink, a regular contributor to *HEALTH AND*

LIFE magazine. She and her husband Stanley farmed organically from the 1940s. Member of SOIL ASSOCIATION Council for most of the 1970s. Mentor to members of the SEVENTIES GENERATION, particularly PATRICK HOLDEN and PETER SEGGER of the West Wales Group of the Soil Association, and mother of Rachel Rowlands, the founder of successful brand Rachel's Dairy.

WILLIAMSON, DR GEORGE SCOTT (1884–1953) With EVE BALFOUR and FRIEND SYKES, one of the three founders of the SOIL ASSOCIATION and a profound influence on Balfour herself, KENNETH BARLOW, MARY LANGMAN and Douglas Trotter. Partner and husband of INNES PEARSE, with whom he established the PIONEER HEALTH CENTRE; both had known SIR CEDRIC STANTON HICKS and SIR ROBERT McCARRISON since the 1920s. As well as establishing a notable experiment in communal health, Scott Williamson developed his own, rather complex, biological philosophy and reflected on the political implications of the Pioneer Health Centre from a standpoint of 'liberal socialism'.

WILSON, C. DONALD (1900–73) Important figure in first two decades of the SOIL ASSOCIATION: EVE BALFOUR described him as the Asociation's main 'think tank'. Scientist who worked in industry before involvement in social welfare schemes during 1930s. Manager of the PIONEER HEALTH CENTRE in post-war years and first secretary of the Soil Association, serving the Association also as Council member and Assistant Treasurer. Worked with DR REGINALD MILTON on observation of the HAUGHLEY EXPERIMENT. Responsible for establishing the WHOLEFOOD shop in London, and the joint working party on municipal composting. Met MICHAEL ALLABY through Morton Whitby's Cancer Prevention Detection Centre and suggested he work for the Soil Association. Taught composting at FINDHORN.

WOOD, MAURICE* (1884–1960) A BIODYNAMIC farmer at Huby in Yorkshire from 1928 onwards, he was also a successful miller whose stoneground, wholemeal 'Huby' flour was much in demand from members of the organic movement. Influenced LAWRENCE HILLS. Founder member of the SOIL ASSOCIATION and a member of its Council from 1946 to 1953.

WOODWARD, LAWRENCE (b.1951) From a South Yorkshire mining background, following a school exchange with Dartington College he married DAVID ASTOR's daughter Alice and became influenced by the views of Astor and his friend E.F. SCHUMACHER on the likely social effects of an energy crisis. Studied agriculture and farmed on the estate which had belonged to ROLF GARDINER, being advised for a time by DAVID STICKLAND. From 1980, director of the ELM FARM RESEARCH CENTRE and played a key role, along with other members of the SEVENTIES GENERATION, in taking the organic movement out into the world of marketing, the media and government policy. Involved in BRITISH ORGANIC FARMERS and *NEW FARMER AND GROWER;* worked with BARRY WOOKEY; member of SOIL ASSOCIATION COUNCIL. Has had particularly close links with DR HARTMUT VOGTMANN.

WOOKEY, BARRY Large-scale arable farmer in Wiltshire whose Rushall estate became a showpiece of organic methods from the mid-1970s onwards. Although differing from them politically, he supported the initiatives of the SEVENTIES GENERATION and was prominent in BRITISH ORGANIC FARMERS. Provided useful contacts with the world of conventional agricultural policy during the period of the Thatcher government in the 1980s. A patron of the SOIL ASSOCIATION.

YELLOWLEES, DR WALTER (b.1917) Perthshire GP with two farming brothers. Served in Royal Army Medical Corps during the Second World War and was greatly struck by reading EVE BALFOUR's *The Living Soil* shortly before demobilization. An early member of the SOIL ASSOCIATION and a practitioner of SIR ALBERT HOWARD's horticultural composting methods. Helped found the McCARRISON SOCIETY and, in 1981, its Scottish branch. Gave the 1978 James Mackenzie Lecture to the Royal College of General Practitioners. An admirer of the work and ideas of SURGEON-CAPTAIN T.L. CLEAVE.

YEOMANS, P. A. (1905–84) New South Wales mining engineer and agricultural machinery designer, who won the Prince Philip Design Award in 1974 for his sub-soil ripper. Became manager of extensive agricultural land following his brother's death in the 1940s, investigating questions of hydrology and water storage. Developed the KEYLINE SYSTEM of conservation. A SOIL ASSOCIATION member, he helped organize EVE BALFOUR's tour of Australia in the late 1950s. An influence on ROBERT HART and on PERMACULTURE.

YOUNG, RICHARD (b.1950) Cotswold farmer who was influenced by Richard Mayall in the 1970s and turned successfully to organic methods. One of the SEVENTIES GENERATION who helped give the organic movement a higher profile in the 1980s. Worked for BRITISH ORGANIC FARMERS and on the editorial board of *NEW FARMER*

AND GROWER. SOIL ASSOCIATION Council member. Brother of farmer and naturalist ROSAMUND YOUNG.

YOUNG, ROSAMUND (b.1953) Sister of RICHARD YOUNG and authority on the health and behaviour of farm animals; friend of JOANNE BOWER. SOIL ASSOCIATION Council member. The Youngs' farm was visited in 1989 by HRH the PRINCE OF WALES.

Appendix B: Groups, institutions, organizations and journals in the organic movement, 1945–1995

Names of people in CAPITALS indicate an entry in Appendix A. Names of groups, institutions, organizations or journals in CAPITALS indicate an entry in Appendix B.

AGRICULTURE ACT 1947 Brought on to the statute book by the Labour government's Minister of Agriculture Tom Williams, the 1947 Act provided much of the context for the organic movement in the post-war decades. Praised at the time for its intention to give British agriculture the security which it had lacked during the 1920s, the Act became a target both of those who thought it 'feather-bedded' farmers at the tax-payers' expense, and of the organic movement, which condemned its emphasis on increasingly large-scale, chemical-based, industrial techniques in the interests of an efficiency measured by output per worker.

ALBERT HOWARD FOUNDATION OF ORGANIC HUSBANDRY Following the death of SIR ALBERT HOWARD, the Foundation was established in 1948 by friends of his in order to continue his work. LOUISE HOWARD was its President and FRANK NEWMAN TURNER its Chairman of Membership. Dr C. Langley Owen of Mayfield in Sussex provided his house and estate as the Foundation's headquarters. The Foundation aimed to create knowledge of the importance of humus to the soil, and to establish demonstration centres. It published booklets on soil science and related matters. It merged with the Soil Association in the early 1950s.

ANTHROPOSOPHY A form of 'spiritual science' developed by RUDOLF STEINER, who defined it as 'a path of knowledge to guide the spiritual in the human being to the spiritual in the universe': a methodology and body of knowledge deriving from study of his own spiritual experiences and available to others not through revelation but through their own efforts. Unlike Theosophy, it attached central importance to the figure of Christ, and, also unlike Theosophy, it aimed to change the world through gathering objective knowledge and through a wide-ranging pedagogical programme. (See Lachman 2007, pp. 171–73.) Influenced by evolutionary thought and organic science, Anthroposophy emphasized the links between all phenomena, and the influence of cosmic forces on planet Earth. This had implications for agriculture which were expressed through the methods of BIODYNAMIC CULTIVATION. JOHN DAVY, for many years the Science Correspondent of *The Observer*, was a leading teacher of Anthroposophy.

BIODYNAMIC AGRICULTURAL ASSOCIATION Originally named the Anthroposophical Agricultural Association, the BAA emerged in the UK in the late 1920s, with Dr Carl Mier and the Yorkshire farmer MAURICE WOOD as key figures. In 1933 DAVID CLEMENT's farm at Clent in Worcestershire became the Association's base for the next fifty years. The precise influence of BIODYNAMIC methods on the wider organic movement is difficult to determine, but leading figures in the BAA such as Clement, Wood and George Corrin, and, later, Vivian Griffiths and Nick and Ana Jones, were active in the organic mainstream. The BAA produced the journal *STAR AND FURROW*. DR ANTHONY DEAVIN was a BAA member, and PATRICK HOLDEN trained as an organic farmer on the BAA course at Emerson College.

BIODYNAMIC CULTIVATION The name given by EHRENFRIED PFEIFFER to the agricultural and horticultural methods outlined by RUDOLF STEINER and practised by his followers. They are based on the idea that the universe is an evolving organism and that every element in it undergoes purposive development towards realization of its spiritual potential. Science, in Steiner's view, had failed to see agriculture as a totality and the farm as an organism; it saw the soil as inert material rather than a sentient force. As with the Indore Process of SIR ALBERT HOWARD, composting forms the basis of the system, but there are significant differences. Biodynamic methods use activators formed from homoeopathic preparations and based on traditional peasant knowledge of herbal medicine; and Steiner believed in a correspondence between the earth, its products and elements, on the one hand, and the heavens on the other: a system of planetary influences which required sowing according to lunar phases, for instance. Biodynamics are of course a gift to those who wish to mock organic farming as 'muck and mysticism', but there is no reason why they should not be scientifically investigated. DR ANTHONY DEAVIN took a scientist's interest in biodynamic methods, and various leading figures in the organic movement have experimented with them, including LAURENCE EASTERBROOK, PATRICK HOLDEN and PETER SEGGER.

BIOLOGICAL AGRICULTURE AND HORTICULTURE Edited by DR DAVID HODGES, assisted by Drs Joe Lopez-Real and Tony Scofield and supported by a wide range of international advisors, including DR HARTMUT VOGTMANN, *BAH* was established late in 1982 and published under the aegis of the INTERNATIONAL INSTITUTE OF BIOLOGICAL HUSBANDRY (though the relationship was somewhat complex). It was the first British scientific journal specifically to cover alternative husbandry systems and, in particular, the field of biological/eco/organic agriculture. Surviving thanks to Herculean efforts by Hodges, *BAH* published work of a sound scientific or economic nature on research into biological techniques in agriculture and horticulture in both temperate and tropical conditions.

BRITISH ORGANIC FARMERS Founded in 1982, BOF was the result of a perceived need for a serious and professional farmers' group within the organic movement. Following a weekend seminar at ELM FARM RESEARCH CENTRE a steering committee was formed whose members included HUGH COATES, PATRICK HOLDEN (who was appointed co-ordinator), LAWRENCE WOODWARD, DAVID CLEMENT and Richard Mayall. BOF acted as a pressure group for further research, worked to improve marketing strategies and provided an information service for organic farmers. It aroused media interest, organized farm walks and worked closely with both Elm Farm and the ORGANIC GROWERS' ASSOCIATION. Along with the OGA it established the journal *NEW FARMER AND GROWER*. Its merger with the OGA in the 1990s was followed by its absorption into the SOIL ASSOCIATION as its Producer Services section.

BRITISH PEOPLE'S PARTY Run by the pro-organic monetary reformer the Duke of Bedford, this was a national-socialist organization proscribed during the war and re-emerging in 1945. Its leading figures included the pro-Nazi eugenicist Anthony Ludovici, a close friend of the EARL OF PORTSMOUTH and ROLF GARDINER. The BPP published a newspaper, *Peoples Post,* which featured DION BYNGHAM as columnist 'Miles Yarrow', an enthusiastic advocate of organic farming.

BRYNGWYN PROJECT This experiment in South Wales was initiated by E.F. SCHUMACHER through his connection with the National Coal Board. It was set up by SOIL ASSOCIATION Secretary Brigadier Bill Vickers, and managed by the Association in conjunction with the University of Wales at Aberystwyth, its research unit being headed by DR VICTOR STEWART. Its aim was to show how organic techniques might effectively bring back to full production land ruined by open cast coal mining.

COUNCIL FOR THE CHURCH AND COUNTRYSIDE In effect a 'front' for the organic movement, the CCC was founded during the Second World War by the Rev David G. Peck, with the help of ROLF GARDINER and the Bishop of Salisbury. It was based on a perception that the Church was neglecting its roots in agriculture, and attempted to relate the Church's social teaching to rural society (though its headquarters was in London's Soho). Belief in the God-given natural order was the key to its teachings. It issued regular pamphlets during the late 1940s and early 1950s, usually edited by JORIAN JENKS. Several of the organic movement's leading figures were members of the CCC or involved with its work: these included H.J. MASSINGHAM, the EARL OF PORTSMOUTH and SIR ALBERT HOWARD. Its work provides one of the clearest examples of the early organic movement's Christian context.

COVENTRY FAMILY HEALTH CLUB HOUSING SOCIETY A post-war initiative by DR KENNETH BARLOW, with the support of DRS GEORGE SCOTT WILLIAMSON and INNES PEARSE. Its aim was to extend the principles of the PIONEER HEALTH CENTRE to a wider community in which agricultural work and the production of fresh organic foodstuffs would be centrally important. Despite the help of distinguished figures such as Lord Lindsay, Barlow's efforts were rendered unsuccessful by the opposition of Labour Party local councillors.

ECOLOGIST, THE Influential environmentalist journal founded in 1970 by EDWARD GOLDSMITH. At first printed on the same presses as the SOIL ASSOCIATION's journal *MOTHER EARTH*, *The Ecologist* included on its editorial board MICHAEL ALLABY, LAWRENCE HILLS and ROBERT WALLER. For many years a serious, demanding and deeply informative publication, *The Ecologist* was considerably more uncompromising in its political and cultural stance than other organicist publications, displaying a sympathy for the values of traditional societies which at times verged on misanthropy towards contemporary Westerners. It favoured organic farming as an antidote to the environmental destruction wreaked by agri-business techniques.

ELM FARM RESEARCH CENTRE Elm Farm was registered as a charity in 1980 and started operating the following year, funded by the Progressive Farming Trust. It was created by DAVID ASTOR, whose then-son-in-law LAWRENCE WOODWARD became its co-ordinator.

The Council included DR HARTMUT VOGTMANN. Its aims were to manage the 232-acre farm as a complete organic holding; to develop applied research of immediate use to farmers and growers; and to investigate European developments in organic farming and adapt them to British conditions. Elm Farm began a soil advisory service in 1983 with the help of money from J.A. PYE. The Centre worked with the SOIL ASSOCIATION and the HENRY DOUBLEDAY RESEARCH ASSOCIATION, and was closely involved in the work of BRITISH ORGANIC FARMERS and the ORGANIC GROWERS ASSOCIATION. Its work has been highly regarded by the leading agricultural scientist Sir Colin Spedding.

FABER AND FABER Publishers whose agricultural and horticultural editor RICHARD DE LA MARE promoted many of the classic texts of the organic canon, through to the 1970s, including works by EVE BALFOUR, DR KENNETH BARLOW, LAWRENCE HILLS, SIR ALBERT HOWARD, JULIETTE DE BAIRACLI LEVY, SIR GEORGE STAPLEDON, FRIEND SYKES and NEWMAN TURNER.

FARM AND FOOD SOCIETY (for Humane, Wholesome and Fair Farming). It was founded in 1966 by JOANNE BOWER, who had been influenced by the evidence of maltreatment of poultry and livestock presented in RUTH HARRISON's book *Animal Machines* (1964). Among its patrons have been the noted conductor Sir John Eliot Gardiner (son of ROLF GARDINER) and JONATHON PORRITT.

FARMER, THE Subtitled *The Journal of Organic Husbandry*, this publication, edited by FRANK NEWMAN TURNER, ran from 1946 to 1956 and contained a wealth of high-quality material putting the case for the organic philosophy of farming and health. It incorporated *The Gardener*, to which F.C. King and LAWRENCE HILLS contributed. Many leading figures wrote for the journal, including JULIETTE DE BAIRACLI LEVY, Philip Oyler, Dr Annie Cunning, DR W.E. SHEWELL-COOPER, DORIS GRANT, Lady Seton of the NATIONAL GARDENS GUILD, LOUISE HOWARD and Dr Cyril Pink.

FINDHORN COMMUNITY Founded in the 1960s by husband and wife Peter and Eileen Caddy, the community, based in north-east Scotland, became a leading centre of 'New Age' practice and a place of pilgrimage for spiritual seekers. The organic movement's connections with Findhorn included the support which RICHARD ST. BARBE

BAKER gave it (the Findhorn Press re-published his autobiography *My Life My Trees* in 1985) and the interest which PROFESSOR R. LINDSAY ROBB took in its remarkable success at growing splendid crops in unpromising, sandy soil. Other 'organic' visitors included George Corrin and C. DONALD WILSON, who taught composting there. Its significance was also noted by John Butler in *SEED* magazine. Community members believed that it was the responsibility of humans to evoke from the earth the fruition of its seed potential.

GOOD GARDENERS' ASSOCIATION Founded in 1963 by DR W.E. SHEWELL-COOPER, the GGA based its approach on biblical teaching and promoted the proper making and use of compost and the value of 'no dig' techniques. Its demonstration ground was at Arkley Manor in Hertfordshire. Shewell-Cooper's son Ramsay continued the work of the horticultural training college following his father's death.

HAUGHLEY EXPERIMENT Taking its name from the Suffolk village where EVE BALFOUR farmed from the 1920s onwards, the Experiment was an important but over-ambitious attempt to investigate what truth there might be in the ideas of SIR ALBERT HOWARD and SIR ROBERT McCARRISON, that there existed a positive link between cultivation based on the return of biological wastes to the soil, and the health of the plants grown in that way and of the animals and humans who consumed those plants. Balfour conceived the idea of the Experiment late in the 1930s, but it could not be put into practice until after the end of the war. The Experiment's strategy involved farming three comparable areas by three different methods. Two plots carried stock, one being cultivated organically and the other with a mixture of wastes and artificial fertilizers. The third was stockless, relying on artificials and crop residues. Howard himself was dubious about the Experiment, considering it unnecessary, too complex and likely to be too expensive. The newly founded SOIL ASSOCIATION took it over, and it indeed proved an enormous financial burden to the organization. FRIEND SYKES and the BIODYNAMIC farmer Deryck Duffy oversaw the Organic section, and in 1952 DR REGINALD MILTON began a twelve-year stint of sampling and analytical work. In 1957 the scientific journal *Nature* praised the importance of the Experiment's long-term ecological research. The Experiment was wound up in 1969, and in 1975 Eve Balfour gave an account of its results in a revised edition of *The Living Soil*. The final verdict must be that the Experiment failed to achieve its stated aim of tracing the relationship between soil treatment and health, the variables being too numerous and complex. Also, Milton's data were regarded with some scepticism by, for instance, DR NORMAN BURMAN and

MICHAEL ALLABY.

HEALTH AND LIFE One of the organic movement's most important organs of communication in the 1930s and '40s, EDGAR SAXON's journal continued to promote the cause in the post-war years, though it began to lose its edge in the 1950s as Saxon entered his mid-seventies. DION BYNGHAM continued to work for it, and James Gathergood succeeded Saxon as editor in the mid-1950s. DORIS GRANT was a contributor.

HEALTH AND THE SOIL This journal was in effect a continuation of SIR ALBERT HOWARD's journal *SOIL AND HEALTH*. Howard's wish was that the title of *Soil and Health* should die with him, and its last issue, a memorial number, appeared in Spring 1948. In February that year LOUISE HOWARD spoke in Edinburgh about the need to carry on her husband's work, and this occasion became in effect the first meeting of the SCOTTISH SOIL AND HEALTH GROUP, which re-named Howard's journal, with his widow's full support. A quarterly journal, *Health and the Soil* first appeared in the Summer of 1948 and ran until the autumn of 1951, also producing a one-off issue for the Highland Show in 1955. The journal worked closely with the ALBERT HOWARD FOUNDATION OF ORGANIC HUSBANDRY. Its contributors were not exclusively Scottish, and it did not differ significantly in its stance from either *Soil and Health* or *MOTHER EARTH*. It featured various figures important in the organic movement's early years, among them the gardener F.C. King; the expert on municipal composting J.C. Wylie; dental scientist E. Brodie Carpenter, and Dr J.W. Scharff. Sir Albert Howard's personal assistant Miss Ellinor Kirkham supervised the journal's printing and production.

HENRY DOUBLEDAY RESEARCH ASSOCIATION Henry Doubleday was a Victorian Quaker who introduced comfrey into Britain in the 1870s. LAWRENCE HILLS so admired the virtues of comfrey that when in 1954 he established a research garden at Bocking in Essex he named it after Doubleday. The HDRA was formally established as a charity in 1958, when it began recruiting members and producing a newsletter. Hills' idea was that gardeners could carry out research into chemical-free methods on their own plots. FRANK NEWMAN TURNER was the HDRA's President; other notable figures who served on its Council included DR ANTHONY DEAVIN, RICHARD DE LA MARE, H. Witham Fogg, DORIS GRANT, DR DAVID HODGES, Earl Kitchener and Charles Wacher. The Association's most important recruits were ALAN and JACKIE GEAR, who began working at

Bocking in the winter of 1973–74 and in the mid-1980s oversaw the move to Ryton. Such was the success of this move, and the interest aroused by the television series *All Muck and Magic?*, that membership increased from 7,500 at the end of 1985 to 18,000 five years later.

HOME FARM Magazine run by KATIE and DAVID THEAR from their Essex smallholding. Founded in 1975 and first titled *Practical Self-Sufficiency*, it incorporated *Goat World* in 1982 and the following year changed its name to *Home Farm*. In 1986 it incorporated the newsletter of the Small Farmers Association and also began to describe itself as 'Britain's Organic Farming and Growing Magazine', despite the existence of *NEW FARMER AND GROWER*. It covered a wide range of topics, practical and legal, and supported the organic movement. Contributors included Francis Blake of the SOIL ASSOCIATION, SIR RICHARD BODY, ALAN GEAR, DR W.E. SHEWELL-COOPER, Sue Stickland of the HENRY DOUBLEDAY RESEARCH ASSOCIATION, and Penny Strange of PERMACULTURE.

INSTITUTE OF ORGANIC HUSBANDRY Founded by FRANK NEWMAN TURNER after the war, the Institute was based at his Somerset farm and described itself as 'the World's first centre of Instruction and advice on Organic Husbandry, and for the Natural Treatment of Diseases of Animals'. The Institute offered advisory services for farmers and landowners wishing to manage their holdings organically, and Turner undertook a certain amount of veterinary work based on herbal treatments. There were courses for a small number of students, and weekend courses for farmers and gardeners. Turner was assisted by, among others, the organic gardener F.C. King.

INTERNATIONAL FEDERATION OF ORGANIC AGRICULTURE MOVEMENTS (IFOAM) IFOAM was founded at a conference of Nature et Progrès held at Versailles in November 1972. Its establishment stemmed from an increasing feeling that a diversity of national movements and organizations promoting organic methods needed to co-ordinate their activities on an international scale. EVE BALFOUR and MARY LANGMAN played a central part in setting it up and the SOIL ASSOCIATION was one of its five founding members. By the early 1980s there were more than 80 member groups drawn from about 30 countries. Its wider aim has been to develop and encourage an agriculture which is ecologically and socially sustainable, and to do this through exchange of information, provision of common platforms for interest groups, continuous revision of standards, and representation of the organic movement in international institutions and agencies. DR HARTMUT VOGTMANN has been a prominent figure in IFOAM, and in 1982 LAWRENCE WOODWARD was elected a member of the Directorate. For more on IFOAM, see Lockeretz 2007, pp. 175–86.

INTERNATIONAL INSTITUTE OF BIOLOGICAL HUSBANDRY (IIBH) The IIBH was founded in 1975, the brainchild of DAVID STICKLAND, who wanted to establish some scientific credentials for organic/biological methods of cultivation. It was in part a response to the oil crisis of the winter of 1973–74. Stickland, like many in the organic movement, felt that conventional agriculture was vulnerable to energy shortages and that biological husbandry could be developed to provide food at a fraction of the energy input required by conventional methods. The IIBH was established to serve as a centre for the co-ordination and scientific development of this alternative, in which many scientists, agriculturalists and farmers were increasingly interested. It would bring together all such people in order to raise the level of credibility of biological agriculture. Its ambitious activities would include setting up channels of communication between those who had previously been isolated; gathering and disseminating information; publishing a journal (*BIOLOGICAL AGRICULTURE AND HORTICULTURE* in effect became this journal); liaising with government and universities to establish research work; taking a particular interest in Third World countries, and co-operating with other relevant bodies. DR DAVID HODGES became its Honorary Director, and HUGH COATES was also involved in its establishment. The SOIL ASSOCIATION regarded the IIBH with some mistrust, believing its establishment to have been secretive. The IIBH held an international symposium at Wye College in August 1980. Its demise in the mid-1980s was largely due to lack of finance.

KEYLINE SYSTEM A land-use system developed in Australia by P. A. YEOMANS. It was much admired by SIR CEDRIC STANTON HICKS, EVE BALFOUR, George Corrin and ROBERT HART, and influenced the principles of PERMACULTURE. In his desire to restore run-down land, Yeomans read various books on soil, among them works by SIR GEORGE STAPLEDON, SIR ALBERT HOWARD and R.H. Elliot, and began applying ecological principles to the treatment of the land. The term 'Keyline' reflects the system's use of contours. It is a complete method of layout and cultivation which respects topography, plants trees and conserves water while acting as a form of flood control. Yeomans' work demonstrated that ecological principles applied to pastoral management could rapidly create a deep, fertile soil. Various articles on

Keyline appeared in *MOTHER EARTH* in 1959–61, and there was a substantial piece on it in the *SOIL ASSOCIATION QUARTERLY REVIEW* in December 1979 (pp. 2–6).

KINSHIP IN HUSBANDRY A group founded in 1941 by ROLF GARDINER. It was at the centre of the emerging organic movement, formulating a philosophy of 'husbandry' of the earth in opposition to the exploitation of natural resources by industrialism. Its members included the poet Edmund Blunden; farmer-writer Adrian Bell; historian Arthur Bryant; seed merchant J.E. Hosking; PHILIP MAIRET; H.J. MASSINGHAM; LORD NORTHBOURNE and the EARL OF PORTSMOUTH. It aimed to 'percolate' various organizations to spread its views, and among those 'percolated' were the COUNCIL FOR THE CHURCH AND COUNTRYSIDE, MEN OF THE TREES, the RURAL RECONSTRUCTION ASSOCIATION and the SOIL ASSOCIATION. It seems to have been dominated by Gardiner, Northbourne and Portsmouth, giving it a distinct radical-Right tinge (its name, after all, carries connotations of 'blood and soil'), but nevertheless there were some serious internal tensions. In 1950 it organized a major meeting in London on European Husbandry, but seems to have faded out following Massingham's death in 1952.

LIVING EARTH The title of the SOIL ASSOCIATION's journal from 1988 onwards, combining references to *MOTHER EARTH* and to EVE BALFOUR's book *The Living Soil*. It was felt that a name rather more exciting than the *SOIL ASSOCIATION QUARTERLY REVIEW* was required. The journal was 'devised and edited' by GEOFFREY CANNON, and the editorial team at the beginning included Francis Blake, Nigel Dudley, GRAHAM HARVEY, PATRICK HOLDEN, ROBERT WALLER and ROSAMUND YOUNG. Its professed aims were 'to preserve the zeal of the *Quarterly Review* and to revive the thought in *Mother Earth*'. (By the middle years of the 21st century's first decade, the latter aim had been thoroughly shelved.)

McCARRISON SOCIETY The McCarrison Society for Nutrition in Health, named after nutritional scientist SIR ROBERT McCARRISON, was founded early in 1966, following a SOIL ASSOCIATION Attingham Park conference the previous year on the theme of whole food. The Society's aim was to try and overcome the medical profession's sceptical attitude towards the concept of positive health through nutrition, and its membership was open to qualified doctors, dentists and veterinary surgeons. The Society would collect and publish evidence relating to nutrition and health, encourage relevant research, liaise with other interested scientific bodies and press for a review of food and health legislation. Many leading figures of the organic movement were members, among them DR KENNETH BARLOW, the dentist E. Brodie Carpenter, the surgeon Arthur Elliot-Smith, the Latto family, DR INNES PEARSE, DR HUGH SINCLAIR, Dr Kenneth Vickery and DR AUBREY WESTLAKE. The first meeting was chaired by DR WALTER YELLOWLEES. The Society organized various conferences in the 1970s and established good relations with the Royal Society of Medicine. In the early 1980s it began publishing the journal *NUTRITION AND HEALTH,* and later produced a series of booklets *The Founders of Modern Nutrition,* under the editorship of GEOFFREY CANNON.

MEN OF THE TREES Founded in 1922 by RICHARD ST. BARBE BAKER, this body was dedicated to the care, protection and maintenance of the world's woodlands and forests. It had close links with the organic movement: figures such as SIR ALBERT HOWARD, PHILIP MAIRET and the EARL OF PORTSMOUTH were involved in its activities. It continues to exist today as the International Tree Foundation, with HRH the PRINCE OF WALES as its Patron.

MOTHER EARTH The SOIL ASSOCIATION journal from its inception in 1946 until the late 1960s, when it became *The Journal of the Soil Association.* During this time, it had just two editors: JORIAN JENKS, until his death in 1963, and then ROBERT WALLER. The title was suggested by GEORGE SCOTT WILLIAMSON, who considered it to be not sentimental, but literally true. Many in the Association were unhappy with the name, and there was much debate about changing it in the mid-1950s, but nothing conclusive emerged at that time. In the 21st century, older members of the Association have looked back nostalgically to *Mother Earth,* and it is not hard to see why. The journal was tasteful and solid in appearance, and contained a wealth of information, not just on the organic movement but on its wider agricultural context, as well as serious scientific research and philosophical debate. It was always ecologically minded, but became more strongly so under Waller's editorship.

NATIONAL GARDENS GUILD Founded in 1927 by Lady Frances Seton, the NGG was by the 1940s a leading encourager of organic horticulture. It produced a monthly journal *The Guild Gardener,* to which various figures in the organic movement contributed, among them RICHARD ST. BARBE BAKER, Roy Bridger and H.E. Witham Fogg, who also wrote for the SOIL ASSOCIATION journal in the 1970s.

NEW ENGLISH WEEKLY Founded in 1932 by A.R. Orage as a

platform for Social Credit ideas, the *NEW* was edited by PHILIP MAIRET from 1935 onwards and became the leading forum for the philosophy of organic husbandry. It survived the end of the war by only four years, but continued during that time to argue the organic case, publish various leading figures in the organic movement, such as H.J. MASSINGHAM, and review important books on health and organic farming. Its importance as a starting-point for the period covered in this book is very considerable.

NEW FARMER AND GROWER Subtitled 'Britain's Journal for Organic Food Production', *NFG,* jointly published by the Organic Growers' Association and British Organic Farmers, first appeared in the summer of 1983 and ran for more than ten years before being re-constituted as *Organic Farming* when BOF/OGA merged with the SOIL ASSOCIATION in the mid-1990s. Chris Mair was its first editor and those who helped to launch it were Douglas Blair, GRAHAM HARVEY, PATRICK HOLDEN, Michael Marriage, PETER SEGGER, Charles Wacher and LAWRENCE WOODWARD. By the end of the 1980s the editor was RICHARD YOUNG, assisted by a team including Tracey Clunies-Ross, Laura Davis, Stuart Donaldson, David Frost, Bill Starling, Chris Stopes and Iain Tolhurst. It was thus the product of the SEVENTIES GENERATION of organic activists, and was produced thanks to their energy and commitment. Its aim was to serve the practical and technical needs of organic producers in the UK. It certainly did so, and included much discussion of the organic movement's principles and strategy. Very much a publication produced 'from the ground up', it was, as so many of the organic publications once were, packed with information and controversy. MARY LANGMAN considered it superior to the SOIL ASSOCIATION's journal of the time.

NUTRITION AND HEALTH Journal published in association with the McCARRISON SOCIETY. It was founded in 1981 with the aim of providing a broad perspective on the subject of food and its effect on the human body, and to serve as a forum for communication of newer knowledge of research into all aspects of nutrition and foods. The journal emphasized the importance of nutrition in preventive medicine. It was edited by DR KENNETH BARLOW, who was succeeded in 1990 by Edward Kirby, and the international editorial board included as members Professor Ross Hume Hall and DR HUGH SINCLAIR. Among the contributors were DR HARTMUT VOGTMANN, DR WALTER YELLOWLEES, Beata Bishop and Michael Crawford.

ORGANIC ADVISORY SERVICE Founded in 1984 and headed by

Mark Measures of the ELM FARM RESEARCH CENTRE, the Organic Advisory Service was established to ensure that organic farmers and growers had somewhere to go for information and guidance, given that the Ministry of Agriculture had minimal interest in organic methods. The Service was developed jointly by the SOIL ASSOCIATION, BRITISH ORGANIC FARMERS, the ORGANIC GROWERS' ASSOCIATION and Elm Farm. It aimed to advise on principles of organic production, conversion to organic systems and the practices necessary for producing to organic standards; it also provided information on marketing and supplies. Services, which were offered on a commercial basis, included farm visits, feasibility studies, conversion plans, soil analysis and a consultancy service. The advisors all had practical experience in organic production.

ORGANIC FARMERS AND GROWERS One of the longest-lasting initiatives in the post-war history of the organic movement, OFG has also been one of the most controversial, with its founder DAVID STICKLAND an object of considerable mistrust for the SEVENTIES GENERATION. OFG's precise origins are lost in dispute, but it stemmed from a desire on the part of the SOIL ASSOCIATION to be more pro-active and well organized in the marketing of organic produce. The Association's journal for September/October 1974 (p. 23) gives an account of the Council meeting at which it was decided that the Soil Association Organic Marketing Company could no longer be supported without additional help, and that the time had come to 'establish an independent financially self-supporting organisation to take over the development of organic production and marketing'. It was agreed that the Council would support the setting up of an independent co-operative to be called Organic Farmers and Growers Ltd, which would trade under the Soil Association mark but be entirely independent of it. Exactly how OFG became Stickland's own business is a matter of debate, but he left the Soil Association in order to run it, and much bad feeling towards him resulted. Nevertheless, there were those in the Association such as Angela Bates and Michael Rust who worked with Stickland as Directors of OFG. The co-operative provided seed, organic fertilizer, advice and a market for organic cereals. By the early 1980s OFG had 80 members, but further controversy was provoked when OFG developed a 'Conservation' grade of produce which was perceived by many in the Soil Association to be a dilution of standards. This led to much dissension when national standards were thrashed out for UKROFS in the 1980s. The most balanced view of OFG, as expressed by a number of those interviewed for this book, is that it did not satisfy organic purists, but was commercially astute and encouraged a number of hesitant farmers to convert to organic.

ORGANIC GROWERS' ASSOCIATION In January 1980, the first national conference of organic growers was held, at the Royal Agricultural College, Cirencester. It was the result of a meeting held the previous year, at which growers had discussed the problems facing both those already practising organic methods and those wishing to convert from conventional methods. The project was supported by the SOIL ASSOCIATION, the INTERNATIONAL INSTITUTE OF BIOLOGICAL HUSBANDRY, and ORGANIC FARMERS AND GROWERS. Speakers at Cirencester included DAVID STICKLAND, DR HARTMUT VOGTMANN, DR KEN GRAY, DR VICTOR STEWART and Lord O'Hagan of the Soil Association. PETER SEGGER became Chairman of the OGA, which was formally established in February 1980, with Charles Wacher as Secretary and Geoff Mutton as Treasurer. A newsletter was produced from the winter of 1980–81 onwards, edited by Chris and Chris Mair, and a second Cirencester conference featured among its speakers Iain Tolhurst and the Soil Association's Agricultural Advisor R.W. Widdowson. The OGA's example inspired the establishment of BRITISH ORGANIC FARMERS, and the two organizations produced *NEW FARMER AND GROWER*. The OGA was not a commercial organization but was concerned to help growers with marketing, and provided packaging and promotional items. In 1982 it published the *Organic Products Directory,* the first booklet of its kind. Its main areas of work were research, information gathering, advisory activity, marketing and consumer information, and its social side was particularly important, enabling hitherto isolated growers to meet and learn from each other. In the 1990s it merged with BOF and so was swallowed up by the Soil Association when BOF/OGA became the Association's Producer Services section.

PERMACULTURE Permaculture has been defined as 'a system of organising the landscape in a self-sustaining way. [It] integrates ideas from the fields of organic farming, renewable energy technology, forest farming and the fund of human experience.' The name implies not just permanent agriculture, but permanent culture: that is, the idea that its principles can be adapted to all areas of activity where sustainable design systems can be developed, such as building, town planning and water supplies. Permaculture's principles, which were influenced in part by the KEYLINE SYSTEM, were developed in Australia during the 1970s by Bill Mollison and David Holmgren. Holmgren's book *Permaculture One* (1978) was reviewed in *THE ECOLOGIST* by LAWRENCE HILLS. In 1983 the Permaculture Association of Great Britain was founded, with Penny Strange its chief begetter, and began publishing *Permaculture Newsletter.* Leading figures in UK Permaculture have included Graham

Bell, Andy Langford, Patrick Whitefield and Helen Woodley. ROBERT HART, ARTHUR HOLLINS and Scottish farmer Bruce Marshall were important influences. Permaculturists have tended to be sceptical about the organic movement, considering it to have been too willing to align itself with an unsustainable commercial system.

PIONEER HEALTH CENTRE A major experiment in social medicine which originated in a family health club established in Peckham, South London by the doctors GEORGE SCOTT WILLIAMSON and INNES PEARSE. In 1935 it opened in a purpose-built, modernist building which attracted widespread interest, its aim being to investigate the conditions which make for health. Closed during the war, it re-opened in 1946 accompanied by much hope that its example could help inspire a new approach to national health. But the NHS could find no place for it, and it was forced to close in 1950. The organic movement has looked back ever since at this tantalizing, unfulfilled hope, and the Pioneer Health Centre Ltd, for which figures such as DR KENNETH BARLOW and DR PETER MANSFIELD worked, has sought to keep alive the PHC's principles and to embody them in a contemporary equivalent of what was undertaken at Peckham.

PRACTICAL SELF-SUFFICIENCY (see *HOME FARM*)

PRODUCER CONSUMER WHOLE FOOD SOCIETY LTD Perhaps the first major attempt to organize the marketing of organic produce, the Society was founded by FRANK NEWMAN TURNER in 1946 and formally established under the above title in 1948. Its aim was to put producers and potential consumers of wholesome foodstuffs in touch with each other, balancing the interests of producer and consumer in a reciprocally beneficial arrangement. Turner was President and his colleague Derek Randal Vice-President. The Society's committee included Hugh Corley, DORIS GRANT, Col. Robert Henriques, F.C. King, Dr Cyril Pink and Stanley Williams, husband of DINAH WILLIAMS. Ambitiously, the Society saw itself as an alternative to the new National Health Service, believing that consumption of healthy food was a form of preventive medicine. The Society was also ahead of its time in its attempt to establish a Whole Food Mark as a safeguard; but despite discussions with the SOIL ASSOCIATION, the BIODYNAMIC AGRICULTURAL ASSOCIATION and the ALBERT HOWARD FOUNDATION OF ORGANIC HUSBANDRY, this was never achieved. The Society also produced a newsletter, *Whole Food,* which provided a list of producers and of the produce they had available; this was eventually incorporated into *THE FARMER*. The initiative petered out in the mid-1950s.

RESURGENCE Environmentalist journal founded in 1966 by John Papworth as a 'peace publication'; from late 1973 edited by SATISH KUMAR. Combining an interest in the arts, spirituality (with a 'New Age' tinge) and ecology, *Resurgence* helped promote the organic movement and regarded itself as a 'soulmate' of *THE ECOLOGIST.* Contributors from the organic movement have been many, including E.F. SCHUMACHER, MICHAEL ALLABY, JONATHON PORRITT and ROBERT WALLER.

RURAL ECONOMY Journal produced during the 1940s and '50s by the Economic Reform Club and Institute in conjunction with the RURAL RECONSTRUCTION ASSOCIATION, and describing itself as a 'A Non-Party Commentary devoted to the development of a Sound National Economy rooted in the Soil'. It was edited by JORIAN JENKS and contributors included RALPH COWARD, H.J. MASSINGHAM and SIR GEORGE STAPLEDON. The journal reported on or advertised the activities of the COUNCIL FOR THE CHURCH AND COUNTRYSIDE, MEN OF THE TREES (whose *New Earth Charter* it published) and the SOIL ASSOCIATION.

RURAL RECONSTRUCTION ASSOCIATION Founded in 1926 by Montague Fordham, the RRA issued various proposals for regenerating British agriculture and saving rural life from decay. During the 1940s and '50s it was closely linked with the Economic Reform Club and Institute and published the journal *RURAL ECONOMY.* In 1955 it issued the report *Feeding the Fifty Million,* with an introduction by LAURENCE EASTERBROOK. Its Research Committee at that time included JORIAN JENKS and Dr Hugh Martin-Leake of the SOIL ASSOCIATION.

SCOTTISH SOIL AND HEALTH GROUP Following SIR ALBERT HOWARD's death in October 1947, his widow LOUISE HOWARD spoke in Edinburgh in February 1948 about the need to carry on his work. This occasion was in effect the inaugural meeting of the Group, whose leading spirits were DR A.G. BADENOCH, Dr Angus Campbell and CDR ROBERT L. STUART. In the Summer of 1948 they established the journal *HEALTH AND THE SOIL,* and in November that year formally established the Scottish Soil and Health Movement, which in 1950 re-named itself the Scottish Soil and Health Society. The body's aims were to demonstrate the connection between the health of mankind and the health of the soil; to collate and publicize information on the cultivation of foodstuffs by natural methods, particularly through the use of composting; and to encourage the increase of the supply of fresh, humus-grown food to the community. The Group/Movement/Society had many contacts outside Scotland, both in the UK and world-wide, and Louise Howard continued to take a close interest in it until it faded out in the mid-1950s.

SEED Counter-culture magazine produced during the 1970s by CRAIG and GREGORY SAMS. An intriguing mixture of hippie philosophy and entrepreneurialism, of serious intent and playful spirit, *Seed* is reminiscent in certain respects of EDGAR SAXON's journal *HEALTH AND LIFE* as it was in the 1930s, but also anticipates the organic movement's later concern with consumerism and celebrities. Topics covered included alternative health treatments, investigations into the food industry, natural living, spiritual awareness, organic gardening, cookery, esoteric philosophy, vegetarianism and macrobiotics. The organic smallholders John and Shirley Butler wrote regularly for it and SUSAN COPPARD worked on it.

SEVENTIES GENERATION A term used in this book to indicate a highly influential group of activists who became involved in the organic movement during the 1970s, largely took control of it in the 1980s and have continued to dominate it ever since, becoming a new establishment in their turn. The phrase does not imply that all who belonged to this 'Generation' held the same views on either philosophy or strategy, let alone that they necessarily liked each other or related harmoniously. What they shared was a feeling that the older generation of the SOIL ASSOCIATION was insufficiently aware of the problems of those who had to make a living from the organic produce which they grew; that it was too much dominated by the wealthy and aristocratic, and by gentleman-farmers; and that it was too introspective, failing to get its message across to the outside world. The generation's leading figures included Francis Blake, Anne Evans, David Frost, ALAN and JACKIE GEAR, PATRICK HOLDEN, DR NIC LAMPKIN, GINNY MAYALL, PETER SEGGER, Carolyn and Charles Wacher, LAWRENCE WOODWARD, and RICHARD and ROSAMUND YOUNG. Their achievements included the establishment of the ORGANIC GROWERS' ASSOCIATION, BRITISH ORGANIC FARMERS, *NEW FARMER AND GROWER* and the ELM FARM RESEARCH CENTRE; the transformation of the *SOIL ASSOCIATION QUARTERLY REVIEW*; involvement of government in the setting of organic standards and of supermarkets in the sale of organic produce; and the arousal of much media interest in the organic message. There were some in the movement who, while sympathizing with their aims, found their methods and manners unnecessarily abrasive. The Seventies Generation received support from some of their elders, most notably MARY LANGMAN, DR VICTOR STEWART and DINAH WILLIAMS.

SOIL AND HEALTH This short-lived quarterly journal (it ran from February 1946 to Spring 1948), edited by SIR ALBERT HOWARD, was in effect a continuation of DR LIONEL PICTON's *News Letter on Compost*. It was in many respects similar to the SOIL ASSOCIATION's journal *MOTHER EARTH* in its coverage of issues relating to humus farming, municipal composting and nutrition. Contributors included EVE BALFOUR, Donald P. Hopkins, WESTON A. PRICE and FRANK NEWMAN TURNER. With the support of LOUISE HOWARD, it was taken over by the SCOTTISH SOIL AND HEALTH GROUP in 1948 and became *HEALTH AND THE SOIL*.

SOIL ASSOCIATION The Association arose from the interest aroused by EVE BALFOUR's book *The Living Soil* (1943). With the support of DR GEORGE SCOTT WILLIAMSON and FRIEND SYKES a founders' meeting was held in June 1945 and an inaugural meeting followed in May 1946. I refer readers to Lockeretz 2007, pp. 187–200, for a brief history of the Association, which, whatever criticisms may have been directed at it, has to be considered the single most significant organization in the post-war history of the organic movement. Even in the 1970s, when the Association was at a low ebb, the SEVENTIES GENERATION felt obliged to join it because it was the one non-esoteric body which was interested in wholefoods and ecological cultivation, and had a tradition of practical experience and writings which could be drawn upon. Association members were responsible for establishing the McCARRISON SOCIETY, the ORGANIC GROWERS' ASSOCIATION and BRITISH ORGANIC FARMERS. The Association succeeded, where NEWMAN TURNER had failed, in establishing a symbol for organic produce. And even its great rival ORGANIC FARMERS AND GROWERS grew out of a Soil Association body. A full history of the Association has yet to be written, but it will need to be critical and independent rather than official if it is to have any value.

SOIL ASSOCIATION QUARTERLY REVIEW This was the title of the SOIL ASSOCIATION's journal from 1975 until 1987. Following the departure of ROBERT WALLER in 1972, DAVID STICKLAND became editor of the Association's journal, radically changing its format and producing it from April 1973 as a monthly simply titled *The Soil Association*. This proved unsustainable, and it was then produced at two-month intervals until the summer of 1975. From September that year onwards it became the *SAQR,* edited successively by A.J. Benn, Joy Griffith-Jones and Kate Walters. Following what was in effect a coup by the SEVENTIES GENERATION in 1982, the journal was re-designed and produced by an editorial group whose members included Francis Blake, GINNY MAYALL, DR DAVID HODGES, Angus Marland and PETER SEGGER. Later, GRAHAM HARVEY was appointed editorial co-ordinator. The journal became both more garish in appearance and more informative, wide-ranging and combative in content.

SPAN The title was a rather awkward acronym of 'Soil-Plant-Animal-Man'. This monthly newsletter of the SOIL ASSOCIATION, edited by MICHAEL ALLABY, ran from March 1967 until February 1973, when DAVID STICKLAND closed it down. Its aim was to provide more information on current affairs and membership news than could be offered by the quarterly *MOTHER EARTH,* and it was hoped that its more informal design — perhaps reminiscent of a university Students' Union paper — would appeal to a younger readership. Despite the amateurishness of its appearance, *Span* was a very informative publication in times when content was more important than design values.

STAR AND FURROW Journal of the BIODYNAMIC AGRICULTURAL ASSOCIATION, its title reflecting the belief in the interplay of cosmic influences and soil fertility at the heart of BIODYNAMIC CULTIVATION. From 1928 to 1931 the Anthroposophical Agricultural Association issued a duplicated newsletter which then became a small printed journal called *Notes and Correspondence.* In 1953 this in turn became *Star and Furrow,* its aim being 'to encourage the free exchange of ideas and experience among those who work with, or are interested in, the agricultural teachings of [RUDOLF STEINER]'. The journal featured writings by Katherine Castelliz, DAVID CLEMENT, Wendy Cook, George Corrin, JOHN DAVY, RUTH HARRISON, ROBERT HART, Dr Carl Mier, JOHN SOPER, ROBERT WALLER and MAURICE WOOD, and for many years kept a close eye on what was happening in the world of conventional agriculture in its 'Signs of the Times' column.

UK REGISTER OF ORGANIC FOOD STANDARDS (UKROFS) A sign that the organic market was expanding and that government was beginning to take organic food and farming seriously, was set up in July 1987 and held its first board meeting in November that year. The Standards were to be supervised by a board of representatives from the organic movement, multiple retailers and consumer interests. Their tasks were to set minimum production standards tied to a UK organic logo; to consider the registration of existing organic sector standards and inspection schemes; to establish a register of producers entitled to use the logo, and to arrange inspection of farms registering without being linked to existing schemes. The SOIL ASSOCIATION at first regarded this development as the most significant the movement had ever faced and as potentially disastrous unless the voices of those with deep understanding of organic principles prevailed; the Association's symbol might have been put at risk. By spring of 1988 it saw this fear as groundless. Prof. Colin Spedding, who chaired the board, and his colleagues, who included PATRICK HOLDEN, Julian Rose and LAWRENCE WOODWARD, were clear that existing schemes and symbols would be respected. Woodward felt that the movement had to enter the mainstream and work with government in order to accept a wider responsibility for bringing about a better way of living. But there was always a danger that the movement would be swamped, taken over and controlled by officialdom, and by 1993 Woodward felt that this had happened, with the main stream changing the organic movement more than the movement was changing the main stream. Francis Blake of the Soil Association, however, considered UKROFS necessary, as self-regulation had not worked, and the Association had ensured that UKROFS adopted its standards.

VOLE Environmentalist magazine edited by *Guardian* journalist and writer Richard Boston, which ran from 1977 until 1981. Although playful like *SEED* magazine rather than sombre like *THE ECOLOGIST*, *Vole* could not really be described as a product of the counter-culture, containing as it did a strong element of specifically English radical traditionalism. Like the wider organic movement, it favoured the local and small-scale, de-centralization, alternative technology and craftsmanship, and supported conservation, self-sufficiency and responsible use of resources. It also anticipated the 'slow' movement. *Vole* had a sense of organic history, featuring articles on Distributism, H.J. MASSINGHAM, Henry George and Henry Williamson. Contributors included MICHAEL ALLABY, LAWRENCE HILLS, JOHN SEYMOUR, Marion Shoard and Colin Ward, and the organic growers Douglas and Penny Blair were featured.

WHOLE FOOD SOCIETY (see **PRODUCER CONSUMER WHOLE FOOD SOCIETY LTD**)

WHOLEFOOD According to MARY LANGMAN, who helped LILIAN SCHOFIELD run the Wholefood shop and was one of its Directors, Wholefood was 'the first retail outlet in London for the produce of organic agriculture'. This may be an exaggerated claim, given the existence of Roy Wilson's Iceni outlet in 1930s, but there is no doubt that Wholefood played an important part in the organic movement's post-war history. It stemmed from C. DONALD WILSON's concern that those wishing to consume organic or whole food did not know where in London they could buy it. Together with Earl Kitchener, and fully supported by the SOIL ASSOCIATION, Wilson established the Organic Food Society in 1959, whose retail outlet, Wholefood, in Baker Street, opened the following year. Its aims were twofold: to bring organic producers into contact with buyers, as NEWMAN TURNER's WHOLE FOOD SOCIETY had done ten years earlier; and to demonstrate that if organic produce was offered to the public then demand would increase and production would prove profitable. The shop took some time to break even, but attracted media attention and provided high-quality produce sourced from the UK and overseas. Mick Stuart, son of CDR ROBERT STUART, worked there for a time, and CRAIG and GREGORY SAMS had a close business relationship with Lilian Schofield. PATRICK HOLDEN found a market for his produce there. Yehudi Menuhin was a strong supporter of the enterprise. The shop moved to nearby Paddington Street in 1983 and a charity, the Wholefood Trust Ltd, was established, whose aim was 'to carry on research in the fields of nutrition and organic growing, with diffusion of information to the public, and to research into methods of storing, processing and distributing foodstuffs in ways that maintain maximum nutritive value'. Mary Langman, who was still actively involved in the work of the shop in the early 1990s, also established a bookshop there, stocking many classics of the organic canon.

WORKING WEEKENDS ON ORGANIC FARMS (WWOOF) The brainchild of SUSAN COPPARD, who founded it in the early 1970s. Its aim was to enable would-be organic farmers or growers to gain practical experience and instruction, or for urbanites like herself to undertake part-time farm work, preferably in the company of like-minded people. She contacted JOHN DAVY at Emerson College, and he arranged for the farm managers to give her and other volunteers a trial weekend. The idea was a success and Coppard was interviewed by *SEED* magazine. She produced a newsletter and a 'fix-it-yourself' list of places welcoming 'WWOOFers'. Veteran organicist RALPH COWARD participated in the scheme. By 1990 about 120 farms or smallholdings were helping cater for around a thousand members.

Appendix C: List of Recorded Interviews

(Capital letters indicate an entry in Appendix A or B)

MICHAEL ALLABY Summer 2000 (interview for BBC radio programme "Organic Roots"). 28.2.2006 (by telephone).

Christopher Badenoch 5.10.2005 (Port Ellen, Islay) Son of A.G. BADENOCH.

Anne Baillie 5/6.2.2007 (Putney, London) Daughter of ROBERT WALLER.

Angela Bates 3.3.2007 (North Witham, Lincs.) Vice-President of the SOIL ASSOCIATION, friend of EVE BALFOUR and director of animal feeds firm Vitrition. Had close links with IFOAM.

Prof. A.J. (Joe) Biddlestone 21.10.2008 (Birmingham) Scientist at Birmingham University who worked with DR KEN GRAY on municipal composting.

Heda Borton 22.5.2006 (Speldhurst, Kent) Daughter of CDR. R.L. STUART and friend of DR INNES PEARSE.

Brenda Brayne 23.10.2002 (Ringwood, Hants.) Daughter of DR LIONEL PICTON.

ALAN BROCKMAN 10.5.2006 (Chartham, Kent)

Pauline Bulcock 11.9.2006 (Ringwood, Hants.) Worked at Haughley in the 1940s.

Jeff Bull 14.4.2005 (London) Active in Epsom Group of the SOIL ASSOCIATION and worked at Haughley in the mid-1980s.

DR NORMAN BURMAN 26.6.2006 (Surbiton, Surrey)

APPENDIX C: LIST OF RECORDED INTERVIEWS

John Butler 4/5.11.2008 (Bakewell, Derbys.) Fenland smallholder who was on SOIL ASSOCIATION Council and wrote regularly for *SEED* magazine.

Peggy Campbell 2.12.2008 (Cambourne, Cambs.) Widow of Douglas Campbell.

SUSAN COPPARD 25.4.2006 (Bradford-on-Avon, Wilts.)

Arthur Darlington 28.6.2006 (by telephone) Organic farmer near Aberystwyth.

Jan and Tim Deane 13.9.2006 (Christow, Devon) Organic growers; pioneers of vegetable box scheme.

DR ANTHONY DEAVIN 6.4.2005 (Ewell, Surrey)

Rosemary Fost 24.3.2006 (Lewes, Sussex) Actively involved with PIONEER HEALTH CENTRE, SOIL ASSOCIATION and National Childbirth Trust.

David Frost 9.6.2005 (Llanrhystud, West Wales) Organic grower and prominent member of the SEVENTIES GENERATION.

ALAN and JACKIE GEAR 10.10.2006 (Snettisham, Norfolk)

Andy Goldring 24.11.2008 (by telephone). Leading figure in PERMACULTURE.

David Gordon 21.2.2006 (Bristol) Grassland farmer and pioneer of city farms.

DR KEN GRAY 21.10.2008 (Birmingham)

Vivian Griffiths 22.10.2008 (Stourbridge, Worcs.) Leading figure in BIODYNAMIC AGRICULTURAL ASSOCIATION.

Simon Harris 15.9.2006 (Chepstow, Gwent) Organic farmer; SOIL ASSOCIATION Council member.

GRAHAM HARVEY 28.6.2000 (nr. Bridgwater, Somerset)
Meg Haver 25.7.2006 (by telephone) Grand-daughter of Fred Birks, long-term organic gardener.

GILES HERON 1.3.2007 (Glaisdale, N. Yorks.)

MARY HERON 27/28.2.2007 (Glaisdale, N. Yorks.)

DR DAVID HODGES 17.6.2008 (Ashford, Kent)

PATRICK HOLDEN 26.4.2006 (Bristol)

Charlotte Hollins 26.4.2005 (Market Drayton, Salop.) Daughter of ARTHUR HOLLINS.

Connie Hollins 26.4.2005 (Market Drayton, Salop.) Widow of ARTHUR HOLLINS.

John E. Hosking 9.5.2006 (Charing, Kent) Son of J. E. Hosking of the KINSHIP IN HUSBANDRY.

"Mac" Jaeger 30.6.2006 (Chichester, Sussex) Knew Prof. Patrick Geddes and DR KENNETH BARLOW.

Nick Jones 9.12.2009 (by telephone) Miller in Cumbria, using organic and biodynamic grain.

DR NIC LAMPKIN 10.6.2005 (Aberystwyth)

MARY LANGMAN 28.6.2000 (North Cadbury, Somerset)

Peter Lanyon 17.9.2008 (by telephone) Worked at Haughley for a short period in the mid-1970s.

Constance Leigh 14.7.2006 (by telephone) Long-term SOIL ASSOCIATION member.

Riccardo Ling 28.3.2006 (Sydenham, London) Long-term SOIL ASSOCIATION member and organic farmer.

Hans Lobstein 9.2.2005 (Brighton, Sussex) Knew DR GEORGE SCOTT WILLIAMSON and DR INNES PEARSE; father of Tim Lobstein of the Food Commission.

Clifton Lovegrove 11.9.2006 (Ringwood, Hants.) Long-term supporter of the organic movement.

APPENDIX C: LIST OF RECORDED INTERVIEWS

DR PETER MANSFIELD 6/7.11.2008 (Newark, Notts.)

GINNY MAYALL 19.11.2008 (Harmer Hill, Salop.)

Richard Mayall 19.11.2008 (Harmer Hill, Salop.) Organic farmer; son of SAM MAYALL and father of GINNY MAYALL.

GEORGE McROBIE 21.5.2008 (Ealing, London)

CHARLOTTE MITCHELL 12.6.2006 (by telephone)

Dennis Nightingale-Smith 30.4.2008 (Malvern, Worcs.) Long-term member of SOIL ASSOCIATION and HDRA; founder of the Organic Living Association.

Lord Northbourne 30.10.2007 (House of Lords, London) Son of the organic pioneer Lord Northbourne.

Steven Nutt 12.9.2008 (by telephone) Leading figure in PERMACULTURE.

Charles Peers 23.10.2008 (Great Milton, Oxon.) Leading light in ORGANIC FARMERS AND GROWERS, and friend of DAVID STICKLAND.

Joan Pepper 13.6.2006 (Peppard Common, Oxon.) Widow of Allan Pepper, long-term activist on behalf of PIONEER HEALTH CENTRE.

Joanna Ray 2.2.2007 (Corhampton, Hants.) Daughter of DR KENNETH BARLOW.

Tony Reid 11.7.2007 (Shere, Surrey) Long-term organic farmer.

Marcus Ridsdill-Smith 23.6.2006 (by telephone) Long-term organic farmer, active in ORGANIC FARMERS AND GROWERS.

Julian Rose 2.6.2000 (Whitchurch-on-Thames, Oxon.) Organic farmer; leading figure in SEVENTIES GENERATION.

Rachel Rowlands 7.6.2005 (Borth, West Wales) Daughter of DINAH WILLIAMS and founder of organic brand Rachel's Dairy.

Siegfried Rudel 18.3.2005 (Forest Row, Sussex) Long-term member of the BIODYNAMIC movement.

Deidre and Michael Rust 19.6.2008 (Hastingleigh, Kent) Organic farmers in Leicestershire; took over Cdr Noel Findlay's farm in Kent. Michael Rust was a director of ORGANIC FARMERS AND GROWERS.

CRAIG SAMS 19.7.2005 (Hastings, Sussex)

GREGORY SAMS 4.11.2005 (North London)

PETER SEGGER 9.6.2005 (Cilcennin, West Wales)

Graham Shepperd 14.9.2006 (Wiveliscombe, Somerset) Organic farmer; first SOIL ASSOCIATION Standards Inspector.

Sir Colin Spedding 21.4.2008 (Hurst, Berks.) Noted agricultural scientist who in the 1980s chaired the debates which established the UKROFS organic standards.

Bill Starling 11.10.2006 (Tivetshall St. Mary, Norfolk) Authority on organic grain trading; Deputy-Chairman of BRITISH ORGANIC FARMERS.

DR VICTOR STEWART 8.6.2005 (Bow Street, West Wales)

DAVID STICKLAND 8.11.2007 (Swaffham, Norfolk)
Mick Stuart 5.10.2005 (Port Ellen, Islay) Son of CDR R.L. STUART.

Linda Theophilus 3.12.2008 (Sudbury, Suffolk) As Linda Girling, she worked in the SOIL ASSOCIATION's editorial department at Haughley for MICHAEL ALLABY.

Richard Thompson 15.12.2009 (by telephone) Organic farmer in East Yorkshire; director of ORGANIC FARMERS AND GROWERS; son of Michael Thompson, SOIL ASSOCIATION Council member in the 1950s/60s.

Iain Tolhurst 20.6.2007 (Whitchurch-on-Thames, Oxon.) Leading figure in SEVENTIES GENERATION and ORGANIC GROWERS' ASSOCIATION; leading organic producer.

Henrietta Trotter 27.6.2006 (Marlborough, Wilts.) Widow of Douglas Trotter. Both she and her late husband had worked at the PIONEER

APPENDIX C: LIST OF RECORDED INTERVIEWS

HEALTH CENTRE in the 1940s and continued to support the work of the Pioneer Health Centre Ltd

Roger Newman Turner 17.7.2006 (Letchworth, Herts.) Son of FRANK NEWMAN TURNER.

Carolyn Wacher 10.6.2005 (Aberystwyth) Organic grower; prominent member of the SEVENTIES GENERATION.

John Weller 13.10.2006 (Bildeston, Suffolk) Agricultural architect and historian who worked for the SOIL ASSOCIATION at Haughley in the 1960s.

John Wheals 12.10.2006 (Bentley, Suffolk) SOIL ASSOCIATION Treasurer in 1970s.

Patrick Whitefield 27.10.2009 (by telephone) Leading figure in PERMACULTURE.

DINAH WILLIAMS 7/8.6.2005 (Borth, West Wales)

Marion Wilson 4.12.2008 (Haughley, Suffolk) Daughter-in-law of C. DONALD WILSON.

LAWRENCE WOODWARD 28.7.2006 (Hamstead Marshall, Berks.)

DR WALTER YELLOWLEES Summer 2000 (by radio link, for BBC Radio 4 Programme *Organic Roots*).

RICHARD YOUNG 29.4.2008 (Snowshill, Worcs.).

Endnotes

Introduction

1. Conford 2008.
2. See for instance the 'Annual List of Publications on Agrarian History, 2007' in *Agricultural History Review* 57/1 (2009), pp. 109–23.
3. Matthew Reed, review of *The Origins of the Organic Movement* in *Environmental Politics* Vol.11, No.1, Spring 2002, pp. 214–15. This is just one of various criticisms which Reed has levelled at me: see Reed 2003, pp. 35–39, 42, 88–89, 92. Judt 2005, p. 399. 'The Men of Old', in Palgrave 1938, p. 356.
4. Reed 2003, pp. 43f.
5. Rhiannon Harries, 'His Dark Materials', *Independent on Sunday Magazine* 27.9.2009, p. 16.
6. On the morning that I wrote this paragraph, an interview with Patrick Holden appeared in the Saturday magazine of the *Financial Times* in which arable farmer Oliver Walston was quoted as considering the organic movement 'no more than a religion', and the interviewer fatuously hinted that Holden's supposed 'fundamentalism' about organic farming might have been due to the fact that one of his four great-grandfathers was a bishop. Michael Wale, 'Back to Earth', *Financial Times* Saturday magazine 10.10.2009, pp. 16–25.

Prologue

1. Burchardt and Conford 2008, p. 34.
2. CCC 1945, p. 1.
3. CCC 1945, p. 8, 9.
4. CCC 1945, pp. 15, 23, 26.
5. CCC 1945, p. 10, 1.
6. Portsmouth and Walston 1947, pp. 14, 17, 15, 18, 16, 8, 4.
7. Blackburn 1949, p. 138.
8. The author was involved in the making of these programmes, and a copy of the letter from Yellowlees is in his possession. On Teviot, see Griffiths 1998; on the close connection between the organic movement and financial reform, see Conford 2002.

Chapter 1. The Context: Agricultural Efficiency and Industrial Food

1. Harvey 1997, p. 23.
2. Williams 1965, p. 158.
3. Stapledon 1946, p. 46, 36.
4. Foreman 1989, p. 49. Williams 1965, p. 155.
5. Bateson 1946, p. viii. Astor and Rowntree 1946, p. 14. Stapledon 1946, p. 94, 96.
6. Self, P. and Storing, H.J. 1961, p. 116. Imperial Chemical Industries 1957, p. 11, 38, 11.
7. Williams 1960, p. 176, 179, 181. Body 1982, p. 2
8. Holderness 1985, p. 15. Williams 1965, p. 151, 179. McAllister 1945, p. 26.
9. McAllister 1945, p. 33.
10. Martin 2000, pp. 130-31. Grigg 1987, p. 185, 187.
11. Wormell 1978, p. 360. Martin 2000, pp. 83-85. Dexter and Barber 1961, p. 243. Donaldson 1972, pp. 151-52, 150-51.
12. Body 1982, p. 13. Carter and Stansfield 1994, p. 97. Grigg 1989, p. 119.
13. Rickard 2000, p. 12.
14. French 1969, p. 89. Burchardt 2002, p. 152. Beresford 1975, p. 190.

Shoard 1980, p. 16.
15. Martin 2000, p. 106. Blaxter and Robertson 1995, pp. 56-57. Holderness 1985, p. 114. Grigg 1989, p. 162.
16. Shoard 1980, p. 15, 21.
17. Harrison 1964, p. 1. Brambell 1965. Martin 2000, pp. 126-27.
18. Martin 2000, pp. 120-25.
19. Astor and Rowntree 1946, p. 134, 64.
20. *FW* 10.8.45, p. 33. *FJ* 26.4.44, p. 200; 24.10.45, p. 634; 26.9.45, p. 563; 10.10.45, p. 596; 30.8.44, p. 416.
21. *FJ* 14.8.46, p. 525; 8.2.50, p. 71.
22. Seddon 1989, p. 22. Cooke 1981, pp. 186-87. Conford 2001, p. 42. Body 1991, p. 39. Harvey 1997, p. 45.
23. Grigg 1989, p. 74, 75.
24. Body 1987, p. 74. Martin 2000, p. 102.
25. Harvey 1997, pp. 42-43. Blaxter and Robertson 1995, p. 115.
26. Blaxter and Robertson 1995, pp. 34–35.
27. Blaxter and Robertson 1995, p. 221.
28. Blaxter and Robertson 1995, p. 274.
29. Blaxter and Robertson 1995, p. 37, 268.
30. Body 1982, p. 124; 1987, p. 30.
31. Blaxter and Robertson 1995, p. 24, 266.
32. See Street 1954, pp. 11–21.
33. Marr 2008, p. xxi, 361–62.
34. Hennessy 2007, p. 9.
35. *ME* April 1953, p. 35. Pyke 1952, pp. 7–8, 11.
36. Lang and Heasman 2004, pp. 18-20, 29, 290, 29.
37. Huxley 1965, p. 35.
38. Pyke 1952, p. viii. Griggs 1986, p. 72. Pollan 2004, p. 36. Mackarness 1980, p. 53.
39. Bicknell 1960, p. 49.
40. Bicknell 1960, p. 15. Geoffrey Cannon tells the story of additives in detail in Cannon 1988, pp. 147–220.
41. Griggs 1986, p. 257.
42. Cannon 1988, pp. 234–38.
43. Mount 1979, p. 9. For the most influential account of the harm done by sugar, see Yudkin 1972.
44. Mount 1979, p. 20. Mackarness 1980, pp. 66–67. Cannon 1988, p. 78. Stitt 1998, p. 207.
45. Pyke 1952, p. 18, 21.
46. Massingham 1945a, pp. 32–33. Girardet 1976, p. 60.
47. Lang and Heasman 2004, p. 237.
48. Hardyment 1995, p. 123, 38.
49. Lang and Heasman 2004, p. 209.
50. Lang and Heasman 2004, p. 31.

Chapter 2. The Organic Alternative: Farming

1. Wookey 1987, p. 42. On Oriental agriculture, see King 1949.
2. Lord Northbourne to Ned Halley, 13.2.1982; photocopy of letter in author's possession. The *Somerset County Herald* of 9.4.1938 (p. 15) carried an item headed 'Organic Methods of Husbandry: Conference at Springhead, Fontmell Magna'. Private information from Will Best, Harper Adams College, 7.1.2009. O'Riordan and Cobb 1996. Balfour 1977. Waddell 1977. Kiley-Worthington 1993. Kiley-Worthington and Rendle 1984. Lampkin 1990, pp. 2–6.
3. Sykes 1951, p. 43, 44.
4. *ME* July 1958, p. 171, 172, 173.
5. *JSA*, Oct. 1972, p. 115, 108.
6. Vine and Bateman 1982, p. 66.
7. Widdowson 1987, p. viii, 8.
8. *NFG* Spring 1988, p. 44. Lampkin 1990, p. 2, 5, 4.
9. Woodward 2002, p. 115.
10. Woodward 2002, pp. 116–17.
11. Lorimer 2003, p. 86.
12. Conford 2001, pp. 65–80, 80. Seifert 1962, p. 105. On Seifert's Nazi connections, see Blackbourn 2007, pp. 285–87, 329–30. In response to an article which Johann Hari wrote for *The Independent* (5.1.2006), in which he attributed the existence of the British organic movement to Steiner and failed to mention Howard, I wrote to provide Hari with several basic pieces of knowledge about the subject on which he had pontificated. Despite

making it clear that I was not a disciple of Steiner, I received the following reply, which I give in its entirety. 'Dear Mr Conford, Thanks for your letter. If you think a man who practiced [sic] Zodiac farming is "a great scientist" [a phrase I did not in fact use], we'll have to agree to differ. Yours sincerely, Johann Hari.' Such is the level of argument deployed by a prominent contemporary 'rationalist'. Lord Taverne (2005, p. 63) similarly identifies Steiner as the key influence on the organic movement in Britain. In 2002, the more historically aware Dave Wood attacked Howard on behalf of the Center for Global Food Issues in a piece entitled 'One Hand Clapping: Organic Farming in India': http://www.cgfi.org/2002/12/12/one-hand-clapping-organic-farming-in-india/ (accessed 14.12.2009).

13. Interview with Michael Rust, 19.6.2008. Griffiths 1999, p. 3. Since Waller was a poet, we can take it that his comment was intended as praise. Ellis' views were reported in *ME* April 1957, p. 900. *SA* Nov. 1974, p. 20.

14. According to Jean Westlake, her father studied Anthroposophy in depth. She herself became active in the BAA in the mid-1980s. Westlake 2000, p. 200. Hills 1989, p. 60. See Hills' letter about biodynamics in *JSA* April 1970, p. 128. Mier's article 'Organic and Biodynamic: Towards an Agricultural Philosophy', was published in *SF* No.6, Spring 1956, pp. 1–8, and referred to in *ME* July 1956, p. 633.

15. *SF* No.2, Spring 1954, inside front cover; No.6, Spring 1956, pp. 8–10, 30–31; No.11, Autumn 1958, p. 21; No.12, Spring 1959, 30–31; No.19, Autumn 1962, pp. 45–47; No. 20, Spring 1963, pp. 32–35; No. 23, Autumn 1964, pp. 30–31; No. 24, Spring 1965, pp. 5–11; No. 25, Autumn 1965, pp. 9–11; No. 30, Spring 1968, pp. 31–33. No. 38, Spring 1972, pp. 1–4, 28–29. *Just Consequences* contained essays by, among others, Surgeon-Captain T.L. Cleave, Laurence Knights, Kenneth Mellanby, Innes Pearse and Hugh Sinclair: Waller 1971.

16. Private communications from Mrs Helen Zipperlen. Heron 2009, pp. 67–73. Interview with Vivian Griffiths, 22.10.2009.

17. The quotation is from notes sent by Helen Zipperlen to the author, undated but from 2008.

18. Helen Zipperlen, letter to the author, 15.11.2008. Howard 1940, p. ix.

19. *SA* Nov. 1973, p. 11. *SF* No.18, Spring 1962, pp. 26–27; No.21, Autumn 1963, p. 26. Interview with Vivian Griffiths, 22.10.2008. See Soper's articles on the Steiner course in *SF* No.29, Autumn 1967, pp. 5–9; No. 30, Spring 1968, pp. 6–12; No. 31, Autumn 1968, pp. 16–20; No. 32, Spring 1969, pp. 18–25, and others in the 1970s.

20. A pencil annotation on the carbon copy of the letter indicates that Wilson gave no answer and made no further reference to the matter. Correspondence in the author's possession: Robert Stuart to C. D. Wilson, 8.2.1955; Wilson to Stuart 11.3.1955; Stuart to Wilson, 19.3.1955. David Clement recalled Duffy as a difficult character in a telephone conversation with the author, 22.8.2002.

21. Griffiths 1999, p. 2. *SF* No.13, Autumn 1959, pp. 1–4. *ME* Oct. 1959, pp. 717–21. David Clement, letter to the author, 30.8.2002.

22. Interview with Alan Brockman, 10.5.2006. Interview with Simon Harris, 15.9.2006.

23. These were Alan Brockman's views as expressed in the interview on 10.5.2006.

24. Waterman 1946, p. 255. On Charles and John Davy at *The Observer*, see Cockett 1991, pp. 157–58 and 167–68. John Davy wrote on the dangers of pesticides three years before Carson's *Silent Spring* was published: *Observer*

4.10.1959.
25. *JSA* July 1970, p. 132. Davy 1985, p. vi. Snow 1959. Waller's copy of Charles Davy's book, now in the author's possession, shows that he had studied it closely. Charles Davy 1961, pp. 121–27. John Davy 1985, pp. 57–58. *JSA* July 1970, p. 132.
26. Barlow and Bunyard 1981, pp. 23–24.
27. *SF* No. 50, Summer 1978, pp. 23–26. *The Organic Grower* No.10, Autumn/Winter 2009, p. 25.
28. Interview with Patrick Holden, 27.4.2006.
29. Burton 2001, 'Democritus to the Reader', p. 107.
30. Interview with Alan Brockman, 10.5.2006. On Steiner education, see Davy 1985, pp. 203–14. Vivian Griffiths (interview 22.10.2008) has described the successful Larchfield community initiative on the outskirts of Middlesborough.
31. See Portsmouth 1965, p. 86 and Louise Howard's comments in *Health and the Soil* Summer 1948, pp. 7–8.
32. *SH* Feb.1946, inside front cover.
33. AHFOH, publication No.1, 1948, p. 1,3,7.
34. *ME* July 1953, pp. 8–9. *AHNS* No.100, Dec. 1964, p. 1.
35. *HS* Summer 1948, p. 10; Summer/Autumn 1950, pp. 27–30
36. Blackburn 1949, p. 141; *F* Summer 1950, p. 11.
37. *F* Summer 1949, p. 4.
38. *Sp* No.1, March 1967, p. 1.
39. *Sp* No.52, June 1971, p. 2. Interview with Riccardo Ling, by telephone, 30.1.2000. Letter from Olive Rose, 10.7.2006.
40. *Sp* No. 64, Dec. 1972, p. 2.
41. *Sp* No.64, Dec. 1972, p. 1.
42. *SAQR* Summer 1983, pp. 1–5; Sept. 1984, p. 15.
43. *LE* April-June 1988, p. 3, 8.
44. *LE* Spring 1993, p. 3.
45. *LE* Autumn 1996, p. 6.
46. Erin Gill, typescript of paper 'Organic Farming and "New Age" Religion', given to the Spring 2008 conference of the British Agricultural History Society at Nottingham University. *SAQR* Winter 1982 supplement, p. 1.
47. *NFG* No.6, Winter/Spring 1984/85, inside front cover.
48. *NFG* No.11, Summer 1986, pp. 12–13.
49. Clunies-Ross and Weisselberg, 1989, p. 6.
50. *NFG* No.44, Autumn 1994, p. 35.
51. Mary Langman, letter to Lawrence Woodward, 12.2.1987. Richard Young, interview, 29.4.2008.
52. *LE* May 1992, p. 4. *NFG* No.47, Summer 1995, p. 34.
53. *PN* No.1, Feb. 1983, p. 1.
54. *NE* Vol.8/No.5, Sept./Oct. 1978, p. 176, 178.
55. *PN* No.9, Feb. 1987, pp. 3–5. Mollison and Holmgren 1978, p. 1, 2. Whitefield interview 27.10.2009. Whitefield 2005, p. 2, 4.
56. Mollison and Holmgren 1978, p. 4, 9, 12, 6. On Marshall, see *PN* No.14, Winter 1988, pp. 6–7, and Simon Pratt's obituary of him in *Permaculture Magazine* No.5, Spring 1994, p. 31.
57. Mollison and Holmgren 1978, pp. 96–97.
58. *PN* No.14, Winter 1988, p. 3; No.1, Feb. 1983, pp. 7–8; No.15, Spring 1989, p. 4; No.24, Autumn 1991, p. 24. Fukuoka 1987, 1989. *PN* No. 3, Summer 1984, pp. 3–5; No.8, Autumn 1986, pp. 6–7; No.16, Midsummer 1989, pp. 34–37, 40–45; No.14, Winter 1988, p. 21. Opposition to the money economy was a central feature of the early organic movement: see Conford 2002.
59. *PN* No.1, Feb. 1983, p. 7, 8; No.3, Summer 1984, p. 3; No.17, Midwinter 1989–90, pp. 8–9. *NFG* No.35, Summer 1992, p. 10; No.49, Winter 1996, p. 26.
60. Whitefield 2004, pp. 10–11, 257.
61. Interviews with Patrick Whitefield (27.10.2009) and Tim and Maddy Harland (29.10.2009). Whitefield 2004, p. 3, 13–37, 36, 5. Penny Strange in *PN* No.1, Feb. 1983, p. 9.
62. *ME* April 1959, pp. 510–18. *PN*

No.5 (n.d. but probably Autumn 1985), pp. 9–11. Hollins 1984, pp. 231–42. *Permaculture Magazine* No.5, Spring 1994, p. 31. *PN* No.5, p. 7. Waller wrote a preface for Gordon's *The Democratic Farm* (Gordon 1996). Also see Gordon's contribution to Conford 2008, pp. 177–84. *PN* No.8, Autumn 1986, pp. 3–4; No.19, Midsummer 1990, pp. 5–6.
63. *PN* No.22, Winter 1991–92, p. 6, 7; No.6, Jan. 1986, pp. 13–15. *Permaculture Magazine* No.7, p. 34.
64. *LE* Autumn 1993, p. 31.
65. *SA* June 1973, p. 2, 3.
66. Interview with Sue Coppard, 25.4.2006. 'A Brief History of WWOOF', 1990.
67. Interview with Nic Lampkin, 10.6.2005. *LE* January 1991, p. 26; *NFG* No. 39/40, Tenth Anniversary Issue, 1993, p. 8. E-mail communication from Nic Lampkin, 10.2.2010.
68. *NFG* No.45, Winter 1995, p. 34.
69. *ME* April 1956, pp. 519–23, April 1961, p. 576. *SA* June 1973, pp. 4–5. *SAQR* December 1979, pp. 2–6.
70. On Del Pelo Pardi, see *ME* April 1956, pp. 507–19; July 1960, p. 244; Oct. 1960, p. 365; April 1965, p. 452. *Sp* May 1968, p. 1.
71. *ME* Summer 1949, pp. 19–20 and April 1962, pp. 199–202. *Sp* No.25, March 1969, p. 2.
72. See for instance Easterbrook's contribution to the 18.7.1959 edition of the *News Chronicle*. On Smarden, see Coleman-Cooke 1965, pp. 120–32 and *ME* April 1964, pp. 87–93.
73. *LE* January 1990, p. 21. *ME* Spring 1947, pp. 23–26.
74. *SAQR* Winter 1983/84, pp. 5–9. Shoard 1980. Body 1982, 1984, 1987, 1991.
75. Iain Tolhurst, a vegan, is undertaking work on stockless systems at Hardwick, west of Reading. On earthworms, see Barrett 1949. Probably the definitive organicist work on seaweed as a fertilizer is Stephenson 1968.
76. *ME* July 1953, p. 7; July 1967, p. 449.
77. Hopkins 1945, p. 235. Woodward interview 28.7.2006. *ME* April 1961, p. 577.
78. *NFG* 48, Autumn 1995, p. 22.

Chapter 3. The Organic Alternative: Gardeners and Growers

1. Hills 1989, p. 165. Conford 1992, p. 14,13. For a detailed and illuminating account of the Gears' work, with lively anecdotes about Lawrence and Cherry Hills and a history of the development of the HDRA from the mid-1970s onwards, Gear 2009 is essential reading.
2. Blackburn 1949, p. 145.
3. Easey 1955, p. 280, 281. *ME* Oct. 1960, p. 344.
4. For Shewell-Cooper's views on the value of the GGA, see *SA* May 1973, p. 11. Interviews with Alan Gear (10.10.2006) and John Wheals (12.10.2006). I am grateful to Matt Adams of the Good Gardeners' Association for information about Dr Shewell-Cooper.
5. Elliot 1943, p. 105. *ME* Oct. 1954, p. 9. *SAQR* June 1980, p. 3 and June 1985, pp,24–25.
6. *ME* April 1957, pp. 845–50. *SAQR* June 1977, pp. 22–23.
7. *HDRA Newsletter* No.2, Nov. 1958, p. 4.
8. Gear 2009 gives a most interesting account of the Gears' work at Bocking and then at Ryton. Interview with Peter Lanyon 17.9.2008.
9. *SAQR* March 1985, p. 22. *LE* Jan. 1989, p. 30.
10. Interview with Alan and Jackie Gear, 10.10.2006. Scott 2006, pp. 48–49.
11. *SAQR* Sept. 1979, p. 6. Hart 1996, p. 23, 22.
12. *OF* March 2008, p. 36.
13 *SAQR* June 1976, p. 18. Butler 2008, p. 28. *S* Vol.5/No.6, p. 26.
14. *The Listener* 14.8.1975, pp. 203–04. Shirley wrote a number of articles for *Seed* magazine under the nom-de-

plume Elizabeth Butler.
15. *S* Vol.4/No.1, p. 22. *SAQR* Spring 1981, p. 13.
16. Interview with Peter Segger, 9.6.2005.
17. Interview with Carolyn Wacher, 10.6.2005. *SAQR* Dec.1986, p. 29.
18. Interview with David Frost, 9.6.2005. Frost 1998.
19. Interview with Iain Tolhurst, 20.6.2007.
20. *OGA Newsletter* April 1981, p. 1; July 1982, p. 1, 2. Interview with Angela Bates, 3.3.2007. For a portrait of Segger by his friend John Humphrys, see *LE* April 2000, pp. 4–5. For Peter Segger's own memories of the OGA, see *The Organic Grower* No.10, Autumn/Winter 2009, pp. 24–27.
21. *NFG* No.8, Autumn 1985, p. 3; No.16, Autumn 1987, p. 3.
22. *NFG* No.17, Winter 1987, p. 3; No.22, Spring 1989, p. 9; No.30, Spring 1991, pp. 13–14.
23. *NFG* No.32, Autumn 1991, p. 5; No.33, Winter 1991-92, p. 5, 10; No.35, Summer 1992, p. 5; No.36, Autumn 1992, p. 5; No.51, Summer 1996, p. 2.
24. *NFG* No.14, Spring 1987, pp. 16–17, 18–19; No.17, Winter 1987, 28–29; No.30, Spring 1991, pp. 38–40.
25. *NFG* No.30, Spring 1991, pp. 29–33; No.43, Summer 1994, p. 5; No.46, Spring 1995, pp. 32–33.
26. Copy of growers' responses supplied to the author by Carolyn Wacher, to whom he is most grateful.

Chapter 4. The Organic Alternative: Health and Nutrition

1. The PHC pamphlet is undated, but it may well have been produced around the time that the Centre opened in 1935. Howard 1940, p. 178.
2. *HL* Jan. 1946, p. 21; Feb. 1946, pp. 57–60.
3. Balfour 1943, p. 20. *ME* April 1958, p. 106. Recording of talk given at Soil Association Week, 1973, by nutritionist Professor Curtis Shears (in the author's possession.)
4. *ME: Introduction to the Soil Association*, p. 2. Howard 1945, p. 26.
5. Balfour 1943, p. 21. Jenks 1959, p. 44, 88. Williamson 1982, p. 2, 4.
6. Pearse 1971, p. 1, 2. Williamson and Pearse 1965, p. 23. Quoted in Westlake 1961, p. 7.
7. Wrench 1938, p. 9, 7.
8. Barlow 1988, p. 11, 12. Barlow's emphasis on the relations between human beings reflects his reading of *Between Man and Man* (1947) by the personalist Jewish philosopher Martin Buber.
9. *HL* Sept. 1962, p. 282; Nov. 1960, p. 402. Westlake 1961, p. ix, 173, 176. Interview with Dr Peter Mansfield, 6/7.11.2008.
10. Balfour 1943, p. 174. Howard 1945, p. 26. Westlake 1961, p. 120.
11. Williamson and Pearse 1951, p. 8, 10.
12. *New Statesman and Society* 29.9.1989, p. 33. *HL* Sept. 1955, p. 325. Stallibrass 1989, p. 109. Pioneer Health Centre Ltd 1993.
13. Barlow n.d.b., p. 69.
14. Letter from Mary Langman to Douglas Trotter, 2.1.1984. Pearse 1979, p. 159.
15. Pioneer Health Centre Ltd 1986, p. 16, 17; Scott-Samuel 1990, back cover.
16. Scott-Samuel 1990, p. 5.
17. Barlow n.d.a., p. 140, 142, 157.
18. Interview with Peter Mansfield, 6–7.11.2008.
19. Griggs 1986, p. 136. Yellowlees 1989, p. 23.
20. Griggs 1986, p. 204. *ME* April 1958, p. 103. On Sinclair, see Ewin 2001.
21. *ME* April 1958, p. 101, 107, 106–07, 106.
22. Griggs 1986, p. 301.
23. *ME* Jan. 1957, p. 747.
24. *ME* April 1967, p. 403, 405. Typed form from the files of Kenneth Barlow.
25. McCarrison Society newsletter, Feb. 1974.
26. McCarrison Society newsletter March 1977, p. 1. Illich 1976. Hall 1974.

27. *NH* Vol.6/No.1, pp. 21–35; Vol.6/No.2, pp. 105–09; Vol.7/No.3, pp. 143–50; Vol.9/No.4, pp. 237–53; Vol.10/No.4, pp. 313–21; Vol.10/No.2, pp. 135–54.
28. Burkitt n.d., p. 11, 12.
29. Burkitt n.d., p. 19.
30. *NH* Vol.7/No.4, pp. 163–69. Heaton n.d., p. [8]. Cleave 1974, p. iv.
31. *ME* Jan. 1958, pp. 71–72; Jan. 1967, pp. 345–51. Yellowlees 2001, p. 63. Cleave 1974, p. iv, v.
32. Cleave 1974, pp. 181–82, 188, 192. Yellowlees 2001, p. 64.
33. *SAQR* Sept. 1981, p. 26.
34. Interview with Christopher Badenoch, 5.10.2005. As a child, from about the age of nine, Christopher Badenoch spent almost every weekend at the Stuarts' estate, and recalls often seeing Eve Balfour there or at his family home in Edinburgh: a very energetic woman, full of ideas, not very encouraging towards children. He found Louise Howard more appealing: quieter, easy-going and slightly eccentric, but always very focused on the cause. On one occasion, they found themselves unlikely partners in a three-legged race. His experiences at the Stuarts' gave him a consuming interest in agriculture, and he subsequently became an ecologist and land-use consultant.
35. Yellowlees 2001, pp. 15–16. Yellowlees letter to author, 24.3.2001. Yellowlees 1989, p. 17, 32.
36. *ME* Oct. 1963, p. 722. Newman Turner's visitors' book is in the possession of his son Roger Newman Turner. Walter Yellowlees, obituary of Gordon Latto, *NH* Vol.13, 1999, pp. 39–40. Private information from Joanna Ray. Interview with Angela Bates, 3.3.2007.
37. *ME: Further Introduction to the Soil Association*, pp. 12–13; Winter 1950–51, p. 13. Hicks 1971, 1975.
38. Interview with Michael Rust, 19.6.2008. *JSA* Oct. 1972, p. 119. Pollan 2008, pp. 94–101.
39. *ME* Oct. 1963, pp. 783–85.
40. *BDJ* 15.9.1944, pp. 178–79; 5.1.1945, pp. 7–8; 18.7.1947, pp. 30–37. *ME* Oct. 1959, p. 698.
41. *ME* July 1953, pp. 49–54; Jan. 1957, p. 751. *SA* June 1974, pp. 9–12. *JSA* March/April 1975, p. 8.
42. *SAQR* Summer 1983, p. 1 *FJ* 31.5.50, p. 338. Griggs 1986, p. 217. *LE* Autumn 1995, pp. 8–9; April 1991, p. 16; Oct. 1989, p. 7; Oct. 1994, p. 10. Booker and North 2007. *HL* July 1960, p. 277; Oct. 1961, p. 397.
43. Hanssen 1984. *Consumer Reports* Sept. 1956, pp. 455–58; Jan. 1958, p. 46. Westlake 1967, pp. 21–29. *Sp* No. 28, June 1969, p. 3. *S* Vol.2/No.7, p. 20. *E* Vol.7/No.3, April 1977, pp. 94–99. *SAQR* Winter 1982, pp. 9–10.
44. *LE* Spring 1993, pp. 14–15. *ME* April 1953, pp. 35–38.
45. Picton 1946, p. 140. *FJ* 27.2.1946, p. 142.
46. *ME* Winter 1950, p. 38. *Sp* No.66, Feb. 1973, p. 6.
47. *NEW* 7.4.49, p. 312. *ME* April 1954, pp. 11–12. Leaflets from among the papers of Cdr Robert Stuart.
48. *SAQR* March 1985, p. 5. *SA* Feb. 1974, p. 13. Grant 1973, p. 85, 86.
49. Grant 1958, p. 5. *ME* July 1952, p. 30. *Sp* 35, Jan. 1970, p. 6.
50. *SA* June 1973, p. 8. *JSA* May/June 1975, pp. 10–11. *SAQR* March 1976, pp. 12–14.
51. *NFG* No.44, Autumn 1994, pp. 28–29.
52. Ashmole 1993, p. 144. Interview with Craig Sams, 19.7.2005.
53. *HL* June 1948, pp. 249–51; Aug. 1948, pp. 70–71; Aug. 1949, pp. 61–65; April 1950, p. 161; June 1950, pp. 251–52; July 1950, pp. 15–18; Jan. 1957, p. 43; Aug. 1948, p. 56.
54. Baker 1979, p. 98.
55. *HL* March 1957, pp. 132–33; Sept. 1961, pp. 334–35, 325, 323; Oct. 1961, p. 389.
56. *Sp* No.18, Aug. 1968, p. 6, 19, 4; No.22, December 1968, p. 12; No. 41, July 1970, p. 10; No. 43, Sept. 1970 p. 6; No.61, July 1972, p. 4;

No.63, Nov.1972, p. 8. *SA* April 1973, p. 25. *Sp* No.66, Feb. 1973, p. 9.
57. *S* Vol.1/No.3, pp. 1–3; Vol.1/No.5, p. 2; Vol.3/No.9, p. 15, 16.
58. *S* Vol.4/No.2, p. 19; Vol.4/No.5, p. 20. Interview with Mary Heron, 28.2./1.3.2007.
59. *SAQR* June 1982, pp. 1–2; March 1985, p. 5; Sept. 1985, p. 30.
60. *LE* Oct. 1990, p. 8; Jan. 1991, pp. 30–31.
61. Levy 1966, pp. 14–15. Leyel 1952, p. 7, 14, 16, 19.
62. Loewenfeld 1964, p. 17. Leyel 1952, p. 16.
63. Loewenfeld 1964, p. 16, 29, 32.
64. Interview with Pauline Bulcock, 11.9.2006. *ME* April 1958, p. 154; Jan. 1962, pp. 75–77; Oct. 1962, pp. 411–13; Oct. 1965, pp. 713–17.
65. *HL* Feb. 1954, p. 85; Feb. 1957, pp. 82–83, Dec. 1957, pp. 497–99; Oct. 1961, p. 389. *SF* No.16, Spring 1961, inside back cover; No.27, pp. 17–20.
66. *Sp* No.47, Jan.1971, p. 8; No.49, March 1971, p. 6, 7; No.58, March 1972, p. 2. *S* Vol.3/No.8, p. 10; Vol.3/No.11, pp. 10–11; Vol.4/No.2, pp. 8–9. *SAQR* Dec.1978, pp. 17–19.
67. *SA* May 1973, p. 18. *SAQR* March 1985, p. 19. *NFG* No.41, Winter 1994, pp. 18–19.
68. Turner 1952, p. 109, 110, 230.
69. Levy 1953, p. 31, 221; http://www.ashtreepublishing.com/bookshop/juliette.php (accessed 10.7.2008); 1953, p. 35, 23, 145.
70. Levy 1952, p. 17; 1966 p. 12, 164; 1973, p. 31, 10. Levy's mentor Professor Szekely was the founder and President of the Essene School of Life, established in Nice in 1928 and by the late 1940s based in California. It was 'dedicated to the modern Essene Renaissance', according to the heading on its notepaper, and it subscribed to the Scottish Soil and Health Group's journal *Health and the Soil*. A letter from the School exists among the papers of Cdr Robert Stuart, in the author's possession.

71. *ME* April 1953, p. 2. Hancock 1952, p. 72, 136, 219.
72. *ME* Spring 1950, p. 36. Hancock 1952, p. 219. *ME* Autumn 1950, p. 36; July 1952, p. 19, 20.
73. *ME* Autumn 1950, p. 35–36.
74. Interviews with Mary Young (27.4.2005) and Richard Young (29.4.2008). *NFG* No.32, Autumn 1991, pp. 25–26.
75. *LE* March 1988, p. 5; Oct. 1989, p. 13; April 1991, pp. 18–19. Young 2005.

Chapter 5. Commerce and Consumers

1. *FG* 6.2.2009, p. 10; 10.4.2009, p. 6. Jenks 1959, p. 54. Letter from R.W. Widdowson to Soil Association group secretary, dated 10.10.1982, in Box 2 of Mary Langman's papers, held by the Soil Association. Angela Bates, letter to Eve Balfour, Sam Mayall and Robert Brighton, 21.11.1978. Interview with Angela Bates, 3.3.2007.
2. *ME* Harvest 1947, p. 3. *SAQR* March 1977, p. 3. *ME* April 1959, pp. 479–83.
3. *WF* Jan.1949, p. 2, 1, 3. *ME* Summer 1951, p. 53.
4. *WF* Aug. 1951, pp. 1–2.
5. *SH* Winter 1947, p. 216.
6. *F* Summer 1955, p. 9.
7. *ME* April 1959, p. 481, 483.
8. Langman typed notes, copy in author's possession. *Evening Standard* 23.6.1960, p. 7. Langman typescript Feb. 1975; copy in author's possession.
9. Obituary by Mary Langman, *LE* October 1989, p. 18. See also the interview with Lilian Schofield in *Seed*, Vol.3/No.6, pp. 4–5.
10. *The Times*, 1.2.1965, p. 15. *ME* Oct. 1966, pp. 282. *Guardian* 7.11.1967, p. 6. *Sp* No. 55, Nov.1971, p. 1.
11. Langman typescript Feb.1975. Langman typescript, possibly from early 1980s, in author's possession.
12. Interviews with Craig Sams (19.7.2005) and Gregory Sams (4.11.2005). Wright and McCrea

2007, p. 109. See also Sams and Fairley 2008, *The Guardian Review* 13.2.1987, and an article on Gregory Sams, 'Biting into the Vegetarian Boom', by Cordell Marks, *Plus Magazine* 24.10.1988, p. 10.
13. Interviews with Gregory Sams (4.11.2005) and David Stickland (8.11.2007).
14. *S* Vol.4/No.9, p. 7; Vol.2/No.4, pp. 10–11; Vol.4/No.9, pp. 14–15; Vol.1/No.1, p. 1; Vol.5/No.9, pp. 18–20; Vol.1/No.6, p. 3; Vol.2/No.3, pp. 9–10; Vol.3/No.4, p. 11; Vol.6/No.3, p. 10; Vol.4/No.5, p. 19; Vol.4/No.3, pp. 10–11; Vol.4/No.10, p. 21, 20.
15. *S* Vol.2/No.2, p. 25; Vol.3/No.2, p. 21; Vol.5/No.8, pp. 18–19; Vol.2/No.1, p. 15. Miller 1941, pp. 31–46. *S* Vol.6/No.2, p. 8.
16. *S* Vol.2/No.5, p. 3.
17. *S* Vol.5/No.6, p. 13. Fukuoka 1987, 1989.
18. *S* Vol.1/No.1, p. 1; Vol.1/No.5, p. 1, 19; Vol.2/No.3, p. 23, 24; Vol.2/No.7, p. 20; Vol.1/No.2, p. 1; Vol.3/No.3, p. 3.
19. *S* Vol.3/No.6, p. 8; Vol.3/No.5, p. 24; Vol.2/No.3, p. 2.
20. *S* Vol.2/No.8, p. 3; Vol.1/No.1, p. 6; Vol.1/No.2, pp. 8–9; Vol.5/No.5, pp. 10–11; Vol.2/No.5, pp. 4–6; Vol.4/No.10, pp. 4–5; Vol.4/No.11, pp. 4–5; Vol.2/No.6, p. 3. *Guardian* 13.2.1987. It would be interesting to know just how many people adopted vegetarian, wholefood or organic diets simply because a small number of stars from the world of pop music had done so.
21. Interview with Angela Bates, 3.3.2007. One prominent member of the Seventies Generation has told me, simply and perhaps with a hint of remorse: 'We hated him'.
22. Interview with David Stickland, 8.11.2007. Copy of Waller's letter in author's possession.
23. *SA* May 1973, p. 3; Jan. 1974, p. 22.
24. Interviews with David Stickland (8.11.2007) and Charles Peers (23.10.2008).
25. On Bill Jordan, see 'How I Made It: Bill Jordan, Founder of Jordans', by Rachel Bridge, *Sunday Times* 29.11.2009.
26. *SAQR* Sept. 1987, p. 3. *LE* July 1988, p. 4; Oct. 1988, pp. 26–27; Jan.1989, p. 5. Interview with David Stickland, 8.11.2007.
27. Interviews with David Stickland (8.11.2007), Richard Thompson (15.12.2009), Charles Peers (23.10.2008), Deidre and Michael Rust (19.6.2008), Marcus Ridsdill-Smith (23.6.2006), Angela Bates (3.3.2007) and David Hodges (8.1.2010).
28. *SA* March 1974, pp. 19–20.
29. *SAQR* June 1978, p. 8.
30. Interview with Graham Shepperd, 14.9.2006.
31. Segger quotation from a draft of his proposals, Winter 1981–82, found among Mary Langman's papers: copy in author's possession. *SAQR* Autumn 1983, p. 29; June 1984, pp. 14–15. Richard Young, private communication. *SAQR* Autumn 1983, p. 30. On the meaning of the Soil Association symbol, see Julian Rose in *SAQR* June 1985, p. 7.
32. *SAQR* Dec. 1986, p. 7; June 1987, p. 3. Interview with Sir Colin Spedding, 21.4.2008. *LE* July 1988, p. 3. See Lockeretz 2007, pp. 152–56 for more on European standards.
33. *NFG* No.8, Autumn 1985, p. 4; No.33, Winter 1991–92, p. 9.
34. *NFG* No.16, Autumn 1987, p. 3; 39/40, Summer 1993 (Tenth Anniversary), p. 30, 29, 10–11, 41.
35. *NFG* No.10, Spring 1986, p. 25; 11, Summer 1986, p. 21, 22; 13, Jan. 1987, p. 4.
36. *NFG* No.12, Autumn 1986, p. 26.
37. *NFG* No.29, Winter 1990/91, p. 11; 30, Spring 1991, p. 13.
38. The limerick concludes: 'They returned from the ride/ With the lady inside,/ And a smile on the face of the tiger.' *LE* Aug. 1994, pp. 19–21. *NFG* No.48, Autumn 1995, pp. 26–27.
39. *NFG* No.49, Winter 1996, p. 34.

40. Interview with Charlotte Mitchell, 12.6.2006.
41. *LE* Spring 1993, pp. 34–35. On the history of Green & Black's see Sams and Fairley 2008.
42. *LE* Oct. 1991, p. 5. For a detailed analysis of the food waste issue at the end of the twenty-first century's first decade, see Stuart 2009.
43. Browning 2000. *LE* July 1988, pp. 28–29; July 1991, p. 7, 19.
44. Interview with Jan and Tim Deane, 13.9.2006. Deane 2001.
45. *OF* No.98, Summer 2008, p. 14. *Country Life* 21.2.2008, pp. 78–81.

Chapter 6. Ecology, Environmentalism and Self-Sufficiency

1. Mairet 1947, p. 88, 85, 94.
2. Baker 1979, p. 97, xii.
3. *ME* first introductory issue, p. 3; Spring 1948, p. 7, 27–28; Autumn 1948, p. 32; Spring 1951, p. 11, 17; April 1952, p. 45; Oct. 1955, pp. 357–60; July 1953, 29–32; July 1954, pp. 7–8. On Darling, see Chisholm 1972, pp. 39-53.
4. *ME* April 1959, pp. 529–33; Oct. 1960, p. 347; Jan. 1961, pp. 475–80.
5. *ME* Jan. 1963, p. 463, 466. *New Scientist* 11.10.1962. *HDRA Newsletter* No.16, June 1963, p. 5,19; No.19, May 1964, p. 6.
6. Carson 1963, p. 47, 243.
7. *Observer Weekend Review* 17.2.1963, p. 21, 22. *ME* July 1964, p. 168.
8. Conford 2008 pp. 162–63. Interview with Michael Allaby, 28.2.2006. Stapledon 1964. *ME* Oct. 1964, pp. 265–74, 283–86, 287–92, 270; April 1966, pp. 107–12.
9. Chisholm 1972, pp. 67–87. Mellanby 1975, p. 49, 32, 49. *SAQR* Sept. 1975, p. 13.
10. *E* Vol.1/No.1, July 1970, p. 47. *New Ecologist* Vol.8/No. 3, May/June 1978, pp. 77–81; Vol.8/No.5, Sept./Oct. 1978, p. 147, 147–54, 179. Interview with Angela Bates, 3.3.2007. *SAQR* Sept. 1981, p. 27.
11. Conford 2008, p. 175, 165. *E* Vol.1/2, Aug. 1970, pp. 20–22.
12. Wood 1985, pp. 118–19, 221. *ME* Oct. 1967, pp. 499–507. Massingham 1947, pp. 229–56. Robert Waller wrote a tribute to Rolt in *Resurgence* No.142, Sept./Oct. 1990, pp. 46–47.
13. George McRobie (interview 21.5.2008) says that Schumacher in fact believed any size could be beautiful, as long as it was appropriate for its purpose. Schumacher 1973, p. 11, 105.
14. North 1976, p. 4. *Sp* No.2, April 1967, p. 6; No.4, June 1967, p. 5. North 1976, pp. 25–28; 109–10.
15. *ME* April 1962, pp. 215–16.
16. *HF* Aug. 1983, pp. 40–42.
17. *HF* Oct. 1984, p. 9. *PSS* June 1982, pp. 14–15; Sept./Oct. 1979, p. 35; Oct. 1981, p. 30. *HF* Dec. 1986, pp. 32–33; Dec. 1983, p. 65; Aug. 1988, p. 29, 31; June 1983, pp. 12–14.
18. Seymour and Girardet 1988, p. 12, 201–02.
19. Reckitt 1945, pp. 70–84, 72, 83. Seymour and Girardet 1988, p. 213, 18.
20. Letter to the author from Sedley Sweeny, 8.2.2006. *ME* Oct. 1964, pp. 320–23. *SAQR* Autumn 1983, pp. 17–19. Sweeny n.d.; letter to author, 8.2.2006.
21. *ME* Oct. 1964, p. 320.
22. An article on Sam Roddick in *The Ecologist* March 2009, p. 55, shows just how far that publication has declined since its inception as a journal uncompromisingly opposed to the consumerist mentality. *E* Vol.1/No.1, July 1970, pp. 3–5.
23. *E* Vol.1/No.1, July 1970, pp. 44–45, 11–13; Vol.1/No.2, Aug.1970, pp. 40–41; Vol.1/No.4, Oct. 1970, pp. 8–11; Vol.1/No.2, Aug.1970, pp. 10–15; Vol.1/No.8, Feb. 1971, pp. 9–12; Vol.1/No.6, Dec.1970, pp. 24–26.
24. Interviews with Michael Allaby, 10.4.2000 and 28.2.2006. *Ecologist* 1972, pp. 33–36. Massingham and Hyams 1953. *Ecologist* 1972, p. 128.
25. *The Ecologist* 1972, pp. 30–31. *E*

Vol.6/No.4, May 1976, pp. 128–31.
26. *NE* Vol.8/No.1, Jan./Feb.1978, pp. 27–29; Vol.9/No.1, Jan./Feb.1979, pp. 18–19. *E* Vol.11/No.2, March/April 1981, pp. 93–99; Vol.9/No.6, Sept./Oct. 1979, pp. 184–87; Vol.11/No.3, May/June 1981, 144–45. *NE* Vol.8/No.3, May/June 1978, pp. 82–83. *E* Vol.9/No.8–9, Nov./Dec.1979, pp. 259–63. *NE* Vol.8/No.4, July/Aug., pp. 121–24. *E* Vol.7/No.3, April 1977, pp. 100–05; Vol.9/No.7, Oct./Nov. 1979, pp. 248–51; Vol.10/No.6–7, July/Aug./Sept. 1980, pp. 186–88.
27. *V* No.1, p. 1, 2; No.4, p. 6; No.5, pp. 32–35; No.11, pp. 20–22; No.12, pp. 27–30; No.1, pp. 54–55; Vol.2/No.1, pp. 42–44;Vol.2/No.2, pp. 43–44; No.2, pp. 19–21.
28. *V* No.1, pp. 18–19; Vol.4/No.5, pp. 10–11; Vol.2/No.7, pp. 28–30; Vol.3/No.2, pp. 24–27; Vol.3/No.5, pp. 26–28; No.3, pp. 36–37; No.10, pp. 26–27; Vol.3/No.2, pp. 32–33; No.9, pp. 8–9; Vol.3/No.13, p. 21; No.3, pp. 40–42; Vol.3/Oct.1980, p. 5, 9; Vol.2/No.2, p. 2.
29. *E* Vol.17/No.1, Jan./Feb. 1987, pp. 21–25, 26–34; Vol.17, No. 4/5, July/Nov. 1987; Vol.15/No.4, pp. 165–76; Vol.17, No.2/3, March/June 1987, pp. 109–15; Vol.19/No.3, May/June 1989, pp. 104–10; Vol.16/No.6, pp. 249–52; Vol.15/No.4, pp. 187–88; Vol.14/No.1, pp. 32–37; Vol.13/No.5, pp. 175–78.
30. *E* Vol.13, No.2/3, pp. 84–87; Vol.12/No.5, pp. 209–16; Vol.16/No.1, pp. 40–41, 29–35; Vol.19/No.3, pp. 94–97; Vol.11/No.4, pp. 169–73; Vol.13, No.2/3, pp. 88–94.
31. *E* Vol.20/No.4, July/Aug. 1990, pp. 122–24.
32. Porritt 1984, p. 244, 179. Bunyard and Morgan-Grenville 1987, p. 74. Porritt 1984, p. 179.
33. *E* Vol.20/No.4, July/Aug. 1990, p. 123. *R* No.129, July/August 1988, pp. 51–52; No.122, May/June 1987, p. 48.
34. Pye-Smith and Hall 1987, third page of Foreword.
35. *LE* July/Sept. 1991, p. 3; May 1992, pp. 8–10; Oct. 1992, pp. 8–9; Nov. 1993, pp. 13–16.
36. Soil Association Annual Report for 1993–94, p. 5.
37. *LE* Oct. 1994, pp. 8, 9.

Chapter 7. The Role of Science

1. Taverne 2005, p. 62.
2. Taverne 2005, p. 64, 79.
3. *ME* Jan.1957, pp. 742–44; Harvest 1947, pp. 9–11; Spring 1948, p. 4, 5; Summer 1949, p. 15; Spring 1950, pp. 33–34.
4. *ME* Autumn 1950, p. 1, 1–7, 5; Winter 1950–51, pp. 23–24.
5. *ME* Spring 1951, pp. 55–57, 57–59.
6. *ME* Jan. 1952, p. 35, 38.
7. *ME* Oct. 1952, p. 39. Griggs 1986, p. 199. *ME* July 1964, pp. 250–51.
8. Balfour 1976, p. 195, 199, 201, 202.
9. Balfour 1976, p. 203, 204, 206–09.
10. *ME* Oct. 1956, p. 665.
11. *JSA* Oct. 1970, p. 230. *ME* Oct. 1956, p. 657, 658.
12. *ME* Oct. 1957, p. 1009, 1010. The October 1957 issue of *ME* contained a long section, 'Present Position of the Haughley Experiment', pp. 1009–26.
13. Soil Association 1962, p. 41–43.
14. *ME* July 1967, pp. 432–37. Allaby, telephone interview 7.3.2005. Interview with Mrs Peggy Campbell, 2.12.2008. Interview with Siegfried Rudel, 18.3.2005. Hills 1989, pp. 207–08.
15. *ME* Oct. 1965, p. 644. Interview with Peggy Campbell, 2.12.2008. *Sp* No.50, April 1971, p. 3. Balfour 1976, p. 259.
16. *Sp* No.56, Dec. 1971, p. 1, 8. *SAQR* Sept. 1976, pp. 4–8.
17. Balfour 2006, p. xxii, xxii–xxiv. Balfour 1976, p. 270, 266–70, 270–73.
18. Interview with Norman Burman, 26.6.2006.
19. *ME* Oct. 1954, pp. 72–73. Interview with Norman Burman, 26.6.2006.
20. This idea is expressed in symbolic form in the short story 'The Wharf' by Walter de la Mare (De la Mare

1974), whose son Richard was a key figure in the organic movement for many years. Less than a year before he died, Howard addressed the Institute of Sewage Purification on the use of sewage sludge in agriculture and horticulture: Howard 1946. On Venables, see *JSA* April 1970, p. 71, and his article 'Angling and Environment', *JSA* Oct. 1968, pp. 191–94.
21. *ME* Jan. 1953, pp. 23–28; Oct. 1954, p. 45. Davies 1961, p. 192. See for instance the *Interim Report of Joint Working Party on Municipal Composting*, published in 1959 by the Institute of Public Cleansing in conjunction with the Soil Association.
22. *Sp* No.1, March 1967, p. 15.
23. Interview with Ken Gray, 21.10.2008.
24. Interview with Professor Biddlestone, 21.10.2008.
25. Dalzell, Gray and Biddlestone 1979. Dalzell, Gray, Biddlestone and Thurairajan 1987. *LE* Jan. 1990, p. 26.
26. G.W. Robinson had addressed his 1937 book *Mother Earth: Letters on Soil* to R.G. Stapledon. Interview with Victor Stewart, 8.6.2005.
27. *SAQR* Sept. 1976, p. 9. 'Project Bryngwyn — Progress Report', typed report among the papers of Dr V.I. Stewart, marked 'Item 6c 4/78'.
28. *SAQR* June 1985, pp. 9–11. 'Once an Opencast Coal Mine — Now a Productive Farm', *Gwlad* No.49, Feb. 2006, pp. 6–7.]
29. Interview with David Hodges, 17.6.2008.
30. Hodges and Arden-Clarke 1986.
31. I am very grateful to Dr Hodges for making available to me his unpublished personal memoir of his involvement in the organic movement.
32. Information drawn from archival material generously provided by Dr Hodges: currently in author's possession, and to be lodged, in due course, with the Museum of English Rural Life, University of Reading.
33. Information in this paragraph is drawn from the archives of Dr Hodges and reflects his own perspective. Interview with David Hodges, 8.1.2010.
34. Interview with Sir Colin Spedding, 21.4.2008. *BAH* Vol.1/No.3, pp. 181–210; Vol.4/No.1, pp. 1–5. Letters from Lampkin and Hill among Hodges' papers. Hodges calculated that the income he gained from royalties, when divided by the amount of time he spent editing *BAH*, came out at about 35 pence an hour. This means that he would have needed to work for 2,857 hours in order to afford a seat at the Soil Association's Feast of Albion in 2008 — had he wanted to attend.
35. Interview with Lawrence Woodward, 28.7.2006. *EFRC Bulletin* Jan. 2002, p. 3. Astor and Rowntree 1946.
36. Elm Farm Research Centre 1981, p. 1.
37. Elm Farm Research Centre 1981, p. 4, 5, 7. Interview with Lawrence Woodward, 28.7.2006.
38. *SAQR* Sept. 1986, p. 22. Clunies-Ross and Weisselberg 1989, p. 45.
39. Stolton and Dudley 1996, p. 15.
40. Elm Farm Research Centre 1981, p. 9. Woodward 2002, p. 114, 117, 118. For more on the relationship between methods of cultivation and food quality, see Lockeretz 2007, pp. 60–61.
41. Deavin interview, 6.4.2005 and personal communications. On Rusch and his associate Hans Muller, see Lockeretz 2007, pp. 242–246. On Muller, see *ME* Oct. 1965, pp. 654–58.
42. Langman letter to Bryn Lewis, 26.1.198[?], in Box 2 of the Langman archives at the Soil Association. Interview with John Wheals (12.10.2006) and with Anthony Deavin (6.4.2005). Kolisko 1978.
43. Interview with Anthony Deavin, 6.4.2005.
44. Mellanby 1975, p. 32. Allaby 1995, pp. 120–43, 125, 127.
45. Allaby 1995, p. 40.

Chapter 8. The Politics of the Organic Movement: An Overview

1. Yellowlees, letter to C. Stott, 5.10.2000. Massingham and Hyams 1953, p. 6. Interview with Mary Langman, 28.6.2000. *F* Autumn 1955, p. 9. *NFG* No.1, Summer 1983, p. 3.
2. *Collins English Dictionary*, third edition 1994. Yellowlees, letter to C. Stott, 5.10.2000.
3. On the far-Right affiliations of the early organic movement, see Griffiths 1983; Matless 1998, 2001; Moore-Colyer 2004; Stone 2004. Twinch 2001, pp. 176–78. Soil Association List of Members for 1951.
4. The literary critic John Middleton Murry refers to Byngham under the pseudonym 'Dillon' in his book *Community Farm*, and was evidently annoyed that Byngham associated his (Murry's) address with 'some petty political newspaper': Murry 1952, p. 101. Conford 2002. *NEW* 4.11.1948, p. 39. *HL* June 1942, pp. 258–59; Jan. 1947, p. 15.
5. Hyams 1951. Massingham and Hyams 1953, p. 6, 70, 6.
6. 'Journeyman Architect' in *Official Architect* May 1944, p. 212. Williamson 1945, p. 132, 133. See Duncan 1949, 1951.
7. *ME* Jan.1952, p. 3. Rural Reconstruction Association 1955, dust-jacket. *ME* July 1957, p. 913.
8. *ME* July 1964, p. 250.
9. Interview with Michael Allaby, 10.4.2000. *ME* July 1964, p. 250, 251; Jan. 1965, p. 440.
10. *ME* April 1964, p. 78.
11. Interview with George McRobie, 21.5.2008. Interview with Michael Allaby, 28.2.2006. Aaronovitch 2002. Ian Cummins wrote an article on Goldsmith entitled 'Prophet of Loss'. I regret that I do not know its provenance, but it is evidently from a Sunday newspaper supplement c.1973/74. See also the review of Goldsmith's *The Stable Society* by Richard Slaughter in the *Ecology Party Newsletter*, March–April 1979, pp. 9–10.
12. *Sp* No.3, May 1967, p. 1; No.4, June 1967, p. 5; No.32, Oct. 1969, p. 8; No.42, Aug. 1970, p. 5.
13. *Sp* No.45, Nov. 1970, p. 1; No.48, Feb. 1971, p. 8. Judt 2009, p. 127. *Sp* No.54, Sept. 1971, p. 6.
14. Conford 2008, p. 166.
15. *S* Vol.2/No.10, p. 3; Vol.6/No.2, pp. 12–13; Vol.4/No.4, p. 21. *HDRA Newsletter* June 1974, p. 3. Woodward interview 28.7.2006.
16. Frost and Wacher 2004. Interview with Angela Bates, 3.3.2007. Interview with Lawrence Woodward, 28.7.2006.
17. On nineteenth-century communal experiments, see Gould 1988 and Marsh 1982. On the communes of the second half of the twentieth century, see Pepper 1991. On the political ambivalence of the counter-culture, see Goldberg 2009, pp. 358–90.
18. Adrian Bell, diary entry for 4.5.1979. I am grateful to Bell's daughter Anthea Bell for letting me see her father's diaries. *V* Vol.2/No.1, Oct. 1978, p. 3, 4; Vol.2/No.3, Dec. 1978, p. 4.
19. *V* Vol.2/No.6, March 1979, pp. 3–4; Vol.2/No.5, Feb.1979, pp. 7–8.
20. OGA/BOF minutes were printed in *GC* March 1985, pp. 10–11; Spring 1990, p. 12; Summer 1992, pp. 4–6; Autumn 1988, pp. 4–5.
21. *NFG* No.1, Summer 1983, p. 3; No.6, Winter–Spring 1985, p. 27; No.18, Spring 1988, p. 8. *GC* Autumn 1989, p. 9.
22. *SAQR* March 1985, p. 4; June 1986, p. 19; Dec. 1986, p. 5.
23. *LE* April 1988, front cover; July 1988, p. 4; Oct. 1988, p. 4.
24. *LE* July 1988, p. 3. This approach reached what was probably its nadir in a *Living Earth* 'editorial' — if that is the right term for a paragraph of 57 words — more than twenty years later, when Tim Young wrote that going shopping was the sure way to achieve 'radical system renewal'. *LE* Summer 2009, p. 3.
25. *LE* May 1994, p. 2.
26. Seymour 1996, p. 217, 146, 220.

27. Peacock 2005, pp. 206–08. Ward 1989. Sams 1997.
28. Noble 2008.

Chapter 9. Earth and Spirit

1. White 1967. Schumacher 1973, pp. 48–56.
2. *ME* Jan. 1964, pp. 41–44; July 1964, p. 250.
3. Conford 2001, pp. 190–209. Balfour 1943, p. 199. Copy of Langman letter, n.d., in author's possession. Erin Gill, private communication. Deavin interview, 6.4.2005.
4. Conford 1998. Conford 2001, pp. 201–04. Reckitt 1945, pp. 70–84. Conford 2001, pp. 203–04. CCC 1950, pp. 16–17. Hutton 1999, pp. 121–22. CCC 1950, pp. 18–19.
5. *F* Spring 1948, p. 9. On Byngham, see Edgell 1992, pp. 182–214 and Hutton 1999, pp. 165–70. *HL* April 1958, pp. 158–59; May 1960, pp. 173–74; July 1960, pp. 254–56.
6. A copy of the essay was given to me by Giles Heron, whose father Tom was a close friend of Mairet and a member of the *NEW*'s editorial board. The essay is undated, but evidence suggests that it was drafted in 1964.
7. Mairet n.d., pp. 1–2, 7, 8, 10, 16.
8. Mairet n.d., p. 19, 21, 22, 25, 26, 29, 30, 34.
9. Mairet n.d., pp. 35–36.
10. Goldsmith 1996, p. 422. Schumacher 1973, pp. 48–56. Spowers 2003, pp. 287–93. Schumacher 1973, p. 47.
11. On 'the limits to growth', see Meadows 1972. Schumacher 1974, p. 7, 6, 7.
12. Schumacher 1974, p. 12, 17, 18, 21, 22.
13. See Heron 1945; n.d. Massingham 1947, pp. 229–56. McRobie 1981, pp. 19–71. Schumacher 1977, p. 135. On the Chandos Group, see Conford 2001, pp. 170–71.
14. Rolt 1988, p. 6. *E* Dec. 2004, p. 66. Peacock 2005, p. 197, 197–200. Seymour 1989, pp. 59–60. Peacock 2005, p. 144. Correspondence between Tim Beaumont and Lord Portsmouth in Wallop archives in the Hampshire County Record Office, Winchester: 15M84/F218.
15. Jenks 1959, pp. 228–29.
16. Veldman 1994, p. 266. Macmurray 1933, pp. 103–21. *E* Vol.10/6–7, July/Aug./Sept. 1980, pp. 224–29. Waller 1973, p. 260. *The Ecologist Quarterly* No.2, Summer 1978, pp. 144–48. Waller, 'Earth and Spirit' in Conford 1992, pp. 205–13. Details of Chirnside Soil Association Weeks from Cdr Stuart's papers, in author's possession.
17. *SAQR* Spring 1981, pp. 1–2. *NEW* 15.10.1942.
18. *E* Vol.14/No.3, pp. 125–33. Hillel 1992, pp. 11–14.
19. Lachman 2007, p. 162, 128–33. *SF* No.22, Spring 1964, p. 21; No. 7, Autumn 1956, p. 13. Lachman 2007, p. 131. Waterman 1946, p. 8. Cook 2006, p. 356. In correspondence with me, Mrs Helen Zipperlen, formerly Helen Murray, has said that although Steiner's religious thought could not strictly be said to belong to the Christian *tradition* (her emphasis), it undoubtedly attaches central significance to the figure of Christ in the spiritual history of the world. Steiner 2006 is the key text here.
20. *ME* April 1961, pp. 611–14. Conford 2001, pp. 76–77. *ME* April 1961, p. 612, 614, 611.
21. *S* Vol.2/No.10, p. 6; Vol.3/No.11, pp. 16–17; Vol.4/No.9, p. 26. On the philosophy of 'as if', see Vaihinger 1935.
22. *S* Vol.3/No.11, p. 17. I have sometimes wondered whether the world of Summer Isle in the 1973 film *The Wicker Man* is intended either as a kind of inverted Findhorn, or as a portrait of the ends to which re-invented pagan cults might logically lead. The film presents a truly 'organic' or 'holistic' feudal community in which there are no dissenters from the laird's murderous nature-based cult, which is posited on his rejection of Christianity. The

austere but resourceful Christian policeman, whose concern for the supposedly missing girl is so ruthlessly exploited, arrives at the island independently, in a sea-plane he pilots himself. This plane may symbolize a capacity for transcendence which enables him to keep a moral perspective on the islanders' surrender of their individual consciences. For more on this sort of issue, see Goldberg 2009, pp. 358-90.

23. *SA* May 1973, p. 23; July 1973, p. 21. *SAQR* March 1979, p. 26. Interview with Anthony Deavin, 6.4.2005.

24. James Webb's book *The Occult Establishment* (1981) is particularly valuable on the links between what he terms 'occult' ideas and the early organic movement. Interview with Lord Northbourne, 30.10.2007. Northbourne 1940, pp. 169–72; 1963, p. 2; 1970. *JSA* April 1969, pp. 321–23; Jan. 1971, pp. 261–64. Northbourne 1970, p. 9, 17, 18.

25. *Sunday Times*, 23.2.1996. 'Welcome to the SES', www.schooleconomicscience.org, accessed 30.12.2008. *SAQR* Dec. 1986, p. 5. Hancock 1952, pp. 223–24. Capra 1982. *SAQR* Sept. 1986, p. 16.

26. *SAQR* Sept. 1986, p. 16.

27. *SAQR* Dec. 1975, p. 13. Massingham 1947, p. 26. Jenks 1950, p. vii; 1959, pp. 228–29. Black 1970, pp. 36–37. Beresford 1975, p. 180, 182–83. Allaby 1972, pp. 11–43. Hicks 1971. Roszak 1995, p. xii, xix. Waller 1962, p. 274.

28. *ME* July 1952, p. 42. Easterbrook 1970, p. 50. Widdowson 1987. Woodward 2002, p. 14. Smuts, quoted by Woodward in Balfour 2006, p. xxi. Spedding 1996, p. 240.

29. Smuts 1927, pp. 120–21, 353.

30. Hopkins 1945, p. 235. *SAQR* Sept. 1986, p. 16.

31. Macmurray 1961, p. 44. On Macmurray's interest in, and reaction to, the philosophy of Smuts, see Costello 2002, p. 126. Mairet 1981, p. 89. For Mairet on Geddes, see Mairet 1957. Julian Rose's article 'A Quest for Wholeness Through Art', *SAQR* Dec. 1978, p. 22, expresses the concerns of this section.

32. Allaby 1995, pp. 28–29. Goldsmith 1996, pp. 403–05, 485, 419, 47.

33. Goldsmith 1996, p. 48, 405, 406, 292.

34. Goldsmith 1996, pp. 420–21, 443, 438.

35. Self, P. and Storing, H.J. 1961, p. 24. See also Blaxter and Robertson 1995, p. 37. Williams 1960, p. 80. Massingham 1945a, p. 8, 9, 10. Massingham and Hyams 1953, p. 58, 132. Goldsmith 1996, pp. 414–21. Jenks 1950, p. 20.

36. *New Ecologist* May/June 1978, pp. 82–83. Body 1987, p. 32. McRobie 1981, p. 86. *SA* March 1975, p. 5. Davy 1961, p. 88.

Select Bibliography

(Except where otherwise indicated, the place of publication is London)

Aaronovitch, David (2002) 'Lunching with the Enemy', *Independent Review*, 2.5.2002.
Albrecht, W.A. (1975) *The Albrecht Papers,* Acres USA, Raytown, Missouri.
Allaby, Michael (1971) *The Eco-Activists,* Charles Knight.
—, (1972) *Who Will Eat?,* Tom Stacey.
—, (1995) *Facing the Future: The Case for Science,* Bloomsbury.
—, and Bunyard, Peter (1980) *The Politics of Self-Sufficiency,* Oxford University Press, Oxford.
Anderson, Jim & Pauline (n.d.) *Land for Biodynamic Farming,* Forest Row.
Ashmole, Anna (1993) *The Organic Values of Agri-Culture,* Ph.D. thesis, University of Edinburgh.
Astor, Viscount & Rowntree, B.S. (1946) *Mixed Farming and Muddled Thinking,* Macdonald.

Badenoch, A.G. (1949) *The Minerals in Plant and Animal Nutrition,* Albert Howard Foundation of Organic Husbandry, Mayfield.
Baker, Richard St. Barbe (1954) *Sahara Challenge,* Lutterworth.
—, (1966) *Sahara Conquest,* Lutterworth.
—, (1979) *My Life, My Trees,* Findhorn Press, Findhorn.
Balfour, E.B. (1943) *The Living Soil,* Faber and Faber.
—, (1976) *The Living Soil and the Haughley Experiment,* Universe Books, New York.
—, (1977) 'Biological Husbandry: An Alternative Agriculture?', *SAQR* Sep.1977, pp. 1–4.
—, (2006) *The Living Soil,* Soil Association Organic Classics, Bristol.
Barlow, Kenneth (1971) *The Discipline of Peace* [second edition], Charles Knight.
—, (1978) *The Law and the Loaf,* Precision Press, Marlow.
—, (1988) *Recognising Health,* Kenneth Barlow/Wholefood.
—, (n.d.a) *The Family Health Club Housing Society (Coventry) Ltd* (unpublished typescript).
—, (n.d.b) *The Diary of a Doomwatcher* (unpublished typescript).
—, & Bunyard, Peter (1981) *Soil, Food and Health in a Changing World,* A B Academic, Berkhamsted.
Barrett, T.J. (1949) *Harnessing the Earthworm,* Faber and Faber.
Bateson, F.W. (1946) *Towards a Socialist Agriculture,* Gollancz.
Bell, Graham (1992a) *The Permaculture Way,* Thorsons, Wellingborough.
—, (1992b) 'The Sustainable Successes of Permaculture' in Conford (1992), pp. 65–72.
Beresford, Tristram (1975) *We Plough the Fields: British Farming Today,* Penguin, Harmondsworth.
Bicknell, Franklin (1960) *Chemicals in Food and in Farm Produce: Their Harmful Effects,* Faber and Faber.
Bircher, Ruth (1961) *Eating Your Way to Health: The Bircher-Benner Approach to Nutrition* (translated and edited by Claire Loewenfeld), Faber and Faber.
Bishop, Beata (1985) *A Time to Heal: Triumph Over Cancer,* Severn House.
Black, John (1970) *The Dominion of Man,* Edinburgh University Press, Edinburgh.
Blackbourn, David (2007) *The Conquest of Nature: Water, Landscape, and the Making of Modern Germany,* Norton.
Blackburn, J.S. (1949) *Organic*

Husbandry: A Symposium, John S. Blackburn, Ben Rhydding.

Blake, Francis (1987) *The Handbook of Organic Husbandry,* Crowood Press, Marlborough.

Blake, Michael (1970) *Down to Earth: Real Principles for Fertiliser Practice,* Crosby Lockwood.

Blaxter, Kenneth & Robertson, Noel (1995) *From Dearth to Plenty,* Cambridge University Press, Cambridge.

Body, Richard (1982) *Agriculture: The Triumph and the Shame,* Temple Smith.

—, (1984) *Farming in the Clouds,* Temple Smith.

—, (1987) *Red or Green for Farmers,* Broad Leys, Saffron Walden.

—, (1991) *Our Food, Our Land,* Rider.

Bonham-Carter, Victor (1952) *The English Village,* Penguin, Harmondsworth.

—, (1971) *The Survival of the English Countryside,* Hodder and Stoughton.

—, (1996) *What Countryman, Sir?,* B-C Press, Milverton.

Booker, Christopher & North, Richard (2007) *Scared to Death,* Continuum.

Bower, Arthur (1969) 'Philosophy out of Practice: Reflections of an Organic Grower', *Journal of the Soil Association* July 1969, pp. 397–405.

Brambell, Prof. R. (1965) *Report on the Welfare of Animals in Intensive Husbandry Systems,* HMSO.

Brander, Michael (2003) *Eve Balfour,* Gleneil Press, Haddington.

Bromfield, Louis (1949) *Malabar Farm,* Cassell.

Browning, Helen (2000) 'The Organic Model — What Relevance Does it Have for the Rest of UK Agriculture?', *Journal of the Royal Agricultural Society of England,* Vol.161, pp. 55–64.

Bruce, Maye E.(1946) *Common-Sense Compost-Making,* Faber and Faber.

Buber, Martin (1947) *Between Man and Man,* Kegan Paul.

Bunyard, Peter and Morgan-Grenville, Fern (1987) *The Green Alternative,* Methuen.

Burchardt, Jeremy (2002) *Paradise Lost: Rural Idyll and Social Change in England Since 1800,* I.B. Tauris.

—, & Conford, Philip (2008) *The Contested Countryside,* I.B. Tauris.

Burkitt, Denis (n.d.) *The Founders of Modern Nutrition: Trowell,* McCarrison Society.

Burton, Robert (2001) *The Anatomy of Melancholy,* New York Review Books, New York.

Butler, John (2008) *Wonders of Spiritual Unfoldment,* Shepheard-Walwyn.

Campbell, G.D. (1996) 'Cleave the Colossus and the History of the "Saccharine Disease" Concept', *Nutrition and Health,* Vol.11, pp. 1–11.

Cannon, Geoffrey (1988) *The Politics of Food,* Century Hutchinson.

Capra, Fritjof (1975) *The Tao of Physics,* Wildwood House.

—, (1982) *The Turning Point,* Wildwood House.

Carrel, Alexis (1935) *Man the Unknown,* Harper & Bros.

Carson, Rachel (1963) *Silent Spring,* Hamish Hamilton.

Carter, E.S. & Stansfield, J.M. (1994) *British Farming: Changing Policies and Production Systems,* Farming Press, Ipswich.

Carter, Vernon Gill & Dale, Tom (1974) *Topsoil and Civilisation,* University of Oklahoma Press, Norman.

Castelliz, Katherine (1980) *Life to the Land,* Lanthorn Press, East Grinstead.

Chase, J.L.H. (1948) *Cloche Gardening,* Faber and Faber.

—, (1952) *Commercial Cloche Gardening,* Faber and Faber.

Chisholm, Anne (1972) *Philosophers of the Earth,* Sidgwick and Jackson.

Cleave, T.L. (1957) *Fat Consumption and Coronary Disease,* John Wright and Sons, Bristol.

—, (1974) *The Saccharine Disease,* John Wright and Sons, Bristol.

—, & Campbell, G.D. (1966) *Diabetes, Coronary Thrombosis, and the Saccharine Disease,* John Wright and Sons, Bristol.

Clunies-Ross, Tracey (1990) *Agricultural Change and the Politics of Organic Farming,* Ph.D. thesis, University of Bath.

—, & Hildyard, Nicholas (1992) *The Politics of Industrial Agriculture,* Earthscan.

—, & Weisselberg, Tim (1989) *Organic Farming: An Option for the Nineties,* BOF/OGA, Bristol.

Cochrane, E. R. (1946) *The Milch Cow in England,* Faber.

Cockett, Richard (1991) *David Astor and The Observer,* Andre Deutsch.

Coleman-Cooke, John (1965) *The Harvest that Kills,* Odhams Books.

Comerford, John (1947) *Health the Unknown,* Hamish Hamilton.

Commoner, Barry (1966) *Science and Survival,* Gollancz.

Conford, Philip (1988) *The Organic Tradition,* Green Books, Hartland.

—, (1992) *A Future for the Land,* Green Books, Hartland.

—, (1998) 'A Forum for Organic Husbandry: The *New English Weekly* and Agricultural Policy, 1939-1949', *Agricultural History Review* Vol.46, pp. 197–210.

—, (2001) *The Origins of the Organic Movement,* Floris Books, Edinburgh.

—, (2002) 'Finance versus Farming: Rural Reconstruction and Economic Reform, 1894–1955', *Rural History* vol.13/2, pp. 225–41.

—, (2008) *The Poet of Ecology,* Norroy Press.

Cook, Wendy E. (2006) *So Farewell Then: The Untold Life of Peter Cook,* HarperCollins.

Cooke, G.W. (1981) *Agricultural Research 1931-1981: A History of the Agricultural Research Council,* Agricultural Research Council.

Coppard, Sue & Mager, Chris (1984) *Directory of Organic Organisations,* WWOOF, Purley.

Corley, Hugh (1957) *Organic Farming,* Faber and Faber.

Costello, John E. (2002) *John Macmurray: A Biography,* Floris Books, Edinburgh.

Council for the Church and Countryside (1946) *Encounter: Agri-Culture or Agri-Industry,* SPCK.

—, (1950) *Man and Nature,* Council for the Church and Countryside.

Cunning, Annie B. & Innes F.R. (1953) *We Are What We Eat,* Salvationist Publishing.

Dalzell, H., Gray, K.R. & Biddlestone, A.J. (1979) *Composting in Tropical Agriculture,* IIBH, Needham Market.

—, & Thurairajan, K. (1987) *Soil Management: Compost Production and Use in Tropical and Subtropical Environments,* FAO, Rome.

Danziger, Renée (1988) *Political Powerlessness: Agricultural Workers in Post-War England,* Manchester University Press.

Darling, Frank Fraser (1955) *West Highland Survey,* Oxford University Press.

—, (1970) *Wilderness and Plenty,* BBC.

Davies, A.G. (1961) *Municipal Composting,* Faber and Faber.

Davy, Charles (1961) *Towards a Third Culture,* Faber and Faber.

Davy, John (1985) *Hope, Evolution and Change,* Hawthorn Press, Stroud.

Deane, Tim (2001) 'A Short History of Northwood Boxes' (unpublished typescript).

De Bunsen, Mary (1960) *Mount Up with Wings,* Hutchinson.

De la Mare, Walter (1974) 'The Wharf', in *Modern Short Stories* (ed. Jim Hunter), Faber and Faber.

Dexter, Keith & Barber, Derek (1961) *Farming for Profits,* Penguin, Harmondsworth.

Dobbs, Geoffrey (1969) 'Concerning the Control, and the Defence, of Environmental Pollution', *Journal of the Soil Association* July 1969, pp. 379–89.

Donaldson, J.G.S. & Frances, in association with Derek Barber (1972)

Farming in Britain Today, Penguin, Harmondsworth.

Douglas, J. Sholto & Robert A. de J. Hart (1976) *Forest Farming,* Watkins.

Dudley, Nigel (ca 1988) *This Poisoned Earth,* Piatkus.

—, (1990) *Nitrates: The Threat to Food and Water,* Green Print.

Duncan, Ronald (1949) *Jan's Journal,* William Campion.

—, (1951) *The Blue Fox,* Museum Press.

—, & Weston-Smith, Miranda (1979) *Lying Truths,* Pergamon Press, Oxford.

Easey, Ben (1955) *Practical Organic Gardening,* Faber and Faber.

Easterbrook, Laurence (1970) *How to be Happy though Civilised,* Spiritualist Association of Great Britain.

Ecologist, The (1972) *Blueprint for Survival,* Penguin, Harmondsworth.

Edgell, Derek (1992) *The Order of Woodcraft Chivalry 1916-1949,* Edwin Mellen Press, Lampeter.

Elkington, John & Hailes, Julia (1988) *Green Consumer Guide,* Gollancz.

Elliot, R.H. (1943) *The Clifton Park System of Farming,* Faber and Faber.

Elm Farm Research Centre (1981) *The Research Needs of Biological Agriculture in Great Britain,* Elm Farm Research Centre Report No.1, Hamstead Marshall.

Ewin, Jeannette (2001) *Fine Wines and Fish Oil: The Life of Hugh Macdonald Sinclair,* Oxford University Press, Oxford.

Fennell, Rosemary (1997) *The Common Agricultural Policy: Continuity and Change,* Clarendon Press, Oxford.

Finney, Brian (2000) 'Farm Mechanisation in the Second Half of the 20th Century', *Journal of the Royal Agricultural Society of England,* pp. 100–09.

Fogg, H.G. Witham (1976) *Vegetables Naturally,* Bartholomew, Edinburgh.

Foreman, Susan (1989) *Loaves and Fishes: An Illustrated History of the Ministry of Agriculture, Fisheries and Food 1889–1989,* MAFF/HMSO.

Franklin, Michael (2000) 'Politics and Post-War Agriculture', *Journal of the Royal Agricultural Society of England,* pp. 64–75.

French, Mary (1969) *Worm in the Wheat,* John Baker.

Frost, Dave (1998) *Welsh Salad Days,* Y Lolfa Cyf., Talybont.

Frost, David & Wacher, Carolyn (2004) 'A New Incarnation: The Role of the Organic Growers Association in Changing the Production and Marketing of Organic Produce', *Organic-Research.com,* January 2004, pp. 5N–14N.

Fukuoka, Masanobu (1987) *The Natural Way of Farming,* Japan Publications, New York.

—, (1989) *The Road Back to Nature,* Japan Publications, New York.

Gardiner, Rolf (1943) *England Herself,* Faber and Faber.

Gear, Alan (1987) *The New Organic Food Guide,* Dent.

—, & Gear, Jackie (2009) *Organic Gardening: The Whole Story,* Watkins.

Geuter, Maria (1962) *Herbs in Nutrition,* Biodynamic Agricultural Association.

Girardet, Herbert (1976) *Land for the People,* Crescent Books.

Gold, Mark (1983) *Assault and Battery,* Pluto Press.

Goldberg, Joshua (2009) *Liberal Fascism,* Penguin.

Goldsmith, Edward (1978) *The Stable Society,* Wadebridge Press, Wadebridge.

—, (1996) *The Way: An Ecological World-View,* Themis Books, Totnes.

Gordon, David (1996) *The Democratic Farm,* Rural Revival Trust.

Gould, Peter C. (1988) *Early Green Politics,* Harvester, Brighton.

Graham, Michael (1951) *Human Needs,* Cresset Press.

Grant, Doris (1944) *Your Daily Bread,* Faber and Faber.

—, (1958) *Housewives Beware,* Faber and Faber.

—, (1973) *Your Daily Food,* Faber and Faber.
—, & Joice, Jean (1984) *Food Combining for Health,* Thorsons, Wellingborough.
Green Party (1987) *Our Borrowed Land: Food, Farming and You,* The Green Party.
Griffiths, Richard (1983) *Fellow Travellers of the Right,* Oxford University Press, Oxford.
—, (1998) *Patriotism Perverted,* Constable.
Griffiths, Vivian (1999) 'David Clement: A Life near the History of the Organic Movement' (unpublished typescript).
Grigg, David (1987) 'Farm Size in England and Wales', *Agricultural History Review* Vol. 35/2, pp. 179–89.
—, (1989) *English Agriculture: An Historical Perspective,* Blackwell.
Griggs, Barbara (1986) *The Food Factor: Why We Are What We Eat,* Penguin/Viking, Harmondsworth.
Guénon, René (1953) *The Reign of Quantity and the Signs of the Times,* Luzac & Co.

Hall, Ross Hume (1974) *Food for Naught,* Harper and Row.
Hamilton, Geoff (1987) *Successful Organic Gardening,* Dorling Kindersley.
Hancock, Reginald (1952) *Memoirs of a Veterinary Surgeon,* Macgibbon and Kee.
Hanssen, Maurice (1984) *E for Additives,* Thorsons, Wellingborough.
Hardyment, Christina (1995) *Slice of Life: The British Way of Eating since 1945,* BBC.
Harrison, Ruth (1964) *Animal Machines,* Vincent Stuart.
Hart, Robert A. de J. (1963) 'The Odyssey of a Twentieth-Century Peasant', *Star and Furrow* 21, Autumn 1963, pp. 1–5.
—, (1968) *The Inviolable Hills,* Stuart and Watkins.
—, (1996) *Beyond the Forest Garden,* Gaia Books.

Harvey, Graham (1997) *The Killing of the Countryside,* Cape.
Hatfield, Audrey Wynne (1969) *How to Enjoy Your Weeds,* Muller.
Heaton, Kenneth (n.d.) *The Founders of Modern Nutrition: Cleave,* McCarrison Society.
Hennessy, Peter (2007) *Having it so Good: Britain in the Fifties,* Penguin.
Heron, Giles (2009) *Farming with Mary,* Authors OnLine, Sandy.
Heron, T.M. (1945) 'Prolegomena to the Methodology of the Study of Unnecessary Sweat', pamphlet of article from *New English Weekly.*
—, (n.d.) *What has Christianity to Say on Leisure in the Modern State?,* Industrial Christian Fellowship.
Hicks, C. Stanton (1951) *Food and Folly,* Soil Association, Haughley.
—, (1971) *Ecology — and Us,* Australian Heritage Society, Melbourne.
—, (1975) *Man and Natural Resources,* Croom Helm.
—, & White, H.F. (1953) *Life from the Soil,* Longmans Green.
Hillel, Daniel (1992) *Out of the Earth,* Aurum Press.
Hills, Lawrence D. (1953) *Russian Comfrey,* Faber and Faber.
—, (1964) *Pest Control Without Poisons,* HDRA, Bocking.
—, (1971) *Grow Your Own Fruit and Vegetables,* Faber and Faber.
—, (1975) *Fertility Without Fertilisers,* HDRA, Bocking.
—, (1977) *Organic Gardening,* Penguin, Harmondsworth.
—, (1989) *Fighting Like the Flowers,* Green Books, Hartland.
Hodges, R.D. (1974) *The Histology of the Fowl,* Academic Press.
—, (1978) 'The Case for Biological Agriculture', *Ecologist Quarterly* No.2, pp. 122–43.
—, (1982) 'Agriculture and Horticulture: The Need for a More Biological Approach', *Biological Agriculture and Horticulture* Vol.1, pp. 1–13.
—, & Arden-Clarke, Charles (1986) *Soil Erosion in Britain,* Soil Association, Bristol.

Hodgkinson, Tom (2007) *How to be Free*, Penguin.
Hogh-Jensen, H. (1998) 'Systems Theory as a Scientific Approach Towards Organic Farming', *Biological Agriculture and Horticulture* Vol.16, pp. 37–52.
Holderness, B.A. (1985) *British Agriculture Since 1945*, Manchester University Press.
Hollins, Arthur (1984) *The Farmer, the Plough and the Devil*, Ashgrove Press, Bath.
Holman, R.A. (1962) 'What is Health?', *Mother Earth* July 1962, pp. 249–51.
Hopkins, D.P. (1945) *Chemicals, Humus, and the Soil*, Faber and Faber.
Hopkins, Rob (2008) *The Transition Handbook*, Green Books, Totnes.
Houriet, Robert (1973) *Getting Back Together*, Sphere Books.
Howard, Albert (1940) *An Agricultural Testament*, Oxford University Press.
—, (1945a) *Farming and Gardening for Health or Disease*, Faber and Faber.
—, (1945b) *Darwin on Humus and the Earthworm*, Faber and Faber.
—, (1946) 'Activated and Digested Sewage Sludge in Agriculture and Horticulture', lecture given to AGM of the Institute of Sewage Purification, 20.11.1946.
—, & Wad, Y.D. (1931) *The Waste Products of Agriculture*, Oxford University Press.
Howard, Louise E. (1953) *Sir Albert Howard in India*, Faber and Faber.
Hugo, Victor (1909) *Les Misérables*, Dent.
Hutton, Ronald (1999) *The Triumph of the Moon: A History of Modern Pagan Witchcraft*, Oxford University Press, Oxford.
Huxley, Elspeth (1965) *Brave New Victuals*, Chatto and Windus.
Hyams, Edward (1951) 'Soil and Socialism', *New Statesman* 8.12.1951, pp. 661–62.
—, (1976) *Soil and Civilization*, John Murray.

Illich, Ivan (1976) *Limits to Medicine*, Marion Boyars.
Imperial Chemical Industries Ltd (1957) *Agriculture in the British Economy*, ICI Ltd

Jacks, G.V. and Whyte, R. O. (1939) *The Rape of the Earth*, Faber and Faber.
Jenks, Jorian (1950) *From the Ground Up*, Hollis & Carter.
—, (1959) *The Stuff Man's Made Of*, Faber and Faber.
Judt, Tony (2005) *Postwar: A History of Europe since 1945*, Heinemann.
—, (2009) *Reappraisals: Reflections on the Forgotten Twentieth Century*, Vintage.

Kennedy, Gordon & Ryan, Kody (n.d.) *Hippie Roots and the Perennial Subculture*, Hippyland Archive, http://www.hippy.com/php/article-243.html, accessed 30.12.2008.
Kiley-Worthington, Marthe (1993) *Eco-Agriculture*, Souvenir Press.
—, & Rendle, C. (1984) 'Ecological Agriculture: A Case Study of an Ecological Farm in the South of England', *BAH* Vol.2, No.2, pp. 101–33.
Kimbrell, Andrew (2002) *Fatal Harvest: The Tragedy of Industrial Agriculture*, Island Press.
King, F.C. (1944) *Gardening with Compost*, Faber and Faber.
—, (1951) *The Weed Problem*, Faber and Faber.
King, F.H. (1949) *Farmers of Forty Centuries*, Cape.
Koepf, Herbert H. (1982) *What is Biodynamic Agriculture?*, Biodynamic Literature, Wyoming, Rhode Island.
Kolisko, E. & L.(1978) *Agriculture of Tomorrow*, Acorn Press, Bournemouth.
Korch, Charles R. (1960) 'William Albrecht Sums Up a Career in Soil Research', *The Farm Quarterly*, Winter 1960.

SELECT BIBLIOGRAPHY

Lacey, Roy (1988) *Organic Gardening,* David and Charles, Newton Abbot.

Lachman, Gary (2007) *Rudolf Steiner: An Introduction to his Life and Work,* Floris Books, Edinburgh.

Lampkin, Nicolas (1990) *Organic Farming,* Farming Press, Ipswich.

Landau, Rom (1935) *God is My Adventure,* Nicholson and Watson.

Lang, Tim and Heasman, Michael (2004) *Food Wars,* Earthscan.

Leopold, Aldo (1949) *A Sand County Almanac,* Oxford University Press, New York.

Lévi-Strauss, Claude (1969) *The Raw and the Cooked,* Harper and Row, New York.

Levy, Juliette de Bairacli (1953) *As Gypsies Wander,* Faber and Faber.

—, (1958) *Wanderers in the New Forest,* Faber and Faber.

—, (1966) *Herbal Handbook for Everyone,* Faber and Faber.

—, (1973) *Herbal Handbook for Farm and Stable,* Faber and Faber.

—, (1974) *The Illustrated Herbal Handbook,* Faber and Faber.

Leyel, Mrs. C.F. (1952) *Green Medicine,* Faber and Faber.

Lockeretz, W. (2007) *Organic Farming: An International History,* CAB International, Wallingford.

Loewenfeld, Claire (1964) *Herb Gardening,* Faber and Faber.

Lorimer, David (2003) *Radical Prince: The Practical Vision of the Prince of Wales,* Floris Books, Edinburgh.

Lovelock, James (1979) *Gaia: A New Look at Life on Earth,* Oxford University Press, Oxford.

Ludovici, Anthony M. (1945) *The Four Pillars of Health,* Heath Cranton.

Lymington, Viscount (1938) *Famine in England,* Right Book Club.

Mabey, David and Gear, Alan and Jackie (1990) *Thorson's Organic Consumer Guide,* Thorson's, Wellingborough.

Mackarness, Richard (1980) *Chemical Victims,* Pan.

Macmurray, John (1933) *Interpreting the Universe,* Faber and Faber.

—, (1961) *Religion, Art, and Science,* Liverpool University Press, Liverpool.

Mairet, Philip (1947) *The Ecological Basis of Civilization,* Men of the Trees, Abbotsbury.

—, (1957) *Pioneer of Sociology: The Life and Letters of Patrick Geddes,* Lund Humphries.

—, (1989) *Autobiographical and Other Papers,* ed. C.H. Sisson, Carcanet, Manchester.

—, (n.d.) *Bailiff for God's Estate on Earth,* Council for the Church and Countryside.

Mansfield, Peter (1987) *Chemical Children,* Century.

—, (1988) *The Good Health Handbook,* Grafton Books.

Marr, Andrew (2008) *A History of Modern Britain,* Pan.

Marsh, Jan (1982) *Back to the Land,* Quartet.

Marshall, Peter (1992) *Nature's Web: An Exploration of Ecological Thinking,* Simon and Schuster.

Martin, John (2000) *The Development of Modern Agriculture: British Farming since 1931,* Macmillan, Basingstoke.

Massingham, H.J. (1944) *This Plot of Earth,* Collins.

—, (1945a) *The Natural Order: Essays in the Return to Husbandry,* Dent.

—, (1945b) *The Wisdom of the Fields,* Collins.

—, (1947) *The Small Farmer,* Collins.

—, (1951) *The Faith of a Fieldsman,* Museum Press.

—, (1988) *A Mirror of England* (ed. Edward Abelson), Green Books, Hartland.

—, & Hyams, Edward (1953) *Prophecy of Famine,* Thames and Hudson.

Matless, David (1998) *Landscape and Englishness,* Reaktion Books.

—, (2001) 'Bodies Made of Grass Made of Earth Made of Bodies: Organicism, Diet and National Health in Mid-Twentieth-Century England', *Journal of Historical Geography* Vol.27/3, pp. 355–76.

McAllister, Gilbert & Glen, Elizabeth

(1945) *Homes, Towns and Countryside,* Batsford.
McCarrison, Robert (1953) *The Work of Sir Robert McCarrison* (edited H.M. Sinclair), Faber and Faber.
—, & Sinclair, H.M. (1961) *Nutrition and Health,* Faber and Faber.
McRobie, George (1981) *Small is Possible,* Cape.
McSheehy, Trevor (1970) 'A Constant Environment Animal House Suitable for Nutritional Research', *Laboratory Animals* Vol.4, pp. 273-87.
—, & Rawlings, J.A. (1973) 'The Influence of Three Different Farming Systems on Organic Matter in the Soils', *Quant. Plant. Mater. Veg.* XXII 3-4, pp. 321-33.
Meadows, Dennis L. (1972) *The Limits to Growth,* Earth Island Ltd
Mellanby, Kenneth (1967) *Pesticides and Pollution,* Collins.
—, (1969) 'Haughley Research', *Journal of the Soil Association* April 1969, pp. 334-36.
—, (1975) *Can Britain Feed Itself?,* Merlin Press.
—, (1981) *Farming and Wildlife,* Collins.
Merrill, Margaret C. (1983) 'EcoAgriculture: A Review of its History and Philosophy', *Biological Agriculture and Horticulture* Vol.1/3, pp. 181-210.
Miller, Henry (1941) *The Wisdom of the Heart,* New Directions, New York.
Milton, R.F. (1957) *Trace Elements in Soil, Plant and Animal,* Middlesex Hospital Medical School.
Mishan, E.J. (1967) *The Costs of Economic Growth,* Staples Press.
Mitchell, Charlotte & Wright, Ian (1987) *The Organic Wine Guide,* Mainstream, Edinburgh.
Mollison, Bill (1987a) 'Permaculture: History and Future Direction', *Permaculture Newsletter* Issue 9, Feb. 1987, pp. 3-5.
—, (1987b) 'Permanent Agriculture' in Woodhouse 1987, pp. 154-58.
—, (1990) *Permaculture: A Practical Guide for a Sustainable Future,* Island Press, Washington D.C.
—, & Holmgren, David (1978) *Permaculture One,* Corgi.
Montague, Diane (2000) *Farming, Food and Politics: The Merchant's Tale,* Irish Agricultural Wholesale Society, Dublin.
Moore-Colyer, Richard J. (2004) 'Towards *Mother Earth:* Jorian Jenks, Organicism, the Right and the British Union of Fascists', *Journal of Contemporary History* Vol.39/3, July 2004, pp. 353-71.
Mount, S.J. Lambert (1979) *The Food and Health of Western Man,* Precision Press, Marlow.
Mumford, Lewis (1946) *Technics and Civilization,* Routledge.
Murry, John Middleton (1952) *Community Farm,* Peter Nevill.

Nicol, Hugh (1967) *The Limits of Man,* Constable.
Noble, Patrick [under pseud. 'Ernest Organic'] (2008) *Notes from the Old Blair and Bush,* Matador, Leicester.
North, Michael (1976) *Time Running Out?: Best of Resurgence,* Prism Press, Dorchester.
North, Richard & Pye-Smith, Charlie (1984) 'Off the Chemical Treadmill', *The Countryman* Summer 1984, pp. 141-46.
Northbourne, Lord (1940) *Look to the Land,* Dent.
—, (1963) *Religion in the Modern World,* Dent.
—, (1970) *Looking Back on Progress,* Perennial Books, Bedfont.

O'Riordan, Tim & Cobb, Dick (1996) 'That Elusive Definition of Sustainable Agriculture', *T&CP* February 1996, pp. 50-51.
Orr, John Boyd (1936) *Food, Health and Income,* Macmillan.
Osborn, Fairfield (1954) *The Limits of the Earth,* Faber and Faber.

Palgrave, F.T. (1938) *The Golden Treasury,* Oxford University Press.
Payne, Virginia (1971) *A History of*

the Soil Association (M.Sc. thesis), Victoria University of Manchester.

Peacock, Paul (2005) *A Good Life: John Seymour and his Self-Sufficiency Legacy,* Farming Books and Videos Ltd, Preston.

Pears, Pauline & Stickland, Sue (1995) *Organic Gardening,* Mitchell Beazley.

Pearse, Innes H. (1971) *Is Health a Suitable Study for Academic Consideration?,* University of St Andrews, St Andrews.

—, (1979) *The Quality of Life,* Scottish Academic Press, Edinburgh.

Peck, David G. (1947) *Earth and Heaven: A Theology of the Countryside* (2nd edition), Council for the Church and Countryside.

Pepper, David (1991) *Communes and the Green Vision,* Green Print.

Pfeiffer, E. (1947a) *Soil Fertility, Renewal and Preservation,* Faber and Faber.

—, (1947b) *The Earth's Face,* Faber and Faber.

Picton, L.J. (1946) *Thoughts on Feeding,* Faber and Faber.

Pioneer Health Centre (n.d.) *C3 ... or A1?,* Pioneer Health Centre.

Pioneer Health Centre Ltd (1971) *Health: Of the Individual, of the Family, of Society,* Pioneer Health Centre, Rotherfield.

—, (1986) *Positive Prospect for Health,* Pioneer Health Centre Ltd

—, (1993) *A Pool of Information: The Search for Positive Health — the Pioneer Health Centre, Peckham, 1935–50* (videotape), Pioneer Health Centre Ltd

Pollan, Michael (2008) *In Defence of Food,* Allen Lane.

Poore, G.V. (1903) *Essays on Rural Hygiene* (3rd edition), Longmans Green.

Poppelbaum, Hermann (1950) *Biodynamic Farming and Gardening,* BDAA.

Porritt, Jonathon (1984) *Seeing Green,* Basil Blackwell, Oxford.

Portsmouth, Earl of (1965) *A Knot of Roots,* Bles.

—, & Walston, H.D. (1947) *Rural England: The Way Ahead,* National Council of Social Service.

Pottenger, F.M. (1983) *Pottenger's Cats: A Study in Nutrition,* Price-Pottenger Foundation, La Mesa, CA.

Pretty, Jules N. (1995) *Regenerating Agriculture,* Earthscan.

Price, Weston A. (1945) *Nutrition and Physical Degeneration* (revised edition), Weston A. Price, Redlands, Ca.

Pye-Smith, Charlie & Hall, Chris (1987) *The Countryside We Want,* Green Books, Hartland.

—, & North, Richard (1984) *Working the Land: A New Plan for a Healthy Agriculture,* Temple Smith.

Pyke, Magnus (1952) *Townsman's Food,* Turnstile Press.

Reckitt, Maurice B. (1945) *Prospect for Christendom,* Faber and Faber.

Reed, Matthew J. (2003) *Rebels for the Soil: The Lonely Furrow of the Soil Association 1943-2000,* doctoral thesis, University of the West of England, Bristol.

Reich, Charles (1971) *The Greening of America,* Allen Lane.

Rickard, Sean (2000) 'British Farming: Time for a New Mindset', *Journal of the Royal Agricultural Society of England,* vol.161, pp. 8–20.

Robb, R. Lindsay (1965) *The Role of the Soil Association and Haughley Research Farms in the Problems of Soil and Health,* Soil Association, Haughley.

Rolt, L.T.C. (1988) *High Horse Riderless,* Green Books, Hartland.

Rose, Walter (1988) *Good Neighbours,* Green Books, Hartland.

Roszak, Theodore (1995) *The Making of a Counter Culture,* University of California Press.

Rural Reconstruction Association (1955) *Feeding the Fifty Million,* Hollis & Carter.

Rural Resettlement Group (1979) *Rural Resettlement Handbook* (2nd edition), Rural Resettlement Group, Diss.

Sams, Craig and Fairley, Josephine (2008) *Sweet Dreams: The Story of Green and Black's*, Random House.
Sams, Gregory (1997) *Uncommon Sense*, Chaos Works.
Sanderson-Wells, T.H. (1939) *Sun Diet, or Live Food for Live Britons*, John Bale.
Schumacher, E.F. (1973) *Small is Beautiful*, Blond and Briggs.
—, (1974) *The Age of Plenty: A Christian View*, The Saint Andrew Press, Edinburgh.
—, (1977) *A Guide for the Perplexed*, Cape.
—, (1979) *Good Work*, Cape.
Scott, Paul (2006) *Tony and Cherie: Behind the Scenes in Downing Street*, Pan.
Scott-Samuel, Alex (1990) *Total Participation, Total Health*, Scottish Academic Press, Edinburgh.
Seddon, Quentin (1989) *The Silent Revolution*, BBC.
Seifert, Alwin (1962) *Compost*, Faber.
Self, Peter & Storing, Herbert J. (1961) *The State and the Farmer*, Allen and Unwin.
Seymour, John (1961) *The Fat of the Land*, Faber and Faber.
—, (1976) *The Complete Book of Self-Sufficiency*, Faber and Faber.
—, (1983) *The Smallholder*, Sidgwick and Jackson.
—, (1989) *The Ultimate Heresy*, Green Books, Hartland.
—, (1996) *Retrieved from the Future*, New European Publications.
—, & Girardet, Herbert (1988) *Far from Paradise*, Green Print.
Sheail, John (2002) *An Environmental History of Twentieth-Century Britain*, Palgrave, Basingstoke.
Shewell-Cooper, W.E. (1945) *Soil, Humus and Health*, John Gifford.
—, (1952) *Royal Gardeners: King George VI and his Queen*, Cassell.
—, (1959) *The ABC of Soils*, English Universities Press.
—, (1966) *The Good Gardeners' Association Guide to Minimum Work Gardening*, New Homes Press.
—, (1972) *Compost Gardening*, David and Charles, Newton Abbot.
—, (1978) *Basic Book of Natural Gardening*, Barrie and Jenkins.
Shoard, Marion (1980) *The Theft of the Countryside*, Temple Smith.
Smethurst, William (1996) The Archers: *The True Story*, Michael O'Mara Books.
Smuts, J.C. (1927) *Holism and Evolution*, Macmillan.
Snow, C.P. (1959) *The Two Cultures and the Scientific Revolution: The Rede Lecture 1959*, Cambridge University Press.
Soil Association (1962) *The Haughley Experiment, 1938–62*, Soil Association, Haughley.
Soper, John (1996) *Biodynamic Gardening*, Souvenir Press.
Soper, Mike (1995) *Years of Change*, Farming Press, Ipswich.
Spedding, Colin R.W. (1996) *Agriculture and the Citizen*, Chapman & Hall.
—, & Reid, Ian G. (1983) Agriculture: The Triumph and the Shame: *An Independent Assessment*, Centre for Agricultural Strategy, University of Reading.
Spowers, Rory (2003) *Rising Tides: The History and Future of the Environmental Movement*, Canongate, Edinburgh.
Stallibrass, Alison (1989) *Being Me and Also Us*, Scottish Academic Press, Edinburgh.
Standen, Anthony (1952) *Science is a Sacred Cow*, Sheed and Ward.
Stapledon, R.G. (1935) *The Land: Now and To-Morrow*, Faber and Faber.
—, (1946) *Farming and Mechanised Agriculture 1946*, Harrap.
—, (1964) *Human Ecology*, Faber and Faber.
Steiner, Rudolf (1949) *An Outline of Occult Science*, Rudolf Steiner Publishing Co.
—, (2006) *Christianity as Mystical Fact*, SteinerBooks, Gt. Barrington, MASS.
Stephenson, W.A. (1968) *Seaweed in Agriculture and Horticulture*, Faber and Faber.
Stitt, Sean (1998) 'Food for Health or Wealth in the 21st Century',

Nutrition and Health Vol.12/No.4, pp. 203–13.
Stolton, Sue & Dudley, Nigel (1996) *Seeking Permanence: 15 Years and into the Future,* Elm Farm Research Centre, Hamstead Marshall.
Stone, Dan (2004) 'The Far Right and the Back-to-the-Land Movement', in J.V. Gottlieb and T.P. Linehan, *The Culture of Fascism,* I.B. Tauris.
Stonehouse, Bernard (1981) *Biological Husbandry: A Scientific Approach to Organic Farming,* Butterworth.
Street, A.G. (1954) *Feather-Bedding,* Faber and Faber.
Stuart, Tristram (2009) *Waste: Uncovering the Global Food Scandal,* Penguin.
Sweeny, Sedley (1985) *The Challenge of Smallholding,* Oxford University Press, Oxford.
—, (n.d.) *An Intuitive View of the Whole,* Sedley Sweeny, Cortes Island, B.C.
Sykes, Friend (1946) *Humus and the Farmer,* Faber and Faber.
—, (1951) *Food, Farming and the Future,* Faber and Faber.

Tannahill, Reay (2002) *Food in History,* Headline.
Tansey, Geoff & Worsley, Tony (1995) *The Food System: A Guide,* Earthscan.
Taverne, Dick (2005) *The March of Unreason,* Oxford University Press, Oxford.
Taylor, John V. (1975) *Enough is Enough,* SCM Press.
Temple, Jack (1986) *The Here's Health Guide to Gardening Without Chemicals,* Thorsons, Wellingborough.
Thear, Katie (1983) *A Kind of Living: A Practical Guide to Home Food Production and Energy Saving,* Hamish Hamilton.
Thompson, Paul B. (1995) *The Spirit of the Soil,* Routledge.
Tompkins, Peter & Bird, Christopher (1992) *Secrets of the Soil,* Arkana.
Trevelyan, Sir George (1987) 'Spiritual Education' in Woodhouse 1987, pp. 207–13.
Trotter, J.D. (2003) *Wholeness and Holiness,* Pioneer Health Foundation, Glasgow.
Trowell, H. (1960) *Non-Infective Disease in Africa,* Edward Arnold.
Trow-Smith, Robert (1953) *Society and the Land,* Cresset Press.
Turner, Newman (1950) *Cure Your Own Cattle, The Farmer* Publications, Bridgwater.
—, (1951) *Fertility Farming,* Faber and Faber.
—, (1952) *Herdsmanship,* Faber and Faber.
—, (1955) *Fertility Pastures,* Faber and Faber.

Vaihinger, H. (1935) *The Philosophy of 'As If',* Kegan Paul, Trench and Trubner.
Van Allen, Judith & Marine, Gene (1972) *Food Pollution: The Violation of Our Inner Ecology,* Holt, Rinehart and Winston, New York.
Van Vuren, J.P. J. (1949) *Soil Fertility and Sewage,* Faber and Faber.
Veldman, Meredith (1994) *Fantasy, the Bomb, and the Greening of Britain,* Cambridge University Press.
Vine, Anne & Bateman, David (1982) 'Some Economic Aspects of Organic Farming in England and Wales', *Biological Agriculture and Horticulture* Vol.1, pp. 65–72.
Voisin, André (1959) *Soil, Grass and Cancer,* Crosby Lockwood.
—, (1962) *Rational Grazing,* Crosby Lockwood.

Waddell, Eric (1977) 'The Return to Traditional Agriculture', *The Ecologist* Vol.7/4, pp. 144–47.
Waksman, Selman A. (1938) *Humus,* Baillière, Tindall & Cox.
Wales, HRH The Prince of, & Clover, Charles (1993) *Highgrove: Portrait of an Estate,* Chapmans.
Waller, Robert (1962) *Prophet of the New Age,* Faber and Faber.
—, (1973) *Be Human or Die,* Charles Knight.
—, (1980) 'Scientific Materialism: The Strait-jacket of Western Culture',

The Ecologist Vol.10/No.6–7, July/Aug./Sept. 1980, pp. 224–29.

—, (1982) *The Agricultural Balance Sheet,* Green Alliance/Conservation Society.

—, (1990) 'Radical Thoughts of L.T.C. Rolt', *Resurgence* 142, pp. 46–47.

—, & Etté, Anthony (1978) 'The Anomaly of a Christian Ecology', *Ecologist Quarterly* No.2, pp. 144–48.

Walters, A. Harry (1967) *The Living Rocks: An Introduction to Biophilosophy for Technological Age Man,* Classic Publications.

—, (1969) 'Whither Farming in the Year 2000?', *Journal of the Soil Association,* April 1969, pp. 313–18.

Ward, Colin (1989) 'Fringe Benefits', *New Statesman and Society* 29.9.1989, p. 33.

Waterman, Charles (1946) *The Three Spheres of Society,* Faber.

Watson, Lyall (1973) *Supernature,* Hodder and Stoughton.

Webb, James (1981) *The Occult Establishment,* Richard Drew, Glasgow.

Welburn, Andrew (2004) *Rudolf Steiner's Philosophy and the Crisis of Contemporary Thought,* Floris Books, Edinburgh.

Westlake, Aubrey (1961) *The Pattern of Health,* Vincent Stuart.

—, (1967) *Life Threatened,* Stuart and Watkins.

Westlake, Jean (2000) *70 Years a-Growing,* Hawthorn Press, Stroud.

Whitby, H. Morton (1961) *The Prevention of Cancer,* Cancer Prevention Centre.

White, Lynn Jr. (1967) 'The Historic Roots of Our Ecologic Crisis', *Science* Vol.155, pp. 1203–07.

Whitefield, Patrick (1996) *How to Make a Forest Garden,* Permanent Publications, Clanfield.

—, (2004) *The Earth Care Manual,* Permanent Publications, East Meon.

—, (2005) *Permaculture in a Nutshell,* Permanent Publications, East Meon.

Whitley, Andrew (2006) *Bread Matters,* Fourth Estate.

Widdowson, R.W. (1987) *Towards Holistic Agriculture: A Scientific Approach,* Pergamon Press, Oxford.

Williams, H.T. (1960) *Principles for British Agricultural Policy,* Nuffield Foundation/ Oxford University Press.

Williams, Lord, of Barnburgh [Tom Williams] (1965) *Digging for Britain,* Hutchinson.

Williamson, G. Scott (1945) *Physician, Heal Thyself,* Faber and Faber.

—, (1982) *Peckham: The First Health Centre* (reprinted from *The Lancet,* 16.3.1946), Pioneer Health Centre Ltd, Little Bookham.

—, & Pearse, Innes H. (1951) *The Passing of Peckham,* Pioneer Health Centre.

—, (1965) *Science, Synthesis and Sanity,* Collins.

Wood, Barbara (1985) *Alias Papa: A Life of Fritz Schumacher,* Oxford University Press, Oxford.

Wood, John M. (1960) 'Reminiscences of My Father [Maurice Wood] and Letters from Friends', *Star and Furrow* Autumn 1960, pp. 1–4.

Woodhouse, Tom (1987) *People and Planet: Alternative Nobel Prize Speeches,* Green Books, Hartland.

Woodward, Lawrence (2002) 'The Scientific Basis of Organic Farming', *Interdisciplinary Science Reviews* Vol.27/2, pp. 114–19.

Wookey, Barry (1987) *Rushall: The Story of an Organic Farm,* Blackwell, Oxford.

Wormell, Peter (1978) *Anatomy of Agriculture,* Harrap/Kluwer.

Worthington, Jim (1969) *Grain, Grass and Green Foods,* Soil Association, Haughley.

Wrench, G.T. (1938) *The Wheel of Health,* C.W. Daniel.

Wright, Simon and McCrea, Diane (2007) *The Handbook of Organic and Fair Trade Food Marketing,* Blackwell.

Wylie, J.C. (1955) *Fertility from Town Wastes,* Faber and Faber.

—, (1959) *The Wastes of Civilization,* Faber and Faber.

Wynne-Tyson, Jon (1975) *Food for a Future,* Davis-Poynter.

Yellowlees, Walter W. (1985) *Food and Health in the Scottish Highlands,* Clunie Press, Perthshire.
—, (1989) 'Ill Fares the Land' in *Soil, Food and Health,* Wholefood Trust Ltd
—, (1999) 'McCarrison, Farmyard Manure and the Future', *Nutrition and Health* Vol.13, pp. 23–29.
—, (2001) *A Doctor in theWilderness,* Pioneer Associates, Aberfeldy.

Yeomans, P. A. (1971) *The City Forest,* Keyline, Sydney.
Young, Rosamund (1991) *Britain's Largest Nature Reserve?* Soil Association, Bristol.
—, (2005) *The Secret Life of Cows,* Farming Books and Videos Ltd, Preston.
Yudkin, John (1967) 'Sugar and Coronary Thrombosis', *New Scientist* 16.3.1967, pp. 542–43.
—, (1972) *Pure, White and Deadly,* Davis-Poynter.

Index

Aaron, Dr Harold 197
Abbotsholme School 88, 386
Abelson, Edward 280, 284
Abercrombie, Patrick 106
Aberystwyth University 74, 126–128, 154, 310–312, 390, 399, 403, 409, 431f, 434
Acts of Parliament
—, AGRICULTURE ACT, 1947 31f, 43, 46, 57, 163, 333, 378
—, Agriculture Act, 1957 46
—, Farm and Garden Chemicals Act, 1967 142, 336
—, Food and Drugs Amendment Act, 1954 62
Adams, Matt 434
Additives 197f, 200, 230, 232, 234
Adler Society 173
Advisory Panel on Waste Materials 306
Agri-business/agri-industry 28–32, 40–58, 69, 87f, 118, 131–133, 216, 220, 269, 347, 378, 379
—, aesthetic case against 48f, 132
—, in California 29f
—, chemical herbicides and pesticides 28, 53f, 55, 60f, 74f, 109, 131, 136, 142, 196, 219, 237, 261f, 264f, 276, 283, 316, 336, 338
—, complexity of 54–56
—, efficiency measured by output per man 28, 31, 43, 47, 267, 359
—, factory farming 31, 49f, 100, 106, 132, 267, 276, 344
—, farm size increases 43, 45–47
—, food scares 113, 133, 197f
—, landscape transformed by 48f, 55, 57, 132, 272, 375
—, livestock husbandry increasingly intensive 49f, 132
—, mechanization central to 32, 48–50
—, productivity of 51, 55–57, 267, 286
—, soil erosion caused by 99, 131, 260, 283, 314
—, vulnerability of 54, 132, 340
—, workforce reduced 47f
Agricultural Development and Advisory Service 11, 127, 148, 344
Agricultural Reform Group 115
Agricultural Research Council 52, 61, 96, 308, 315, 319
Agricultural Training Board 126
Agriculture
—, Asian 118, 203, 205, 237, 282f
—, Roman 71, 339
ALBERT HOWARD FOUNDATION OF ORGANIC HUSBANDRY 95f, 98f, 137, 189, 225, 228, 306
Albert Howard Memorial Lectures 98
Albert Howard News Sheet 98, 388
Albon, Alan 203
Albrecht, W.A. 128, 262, 392
Alford, Sidney 197
ALLABY, MICHAEL 24, 66, 89, 91, 104, 107, 122, 125f, 143, 263f, 266, 269, 276f, 279f, 283, 300, 325f, 335, 337, 339, 342, 376, 377
Alliance Breweries 235
Allinson, Bertrand 205
Allinson, C.P. 199
Allinson Dr T.R. 199
Allinson's bakery 91, 199
All Muck and Magic? 136
Anarchism 123, 170, 327, 342, 349f
Anderson, Jimmy 90, 92
Anderson, Pauline 90, 92
Animals, treatment and health of 31, 35f, 49f, 54, 57, 61, 72, 76, 87, 96, 98, 101, 106, 132, 134, 137, 142, 155f, 166, 169, 196, 207f, 209, 212–221, 234, 285, 296, 302–304, 320, 325, 375
Antaeus, myth of 192

458

INDEX

Anthroposophical Agricultural Foundation 33, 77
Anthroposophical Society 88
ANTHROPOSOPHY 79, 82f, 90, 127, 244, 356, 366f
Archers, The (radio soap opera) 53, 110, 114, 146
Arden-Clarke, Charles 314
Arkley Manor 138
Arlott, John 215
ARM Ltd, Rugeley 309
Army Dental Corps 195
Art/aesthetics 88f, 91, 94, 114, 119, 132, 134f, 169, 190, 237, 252, 281, 302, 368, 375–379
Asda supermarkets 255
Ashby, A.W. 42
Ashmole, Anna 202
Aspalls 86, 231, 243
Associated British Foods 64
Association for the Study of Human Nutrition 180
Association of Public Health Inspectors 307
Astor, Alice 318
ASTOR, DAVID 88, 141, 317–319, 337, 340
Astor, Jacob 319
Astor, Viscount 42
Attingham Park conferences 80, 82, 129, 181, 195, 275, 338
Attlee, Clement 170
Aubert, Claude 109, 157
Australian Heritage Society 192

Bach, Edward 168, 231
Bach flower remedies 212, 231f
BADENOCH, DR A.G. 97f, 188f, 352, 355
Badenoch, Christopher 189, 436
Baha'i faith 355
BAKER, RICHARD ST. BARBE 14, 33, 38, 112, 137, 184, 204, 225, 258f, 277, 279, 282, 310, 355, 378
BALFOUR, EVE 14, 15, 17, 29, 34, 74f, 78f, 81f, 86, 101, 103f, 107, 109f, 115, 122f, 125, 128f, 139, 163, 177, 189, 192, 194, 201f, 211, 223, 229, 237f, 259, 264, 266, 268, 272, 274, 276, 284, 288, 290, 293, 295f, 301f, 311, 324, 330, 335, 339, 341f, 344, 352f, 371, 374f
Bangor University 127, 219, 310
Barber, Derek 46

Barclay, Peter 207
BARLOW, DR KENNETH 168, 171f, 173–175, 177, 183f, 191, 258, 279, 282, 332f, 349, 378
Barran, Mary (see HERON, MARY)
Barritt, Arthur 147f
Barth, Karl 355
Bastable, Tim 123
Bateman, David 73
Bates, Angela 110, 129f, 158, 191, 222f, 236, 242, 265, 320, 341
Bateson, F.W. 42, 44, 332
Batsford, B.T. 51
Bauer, Dirk 252
Beaumont, Timothy 363
Beckett, J.L. 106, 306
BEDFORD, DUKE OF 34, 95, 165, 331
Bell, Adrian 38, 342
Bell, Anthea 442
Bell, Graham 119, 120
Bellamy, David 287
Bennett, J.G. 233
Bennett, Sir Norman 38, 194
Beresford, Tristram 373
Bernal, Prof. J.D. 291
Best, Will 72
Beveridge Report 163
Bicknell, Dr Franklin 62, 65, 197
Biddlestone, Dr A.J. 307–310
BIODYNAMIC AGRICULTURAL ASSOCIATION 77–79, 101, 109, 113, 225, 243, 246, 316, 324, 365
BIODYNAMIC CULTIVATION 26, 33, 39, 72, 77–94, 124, 135, 212, 243, 277, 324, 366–368
BIOLOGICAL AGRICULTURE AND HORTICULTURE 72, 314, 317f
Bio-technology 40, 160
Bircher-Benner, Max 191
Bishop, Beata 184, 352
Black, Prof. John 373
Blackburn, J.S 36
Blackfriars 355
Blair, Cherie 149
Blair, Douglas 114, 156–158, 280
Blair, Penny 156, 280
Blake, Francis 109, 127, 154–156, 161, 248, 271, 344
Blake, John 155f, 254
Blake, William 192
Blaxter, Sir Kenneth 53, 55–57, 108
Bledisloe, Lord 38
Blond, Anthony 268

459

Blyth, Dr Bill 144
Blythe-Currie sisters 239
Bocking, HDRA trial ground 142–144, 147
BODY, SIR RICHARD 43f, 46, 54, 56, 109f, 131f, 271, 285, 346f, 379
Bolan, Marc 236
Bonham-Carter, Victor 100, 153, 263
Booker, Christopher 197
Booker-McConnell 235
Borton, Heda 177
Boston, Richard 279f, 342
Botton Village 81f, 90
Bower, Arthur 140
BOWER, JOANNE 220f, 229, 276, 279, 352, 379
Bowers, Pam 160, 255
Bowers, Rick 161, 255
Box schemes 36, 251, 255, 257
BRADFORD, LORD 102, 139, 335
BRADY, MARGARET 105, 182, 200f, 204, 211, 352
Brambell Report on animal husbandry 49, 132
Bread
—, Brady's demonstrations and film 105, 201
—, Grant Loaf 200
—, Huby Wholemeal Loaf 199
—, and national fertility 164
—, Rearsby Loaf 106, 199
—, Springhill wholemeal 201f
—, white 63f, 200
—, wholemeal 63f, 105, 176, 183, 198–202, 261
BREEN, DR G.E. 59, 186, 188, 192, 198, 304
Bridger, Roy 137, 203, 276
Brighton, Robert 126, 128, 243, 245
British Association for the Advancement of Science 260, 264, 292
British Dental Association 194
British Dental Journal 181, 194
British Herbalist Union 212
British Humanist Association 289
British Medical Association 179, 185
British Medical Journal 178, 190
British National Party 337
British Nutrition Foundation 64
BRITISH ORGANIC FARMERS 85, 113–116, 126, 150, 156, 158, 161, 208, 238, 241, 244, 246, 253, 321, 344, 372
British Organic Milk Producers 255

British Organic Standards Committee 85, 245, 322
BRITISH PEOPLE'S PARTY 34, 95, 331
British Permaculture Association 119f
British Union of Fascists 329
British Vegetarian Society 191
Broad Leys Publishing 271
BROCKMAN, ALAN 86f, 92–94, 243
Bromfield, Louis 97, 123, 128, 216
BROOKE, HILDA CHERRY (see HILLS, HILDA CHERRY)
Broomfield College 128
Brown, Evelyn Scott 205
Brown, Lester 283
Browne, Sir Thomas 353
BRUCE, MAYE E. 79, 85, 203, 225
Brunt, L.P. 306
Bryce-Smith, Prof. Derek 277
BRYNGWYN PROJECT 108, 129, 310–312
BSE 344
Buber, Martin 435
Buddhism 91, 97, 233, 357, 359f
—, Buddhist economics 359f
Budgens supermarkets 255
Bugler, Jeremy 280
Bulcock, Pauline 211
Bunday, Sally 197f
Bunyard, Peter 175, 284, 342
Burkitt, Denis 182, 185–187, 290, 352
BURMAN, DR NORMAN 98, 106, 188, 303–306
Burton, Robert 93
Busses Farm 90, 255
Butler, Elizabeth (nom de plume of Butler, Shirley) 435
Butler, John 140, 150–152, 158, 176, 244, 352, 371
Butler, Joyce, MP 142, 336
Butler, Shirley 150–152
Butter, John 280
BYNGHAM, DION 165, 203f, 285, 331, 356

Cadbury-Schweppes 235
Caddy, Eileen 368
Caddy, Peter 259, 368
Campaign for Nuclear Disarmament (CND) 327, 335, 341
Campbell, Angus 97f, 189
Campbell, Douglas 242, 300f
Campbell, Peggy 301
Camphill Community/Village Trust 81f, 94, 324

Canadian Agriculture and Horticulture 237
CANNON, GEOFFREY 64, 66, 68, 111, 177, 184, 346f
Canter, David and Kay 229
Capitalism 23, 37, 166, 234f, 281, 329, 332, 337, 339, 341, 346, 349
—, destructive effects of 166, 329, 339
—, and profit motive 346
Capra, Fritjof 369, 371f
Carbohydrates, refined 182, 186f
Carling, Sir Ernest Rock 170f, 294
Carmarthenshire College 127
Carnley, Kathleen 211
Carpenter, E. Brodie 96, 182, 194
Carrel, Dr Alexis 164, 170
Carrington, Lord 42
Carson, Rachel 131f, 258f, 261f, 352
Case, Howard 371
Castelliz, Katherine 90, 93
Castro, Dr Fidel 129
Catholicism
—, Anglo- 34, 362
—, Roman 232, 349, 355, 359, 362f, 364
Cawdrey, E.G. 205
Celebrities 23f, 68f, 147f, 152, 228, 235f, 250, 347, 375
Center for Global Food Issues 432
Centre for Alternative Technology 121, 278
Centre for Organic Husbandry and Agroecology 127
Ceres Grain Shop 230
Chaboussou, Francis 283
Chance, Lady 79
Chandos Group 33, 331, 362
Chapman, Guy B. 193f
Charles Knight (publishers) 175
Chase, Dr Alice 206
CHASE, J.L.H. 139, 304
Chase Organics 139
Chase Protected Cultivation Ltd 139
Cheke, T.W. 204
Chelsea Arts Club 188
Chelsea Flower Show 105, 146, 324
Chemicals (see Agri-business)
—, in food 61–63, 67, 197f, 200, 232, 234
Cherrington, John 42f
Cheshire doctors 163f, 178
Cheveley, Stephen 42f
Chiltern Herb Farms 210
Chipko movement 282
Chisholm, Anne 264
Chorleywood Process 64

Christ 166, 215, 352, 366f
—, as example of radiant health 166
Christendom 34
Christendom Group 34
Christian Community 88
Christian Ecology Link 363
Christianity 25f, 33, 36, 149, 189, 204f, 215, 331, 334, 351–366
—, esoteric 352, 356, 366–368
Church of England 38, 355, 367
City farms 120, 123, 281, 375
Clarke, Eric 110, 243
Clarkson, Jeremy 56
Clayton, Robert 227–229
CLEAVE, SURGEON-CAPTAIN T. L. 186f, 190, 290
CLEMENT, DAVID 78, 80, 84f, 93, 113, 309
Clifton Park 139
Clive, Edward 335
Clunies-Ross, Tracey 341
COATES, HUGH 86, 113, 182, 201, 229, 238f, 314, 324
Coates, Lance 79
Cobbett, William 349, 377
Cochrane, E.R. 81
Colby, Robert 205
Coleman, Eliot 157
Coleman-Cooke, John 131
Comerford, John 169
Comfrey 141f, 176, 212, 233, 271, 336
Common Agricultural Policy 43, 47, 112
Common Ground 112
Common Market 43, 46
Commonwealth Scientific and Industrial Research Organisation 116
Communes 94, 278, 343, 349
Community Supported Agriculture 23, 112, 159, 249f
Compost Engineers Ltd 306
Compost Science 108
Composting 72, 84, 86, 93f, 96–98, 134, 137f, 140, 145, 151, 160f, 190, 194, 203, 211–213, 217, 271, 304, 318, 322, 369
—, Dano system 106
—, Indore Process of 97f, 133, 137, 188
—, municipal 133, 142, 278, 295, 305–310
—, Quick Return 85
Connolly, Geoff 271
Conservation Corps 276, 300
Conservation Grade 92, 241

Conservative Party/governments 337f, 340f, 346f
Consumer Reports 197
Consumerism 23, 111f, Ch.5 passim, 280, 290, 334, 341, 347f, 357, 360, 373
—, 'Green' consumerism 140, 146, 250, 252
Consumers' Union of the United States 197
Cook, Wendy 367
Cooke, Bernard 195f
Cooke, G.W. 52, 103, 315
Coolidge Research Centre 157
Coomaraswamy, Ananda 281, 361
Cooper, Donald 158
Cooper, Frank 199
Co-operative marketing 23, 36f, 114, 118, 159, 174, 224, 226, 237, 239, 244f, 249, 254–256, 274, 345, 349
Copernicus 357
COPPARD, SUE 125f, 212
Corley, Hugh 224
Cornish Organic Growers 254
Corrin, George 80, 83, 125, 129, 368
Council for Nature 276
COUNCIL FOR THE CHURCH AND COUNTRYSIDE 28–31, 34, 36, 38f, 95, 351, 354f, 377
Council for the Preservation of Rural England 106
'Counter-culture' 212, 339, 341, 357
Country College 118
Country Landowners' Association 129
Country Living Books 36, 276
Coutenceau, Professor 139
COVENTRY FAMILY HEALTH CLUB HOUSING SOCIETY 173–175
Coventry Polytechnic/University 145
Cowan, Mark 302
COWARD, RALPH 72, 106, 114, 352
Cowley Wood Conservation Centre 280f
Cranks restaurant 91
Crawford, Dr Michael 177, 183f
Crick, F.H. 88
Cripps, Sir Stafford 330, 333
Croal, Alexander 208
Crocker, Lucy 176, 332
Crystallization experiments 87, 94
Cuba 129, 335
Cunning, Dr Annie B. 192
Cycle of Life, The (film) 105

Daily Mirror (newspaper) 151
Dalai Lama 274
Dalby, John 160
Dalzell, Howard 309f
Daniel, Charles W. 32, 231
Darley, Gillian 280
Darling, Frank Fraser 154, 260, 277, 290
Darlington, Arthur 27
Dartington Hall 153, 318
Darwin, Charles 186, 262
Davies, A.G. 307
Davies, J.L. 306
Davies, Ron, MP 347
Davies, Rupert 228
Davis, Benjamin 362
Davis, Laura 128, 160, 254
Davy, Charles 87, 89, 367, 379
Davy, Gudrun 88
DAVY, JOHN 87–89, 104, 107, 125, 131, 262f, 291, 338, 367
Deane, Jan 156, 159, 255–257
Deane, Tim 156, 159, 255–257
DEAVIN, DR ANTHONY 83, 89, 110, 124f, 143, 147, 271, 311, 314f, 323–325, 354, 370f
De Gaulle, General Charles 336
De la Mare, Rev Ben 363
DE LA MARE, RICHARD 34, 38, 101, 112, 141, 147, 177, 352, 363
De la Mare, Walter 440
Demeter symbol 87
Dentistry/dental health 38, 65, 164, 181–183, 186, 193–196
Department of the Environment 145
Derbyshire College of Agriculture and Horticulture 126
Desai, Pooran 120
Descartes, René 357
Deuteronomy 353
Dexter, Keith 46
Diet, national 58–60, 65, 163–165, 166–169, 178–181, 184f, 198, 234, 285, 337
Dimbleby, Prof. G.W. 264, 277
Dimbleby, Jonathan 288
Disease 30, 65, 123, 166–169, 178, 182, 185, 188–190, 194–196, 224, 232, 234, 266, 276
—, degenerative 179f, 186f, 230
—, result of human interference 214
—, Saccharine Disease 186f
Distributism 280, 349, 362
Dodd, Digby 120
Dodds, Sir Charles 64
Dogra, Bharat 282

Don, Monty 148
Donaldson, Frances 46
Donaldson, Gaye 247
Donaldson, J.G.S. 176f, 352
Donaldson, Stuart 113
Dorwest Growers 213
Doubleday, Henry 141
Douglas, Major C.H. 280
Douglas, F.C.R. (see DOUGLAS OF BARLOCH, LORD)
Douglas, J. Sholto 149f
DOUGLAS OF BARLOCH, LORD 37, 97, 307, 330
Doves Farm 202, 229
Dowding, Charles 156, 158, 256
Driesch, Hans 76
Drought Defeaters 146
Drugs 57, 62, 176, 179, 210, 365
—, sulphonamide 178
Drummond, Sir John 58
Dudley, Nigel 131, 253
Duff, Gail 109, 280
Duffy, Deryck 79f, 84, 297
Dumonteil, Caroline 251f
Duncan, Ronald 36, 38, 276, 281, 333
Dundee College of Further Education 128
Dunn's Farm Seeds 237
Durham University 146, 363
Duveen, Michael 255

Earth Summit, 1992 287
Easey, Ben 80, 137–139, 148
Eason, Dr E.H. 346
East Kent Packers 237
EASTERBROOK, LAURENCE 34, 36, 38, 44, 79f, 131, 139, 240, 276, 352, 367f, 374
Eastern Counties Organic Producers 254
Eastern philosophy 91, 120, 233, 352, 357, 359, 361, 369, 376f
EC Regulation 2092/91 114, 159, 247
Echlin, Dr Edward 351
Eco-fascism 275, 342f, 376
Eco-Logic Books 123
Ecological agriculture 120
Ecological humanism 363–65
Ecological Research Foundation 260
ECOLOGIST, THE 36, 80, 91, 107, 197, 200, 204, 232, 259, 265f, 275–84, 286, 325, 342, 358, 360, 362, 365f
Ecology 117f, 122, 249f, 258–266, 273, 283, 293, 318, 335, 357f
—, human ecology 263, 268, 272, 274, 285f, 364f
—, versus economics 260, 273
Ecology Party 281, 342f, 345
Economic Reform Club and Institute 33, 277, 330
Economics/money system 120, 235, 270, 354, 362
—, versus ecology 263
Edwards, Alexis 368
Eliot, T.S. 365
Elizabeth II 146
Elliot, R.H. 139, 308
Elliot-Smith, Arthur 108, 182, 187
Ellis, Charles 78
ELM FARM RESEARCH CENTRE 75, 89, 113, 129, 158, 202, 208f, 244, 246, 271, 302, 318–23, 372
Emerson College 88, 90, 92, 124f, 126, 316
English Array 330, 384
Enlightenment humanism 357
Entwistle, David 312
Environment/environmentalism 33, 46, 66, 76f, 88, 102, 104–107, 113–119, 151, 153f, 167, 169, 174, 192, 195, 204, 234f, 248, 250f, 253, 255f, Ch.6 passim, 298, 302, 308f, 312f, 317f, 325, 328f, 331, 335f, 337–339, 342–345, 347f, 350, 352, 354–61, 364–366, 370, 374
Eskimos 180
Esoteric thought 25, 39, 72, 77, 83, 93, 169, 233, 244, 323, 352f, 356, 361f, 366–372
Essene School of Life 215, 437
Essential fatty acids 180
Etté, Anthony 279, 365, 379
EU Regulation (2092/91) 114, 159, 247
Eugenics 164f, 183
European Economic Community 62
European Union 47
Evans, Anne 419
Evans, George Ewart 105
Evans, Stanley 57
Evening Standard (newspaper) 333
Everett, Rod 119
Evetts, E. and A. 212
Ewell Technical College (husbandry courses) 110, 124–126, 128, 149, 205, 271, 311, 323–325

FABER AND FABER 34, 38, 79, 101, 112, 138f, 141, 147, 178, 181, 233, 262f, 305, 363

Factory farming (see under Agri-business)
Farm Amalgamation Scheme 46
FARM AND FOOD SOCIETY 221, 276
FARMER, THE 36, 95, 99–101, 137, 199f, 212, 214, 223, 226, 327
Farmer and Stockbreeder 49
Farmers' markets 23, 255, 349
Farmers' Weekly 34f, 99, 214
Farming, mixed 51, 96, 133, 190, 203, 207, 286, 375
Feast of Albion 147, 441
'Feather-bedding' 43, 57
Fertiliser, Feeding Stuffs and Farm Supplies Journal 34, 52
Fertiliser Manufacturers' Association 52
Fertilizers, chemical/artificial 28, 35, 40, 44, 51–55, 61, 69, 71f, 74, 131f, 134, 136, 138f, 148, 184, 196, 203, 210f, 217, 240, 246, 264f, 277, 296f, 300, 312, 316, 338
—, 'National Growmore' 292
Festival of Mind, Body, Spirit 149
Field, The 333
Fillingham, Jolyon 119
FINDHORN 80, 233, 259, 307, 368f, 443
Findlay, Cdr Noel 86, 106, 125, 313
Finlayson, Henry 259
Finney, Albert 228
Fisher, Colin 302
Fisons 34, 40, 51, 142
Fitzwilliams, James 282
Fleming, Sir Alexander 187
Fletcher, Eileen 196
Fletcher, John 340
Flowerdew, Bob 148
Fluoridation of water supply 142, 180, 261, 282
Fogg, H.E. Witham 137, 147f
Foley, Grover 282
Food
—, faddism 180, 202, 261
—, health foods 223, 228, 235, 238, 241, 344
—, 'honest' 33, 60, 66, 169, 231, 245
—, industrial 58–66, 69, 190, 193, 196–199, 201, 210, 234
—, natural 65, 67f, 206, 210, 230, 232, 234–236
—, organically grown 146, 170, 177, 184, 223, 226–228, 230, 235, 244–247, 252f, 255, 321, 344, 347, 350
—, scares 113, 133, 197f
—, slow 281
—, traditional diets 164f, 166f, 178, 193f
—, whole 60, 63, 100, 103, 195f, 223–231, 251
Food and Agriculture Organization 298, 310
Food Commission 111, 253
Food for Britain campaign 110, 254
Food Magazine 111, 130, 196, 198, 348
Food miles 66, 249, 268
Food Programme 146
Food Standards Agency 179
Fordham, Montague 38
Forestry/forest farming and gardening 30, 33, 80, 112, 116f, 119f, 122f, 129, 149–151, 208, 259f, 272, 274, 279, 281–283, 286f, 310, 317
—, Responsible Forestry Programme 287
Forestry Commission 286
Forster, Steve 160
Fost, Rosemary 177
Four Quartets 366
Fox, Claire 328f, 358
Freeman, Noel 233
French, Mary 338
Frost, Anne 154
Frost, David 154, 160, 341
Fruitarianism 207
Fukuoka, Masanobu 120, 233
Fullerton, Dr Helen 125
Furner, Brian 104, 147f, 212

Gaia Theory 371, 376
Galileo 357
Gallimore, Patricia 146
Game Conservancy 315
Gandhi, M.K. 149
Gardener, The 100, 137
Gardening 33, 35, 80, 85f, 97, 103f, 109, 115, 123, 131, 133f, Ch.3 passim, 176, 190, 195, 203, 210–212, 228, 231, 234, 238f, 259, 267, 280, 284, 292, 304, 307, 309, 316, 336, 348, 359, 362, 367–369
—, cloches 139, 157
—, forest 143, 149
—, guerrilla 137
—, no-dig 138, 142
GARDINER, ROLF 29–31, 34f, 39, 71f, 77, 79, 85, 94, 106, 164, 182, 260, 294, 319, 330, 331–333, 335, 349, 352f, 356, 358
Gardyne, Maryel 369f
Gathergood, James 168, 356
Gavin, Sir William 40

Gay Hussars Dining Club 115
GEAR, ALAN 136, 139, 143–147, 149, 271, 352, 363
GEAR, JACKIE 136, 143–147, 149, 352, 363
Geddes, Sir Patrick 76, 173, 258, 282, 376
Geest Industries 255
Gemmett, John 157
General Agreement on Tariffs and Trade (GATT) 198
General Strike 137
Genesis 192, 353, 366
George, Henry 280, 323
George VI 138
Gerson, Max 103, 184, 196
Geuter, Maria 212
Ghebremeskel, K. 184
Gill, Eric 281f
Gill, Erin 113, 330
Girardet, Herbert 272f
Girling, Linda (see Theophilus, Linda)
Gleadells (grain traders) 92
Glen, Dr 157
Glentanar, Lord 84, 174, 331
GM crops 55, 68f, 112, 363
Goat World 271
Goethe 85, 89, 351
Goff, Ed 116
GOLDSMITH, EDWARD 24, 264–266, 275–276, 279, 283, 325, 337f, 359, 376–379
Good Food Guide 146
GOOD GARDENERS' ASSOCIATION 138
Goodman, Peggy 129, 205
Goodwin, Prof. Brian 173
Gordon, David 123, 134
Gordon-Canning, Captain R. 330
Goulden, Henry 86
Gourlay, Sir Simon 115
Graham-Little, Sir Ernest 97
Grant, Sir Alistair 247
Grant, Diane 184
GRANT, DORIS 147, 200, 202, 204, 224, 229, 282, 308
Gray, Cynthia 308f
GRAY, DR KEN 85, 157, 307–310, 317, 352
Great Chain of Being 209
Green Alliance 343
Green Books 284–286, 362, 365
Green Party 342–345

Greenall, Joy 135
Greig, David 109
Grier, D.B. 241
Griffith-Jones, Joy 106, 108, 199, 212
Griffiths, Richard 329
Griffiths, Vivian 78, 82f, 85, 365
Grigg, David 45, 47, 53
Griggs, Barbara 61, 64, 178, 227
Grigson, Sophie 146
Groundnuts Scheme 96
Grow Show Conference 158
Grower, The 161
Growing Concerns 343f
Grussendorf, Dr Werner 98
Guénon, René 362
Guild, Chevallier 86
Guild Gardener, The 33, 137
Gummer, John Selwyn 111, 159, 287, 346f
Gurdjieff, G. 233, 361, 371
Gypsies 215f, 270

Hahnemann, Samuel 218
Halik, C. 334, 353
Hall, Christopher 280, 285
Hall, Sir Daniel 28, 44, 48
Hall, Prof. Ross Hume 183, 279
Halley, Ned 71
Hamilton, Geoff 148
Hamilton, Lady 101
Hancock, Reginald 217–220, 371
Hardyment, Christina 67
Hari, Johann 431f
Harmony Foods 206, 230
Harper Adams College 19
Harper, Peter 121
Harris, Phil 160
Harris, Simon 27, 86, 100, 242
HARRISON, RUTH 49, 106, 132, 267, 274, 352
HART, ROBERT 123, 129, 149f, 345
HARVEY, GRAHAM 44, 53f, 57, 111, 114
Harwell Atomic Energy Establishment 313
Hasson, Lawrence 254
Hatfield, Audrey Wynne 213
HAUGHLEY EXPERIMENT 34, 80, 84, 95, 102, 107, 192, 194, 201, 290, 295–304, 316, 321, 323f, 326
Haworth, Don 152
Hawthorn, Prof. John 197, 277
Hay Diet 200
Hazell, Ethelyn 103, 177

Health Ch.4 passim
—, environmental factors affecting 155
—, health fascism, so-called 163f
—, problems in defining 166–169
—, process, not state 167, 173
—, as wholeness 122, 157, 165–168, 224, 296
HEALTH AND LIFE 32f, 39, 100, 106, 164f, 168, 197, 199, 203–205, 212, 227, 231, 276, 285, 331, 355
HEALTH AND THE SOIL 98f, 164, 189, 355
Health for All 103
Health Show Olympia 146
Healthy Life, The 32, 227
Heasman, Michael 60, 67, 69
Heaton, Kenneth 182f, 186
Hedges, value of 105, 131, 218, 220
Hennessy, Peter 59
Henriques, David 293
Henriques, Col. Robert D.Q. 100, 224f, 293, 366
HENRY DOUBLEDAY RESEARCH ASSOCIATION 83, 103, 110, 113, 136–139, 141–147, 154f, 160, 200, 212, 246f, 261, 274, 277, 279, 314, 324, 336, 340, 344, 372
Henry VIII 363
Herbalism 35, 85, 100f, 139, 157, 191, 205, 209–216, 220, 233, 271, 280, 318, 320, 324
—, School of Herbal Medicine 324
Herbert, Julian 293f
Herbicides (see under Agri-business)
Herb-royal Animal Health Association 214
Here's Health 148f
Heritage Seed Programme 145, 160
HERON, GILES 14, 27, 82, 352
HERON, MARY 14, 27, 82, 207, 324, 352
Heron, Tom 14, 362
Herriot, James 217
HICKS, SIR CEDRIC STANTON 97, 128, 191–193, 290, 293, 305
Hill, Dr Stuart 318
Hill, W.A. 29
Hillel, Daniel 366
HILLS, HILDA CHERRY 143–145, 252
HILLS, LAWRENCE 79, 117, 119, 136, 138, 141–147, 153f, 176, 229, 232, 242, 266, 271, 276, 279f, 282, 285, 305, 308, 317, 336, 340, 352

Hines, Colin 286
Hippocrates 218
HODGES, DR DAVID 72, 109, 131, 147, 197, 242, 312–318, 324, 346, 352
Hodges, Ursula 313
Hodgkinson, Alderman 175
HOLDEN, PATRICK 80, 90, 94, 110, 112f, 115, 123f, 154, 202, 229f, 241, 244, 246f, 250–252, 254, 287, 341, 345, 371
Holderness, Col. Hardwick 330
Holism 74, 76, 122, 322f, 325, 372–376
Hollins, Arthur 123, 134, 255, 344
Holmgren, David 116–119
HOME FARM (see *PRACTICAL SELF-SUFFICIENCY*)
Homoeopathy 218, 233, 243, 323
Honey, Bernard 119
Hony, Henrietta (see Trotter, Henrietta)
Hopes Compost Club 81, 99
Hopkins, Donald P. 134, 299, 375
Hopkins, Rob 123
Horace 187
Horder, Dr John 176
Horder, Lord 179
Horne, Malcolm 207
Horticultural Advisory Bureau and Training Centre 138
Horticulture (see Gardening)
Hosking, J.E. 16, 29f, 47
Houghton, Lord 20
Houriet, Robert 154
HOWARD, SIR ALBERT 30, 33–35, 54, 72f, 83, 95–99, 122, 124, 128, 133f, 137, 140f, 148, 165, 178, 189, 192–194, 209, 213, 215, 217, 225, 228, 262, 268, 276, 282, 290, 298, 304–306, 324, 333, 344, 352, 368
HOWARD, LOUISE E. 97f, 225, 305f
Howe, Dr E. Graham 233
Howell, Sir Evelyn 98
Hudson, G.N. 208
Hugh Sinclair Unit of Human Nutrition 180
Hughes, Grahame 255
Hughes, J.R. 369
Hugo, Victor 277
Hull, Geoffrey 263
Humphreys, Christmas 97
Humphrys, John 287, 371
Humus 133f, 140, 163, 196, 214, 219, 292, 299, 305, 310, 324, 374
Hunter, James 139

INDEX

Hunter, Kathleen 213
Hunter's of Chester 139
Hunza people 100, 133, 164, 166f, 178, 216
Husbandry 114, 116, 120, 125f, 130, 137, 149, 203, 205, 228, 260, 268, 271, 274f, 277, 279, 284f, 291, 293, 310, 314f, 328, 331f, 354, 357f, 365, 373, 378f
Huxley, Elspeth 100
Huxley, Sir Julian 277
Hyams, Edward 123, 192, 260, 278, 284, 327, 332f
Hyperactive Children's Support Group 197

Illich, Ivan 183, 202, 373
Imperial Chemical Industries (ICI) 34, 40, 42, 51,f, 298, 308
Imperial Tobacco 235
Indore Process of composting 97, 133, 137, 188
Infinity Foods 240, 245
INSTITUTE OF ORGANIC HUSBANDRY 35, 95, 99f
Institute of Public Cleansing 305, 307
Institute of Sewage Purification 307
Institution of Municipal Engineers 306
Intermediate Technology 267, 278, 362
INTERNATIONAL FEDERATION OF ORGANIC AGRICULTURE MOVEMENTS 75, 109, 130, 140, 157, 246f, 317, 322
INTERNATIONAL INSTITUTE OF BIOLOGICAL HUSBANDRY 72, 157, 240, 242, 309, 314–318
International Institute of Human Nutrition 180
International Nutrition Foundation 180
International Society for Fluoride Research 180
International Vegetarian Society 191
Iona Community 172
Irving, Charles, MP 347
Islam 360

Jacks, G.V. 123, 131, 258, 272, 283
Jackson, Sego 123, 344
Jacob and Esau (Old Testament) 68
Jarvis, Horace 339
Jenkins, Robin 109, 124
JENKS, JORIAN 24, 29, 31, 33, 37, 39, 73, 80, 83, 85, 99, 135, 166, 211, 215, 222f, 240, 244, 259, 261, 270, 276, 291f, 329f, 333f, 345, 348, 352, 354f, 363f, 373, 379
John, Augustus 215
Johnson, Colin 248f
Joint Working Party on Municipal Compsting 307
Jones, Ana 90–93
Jones, Sir Emrys 315
Jones, H.H. 203, 205
Jones, Nick 90–93
Jones, Paul 235
Jordan, Bill 241
Jordans 231, 241
Journal of the Soil Association 103, 295
Judaeo-Christian tradition 215, 352, 356f, 366
Judt, Tony 20, 339
Jungian psychology 305

Kagawa 149
Kassel University 129
Keeble, Brian 282, 284
Kennedy, President J.F. 335
Kenton, Leslie 347
Kerouac, Jack 106
KEYLINE SYSTEM 122, 128f, 149, 192, 283, 311, 344
Kiley-Worthington, Marthe 120
Kilmartin, Joanna 281
Kindersley, Peter 270
King, F.C. 97, 100, 137, 224
King, F.H. 118, 124, 133, 148, 237
Kingston Clinic, Edinburgh 228
Kingston Lacy 144
KINSHIP IN HUSBANDRY 16, 29, 34f, 38, 79, 286, 331, 336, 377f
Kirkham, Ellinor 97
Kitchener, Earl 147, 227, 229
Knapp, V.J. 184
Knights, Laurence 106, 187, 195f, 229
Knowles, G.K. 29f
Koepf, Dr Herbert 90, 317
Kolisko, E. and L. 324
Kowalski, Dr Robert 302
Krebs, Barbara (see Latto, Barbara)
Krebs, Hans 179f
Krebs, Dr John 179
Krishnamurti 371
Kropotkin, Peter 118, 349
KUMAR, SATISH 284f
Kwashiorkor (disease) 185

467

Laboratory of Human Nutrition 179
Labour Party 174f, 333, 335, 346
Lacey, Roy 109, 147f
Lachman, Gary 366f
Lackham College 128
Lady, The 200
Laing, R.D. 88
LAMPKIN, DR NIC 72, 74f, 126–128, 241, 311, 318, 341
Lancet, The 62, 180f, 190
Landau, Rom 83
Lang, Tim 60, 67, 69, 250f, 287
Langford, Andy 119
LANGMAN, MARY 86, 115, 129f, 170, 176f, 227–229, 231, 240, 324, 327, 329, 341, 354
Lao Tzu 359, 376
Larkcom, Joy 160
Latto, Barbara 182, 191
—, Conrad 182, 191
—, Douglas 182, 191, 204f
—, Gordon 176, 182, 191, 204
Lawrence, D.H. 30
Le Huray, Alfred 204
Lee, Ron 123, 134
Leneman, Leah 207
Lennon, John 236
Leopold, Aldo 128
Lethbridge, T.C. 369
Levens Hall 97, 137
Lévi-Strauss, Claude 68
LEVY, JULIETTE DE BAIRACLI 100, 139, 202, 209, 213–216, 218, 220, 232, 352, 366
Lewin, Miche Fabre 351
Lewis, Bryn 324
Lewis, Shirley 229
Lewis, Vincent 53
Leyel, Mrs C.F. 209–211
Liberal Ecology Group 343
Lightowler, Ronald 205
Limits 21, 73, 94, 187, 268, 277, 280, 360, 364, 372–374, 378
Lindsay, Lord 174
Ling, Riccardo 27, 106, 352
Linnean Society 147
Lister, Joseph 187
Little Gidding community 365f
LIVING EARTH 24, 104, 108, 111f, 124, 127, 130, 196, 220, 241, 250f, 287, 327, 346f, 348, 364
Living Marxism 290
Lobstein, H.E. 15, 260

Lobstein, Tim 111, 253
Local Exchange Trading Systems 120, 251
Lockeretz, Dr Willie 317
Loewenfeld, Claire 205, 210f, 213
Loftus, P. C. 333
London Medical Group 185
London Vegetarian Society 205
Londonderry, Marchioness Dowager of 330
Long Ashton Research Station 157
Lopez-Real, Joe 315, 317
Lorimer, David 76f
Loseley Park 125
Lost Knowledge 369
Lovelock, James 372, 376
Ludovici, Anthony 164f
Lymington, Viscount (see PORTSMOUTH, EARL OF)
Lynch, G.R. 62

Macdonald, Dr W.G. 292
Macfarlane, Helen 369
Mackarness, Richard 65f
MacKinnon, Prof. D.M. 171
Macleod, George 172
Macmurray, Prof. John 338, 364, 376
Maddern, Ralph 346
Mager, Chris 126
Mair, Chris 114, 157
MAIRET, PHILIP 24, 33, 39, 76, 258f, 272f, 277, 282, 331f, 352, 354, 356–363, 363, 376–378
Malnutrition 163, 185
Manpower Services Commission 145, 344
MANSFIELD, DR PETER 109, 168, 175–177, 244
Maoris 193f
Mapperson, Mr and Mrs 199
Marian, Dr Siegfried 95, 165, 331
Marketing 23, 36, 84, 87, 110, 114, 121, 124, 152, 157–160, 201, 223, 227, 237–241, 244–257, 347
Marks and Spencer 253
Marr, Andrew 58
Marriage, Clare 202, 229, 285
Marriage, Michael 113, 202
Mars Bar, notional organic 21, 253
Marshall, Bruce 119, 123
Martin, John 45, 53
Martin-Leake, Hugh 96, 97
Marxism 329, 336, 349, 378

468

INDEX

Marylebone Cricket Club (MCC) 311
MASSINGHAM, H.J. 36, 38, 66, 122, 124, 149, 276, 278, 280f, 284f, 327, 331–333, 345, 349, 352, 355, 363, 377–379
Massingham, Penelope 332
Matless, David 329
Matravers, Graeme 150
MAYALL, GINNY 109, 113, 245
Mayall, Richard 113, 219
MAYALL, SAM 73, 113, 125, 133, 202, 219, 274f, 308, 314,335
Maynard, Robin 111
McCARRISON, SIR ROBERT 38, 105, 149, 167–169, 178–184, 187, 189f, 193, 216, 232, 236, 290, 295, 300, 308, 352
McCARRISON SOCIETY 125, 175–177, 181–184, 190, 195, 200, 217, 282, 372
McCartney, Paul 236
McColl, Prof. Ian 182
McDonagh, J.E.R. 97
McROBIE, GEORGE 122, 314, 317, 337, 379
McSheehy, Trevor 300
Mead, Tim 112
Measures, Mark 321
Medawar, Sir Peter 277
Medical News 188
Medical Press and Circular 188
Medical Research Council 170, 182, 186, 277, 296
Medical Testament 163, 178
Medicine (see also Herbalism) 62f, 99, 132, 168, 170, 173, 182f, 185, 188, 190, 217, 228, 233, 276, 324
—, community 174f
—, preventive 163, 165, 176, 190, 328
—, vested interests of 216
Medworth, Sally (see Seymour, Sally)
Melchett, Peter 285f
MELLANBY, DR KENNETH 107, 131f, 264f, 281, 290, 325
Melton Lodge Farm 316
Melville, Dr Arabella 233, 248–249
MEN OF THE TREES 33, 39, 112, 232, 258f
Menuhin, Yehudi 91, 229
Merrill, Margaret 318
Metropolitan Water Board 303f
Michell, Keith 236
Mier, Dr Carl 79, 81, 86, 225, 367
Mier, Gertrude 367

Milk Marketing Board 127
Millen, Colin 346
Miller, Mrs Constance 335
Miller, Henry 233
Miller, Sylvia 119
Milligan, Spike 236
Millington, Dr Susan 209
Milnes-Coates, Robert 91, 134, 238
MILTON, DR R.F. 261, 297f, 325
Ministry of Agriculture 40, 42, 64, 110, 246, 270, 274, 296, 321f, 367
Ministry of Food 179
Mishan, Dr E.J. 277
MITCHELL, CHARLOTTE 252
Mollison, Bill 116–119, 122f, 283, 344
Monbrison, Christian de 129
Monks Wood Experimental Station 107, 264
Monro, Dr Jean 109
Moore, Elizabeth 106
Moore, Dr I.H. 99
Moore-Colyer, Prof. Richard 329
Morgan-Grenville, Fern 284
MOTHER EARTH 24, 33, 59, 73, 81, 96, 98, 100, 102–104, 106, 111, 132, 138, 151, 165, 186, 189, 192, 194f, 197, 199, 201, 203, 211, 214, 223, 228, 244, 259, 263, 267, 270, 277, 279, 291–295, 300f, 304, 323, 330, 333–336, 353, 356, 377
—, debate over name 291
—, 'scurrilous' journal 293
Moubray, J.M. 196
Mount, James Lambert 65, 125
Mount, Dr S.J. 181
'Muck and mysticism/magic' 81, 139f, 289, 369, 376
Muggeridge, Malcolm 279
Muller, Hans 441
Mumford, Lewis 272, 282, 357, 378
Murray, Charles 80f
Murray, Elizabeth 80f, 130, 229
Murray, Helen 80–83, 90, 92, 94
Murry, John Middleton 36, 442
Museum of English Rural Life (MERL) 27
Mutton, Geoff 157, 213
Myers, Adrian 123
Mysticism 139, 233, 289, 293, 369, 375f

Nabham, Gary 279
National Agricultural Advisory Service 44
National Agricultural Centre, Stoneleigh 180

National Coal Board 266, 311, 337
National Farmers Union 29, 85, 115
National Federation of City Farms and Community Gardens 123
National Front 337
NATIONAL GARDENS GUILD 33, 37, 137
National Health Service 37f, 58, 163, 165, 170, 173, 175, 178f, 187, 189–191, 217, 224, 333
National Institute of Agricultural Botany 29, 146, 157
National Institute of Medical Herbalists 212
National Organic Wine Fair 247
National Union of Agricultural Workers 29
Natural Food Associates 128
Natural Order 25, 36, 72, 96, 165, 167, 218, 235, 260, 354, 369, 370, 374, 376–378
Natural Rearing Products 215
Nature 29, 72f, 75, 119f, 168, 190, 204, 210, 214, 220, 232f, 291f
—, balance of 77, 99, 135, 150f, 165, 213, 261f, 299, 350
—, humanity's relationship with 24f, 78, 88, 134, 204, 259, 261, 268, 270, 273, 288, 297, 323, 351, 354f, 357–359, 361–363, 365f, 368–370, 373, 379
—, instrumentalist view of 30, 32, 87–89, 155, 216, 268, 378
—, remedies provided by (see Herbalism)
—, unconquerable 187, 359, 370
—, wholeness of 98, 165, 213
Nature (journal) 291
Nature Conservancy 260, 264
Nature et Progrès 109, 130, 157
'New Age' 191, 207, 232, 285, 325, 350, 354, 368f
New Earth Charter 204, 259, 355
New Ecologist 117
NEW ENGLISH WEEKLY 14, 33–35, 38f, 76, 95, 114, 122, 285, 331, 340, 354, 366
NEW FARMER AND GROWER 74, 113–116, 119, 127, 135, 150, 158–161, 202, 213, 220, 247f, 251, 271, 327, 344f, 371
New Left Review 341
New Pioneer 285
New Scientist 261f
New Statesman 332

New Zealand Humic Compost Club/ New Zealand Organic Compost Society 193
Newport, Viscount (see BRADFORD, LORD)
News Chronicle 34, 131
News-Letter on Compost 34, 95
Nicholson, E.M. 260f
Nicol, Dr Hugh 198f, 338, 373
Nightingale-Smith, Dennis 27
1999 Committee 285f
Nitrogen/nitrates 52, 60f, 89, 131f, 134, 136, 196, 240, 282, 320, 322, 324
Noble, Patrick 23, 350
North, Richard 281
North, Richard A.E. 197
Northbourne, Lord (4th Baron, 1896–1982) 71, 362, 370
Northbourne, Lord (5th Baron, b. 1926) 370
Northern Organic Farmers' Group 238
Northwood Boxes 256
Notes and Correspondence 79
'NPK mentality' 52, 96, 292
Nuffield Foundation 43
Nutrition 35f, 38, 42, 58–66, 68, 80, 86, 89, 94, 96, 98, 100, 106, 109, 111, 114, 143, Ch.4 passim, 223, 229, 233, 235, 252, 269, 272, 279, 290, 293, 295, 300, 302, 320, 328, 334, 355, 374
—, as a cycle 296
—, as a form of eugenics 164f
NUTRITION AND HEALTH 175, 183f

Obesity 185 197
Observer, The 87f, 107, 131, 137, 141, 262, 291, 337
O'Connell, Ciara 184
Official Architect 332
O'Hagan, Lord 268
Oil 54, 60, 67, 108, 125, 133, 142, 268, 272, 281, 284, 318, 340f, 348–350, 360, 365, 373
Oldfield, Lady 112
'On Your Farm' (Radio 4 programme) 110
Ono, Yoko 236
Operative Ignorance, Law of 373
Optimism 30f, 55f, 59, 69, 159f, 183, 235, 248, 273, 279, 317, 339, 353
ORGANIC ADVISORY SERVICE 126, 254, 321

'Organic analogy' 338
Organic Farm Foods 154, 245, 249
ORGANIC FARMERS AND GROWERS 92, 108, 157, 238–242, 246, 309, 314, 316, 347
Organic farming
—, aesthetic appeal of 114, 134f, 302, 375f
—, alternative to agri-business 31, 54f, 72, 76f, 114, 130, 133, 220, 281, 344, 347, 378f
—, biological efficiency of 135, 260
—, ecological approach of 260, 297f, 318
—, financial problems of 34, 92, 102, 107, 110, 114f, 135, 174, 225f, 237, 239, 253f, 256, 266, 295, 297, 300–302, 316f
—, preventive medicine, a form of 163
—, virtuous circle created by 135, 307
Organic Farming (journal) 116, 150
Organic Food and Farming Centre 247
Organic Grower 150
Organic Growers' Alliance 150, 156
ORGANIC GROWERS' ASSOCIATION 113, 115, 126, 150, 155–162, 238, 241, 245, 249, 253, 321, 341, 344, 372
Organic Growers' Co-operative 254
Organic Living Association 27
Organic Living Party (need for) 340
Organic Marketing Committee 238f
Organic Marketing Company 238
Organic movement
—, 'beautiful people' (see also Celebrities) 23
—, commercialism of 17, 90, 222f, 247–254
—, complexity of 19–21, 25, 341, 351
—, continuity of 14, 22, 85, 134, 140, 189, 200, 202, 252, 266, 285
—, education and training 124–128
—, existence debatable 21
—, lapsarian mythology of 165f
—, local/regional, preference for 32, 45, 116, 120, 122, 247f, 251, 255–257, 268, 274f, 278, 349f, 378
—, organic living, joyfulness of 232
—, philosophy of 356–365, 372–379
—, politics of 325–351
—, religious and spiritual dimension of 351–379
—, science, attitude to (see Science/scientific method)

—, variety of views within 21f, 327–329, 351f, 372
Organic produce (see also Marketing)
—, curing cancer, role in 184
—, health benefits of 199, 206, 234
Organic Standards Committee 85, 201, 243, 245, 322
Organophosphates (OPs) 156, 262
Orr, Sir John Boyd 58
Orwin, C.S. 28, 44f
Otley College 128
Our Daily Bread (film) 105, 201
Ouspensky, P. D. 361, 369
Owen, C. Langley 97
Oxfam 346
Oxford Farming Conference 104, 108
Oxford University Plough Club 363
Oyler, Philip 203, 205, 352

Paganism 232, 355f, 379, 443
Papworth, John 269, 279, 282, 352
Paracelsus 215
Pardi, Tommaso del Pelo 129
Parsons, Robert H. 377
Pasteur, Louis 187
Patten, Marguerite 146
Paulin, Dorothy 373
Payne, Alan 160
Payne, Sandra 160
Peacock, Paul 362
Peake's Farms 226
Pearmain, Gabriel 206
Pears, Pauline 144–146
Peasantry 118, 124, 215f, 237, 251f, 270, 310, 349, 379
Pease, Mrs 367
Peck, Rev David G. 354
Peel, John 236
Peers, Charles 240, 242
Pegson-Marlow Pump 41
Peoples Post 34, 165, 331
Pepper, Allan 171, 354
Peredur School 86
PERMACULTURE 116–124, 129, 135, 149, 208, 272, 283f, 344, 375
Permaculture Association of Great Britain 116
Permaculture News 119f, 123
Pershore College of Horticulture 128, 159
Pest control, biological 75, 99, 127, 142, 144, 148, 157, 160, 271, 320
Pesticide Action Network 131

Pesticides (see under Agri-business)
PFEIFFER, EHRENFRIED 34, 79, 83, 85, 139
Pickering, Pamela 200
PICTON, DR LIONEL 34, 38, 95, 100, 163, 177, 198, 200, 308, 352
Picture Post 100
Pink, Dr Cyril 100, 224
PIONEER HEALTH CENTRE 33, 35, 38, 81, 86f, 94, 101, 161, 164, 166, 169–173, 177, 221, 294, 332, 350
Pioneer Health Centre Ltd 171–173, 175, 177
Planners/planning 31, 43–45, 163, 263, 278, 337
Plato 376
Polarity Therapy 324
Poliakoff, V. 97
Politics 25, 37, 174, Ch. 8 passim
—, bio-politics/eco-politics 283f, 337–340
Pollan, Michael 61, 193
Pollard, Nigel 366
Pollitt, G.P. 264
Polyculture 118
Poore, Dr G.V. 209, 277, 305
Pope, Alexander 353
Pope John XXIII 358
Population control 204, 208, 230, 276, 278, 338
PORRITT, JONATHON 115, 123, 253, 281, 283–285, 342f, 352
PORTSMOUTH, EARL OF 28–33, 36–38, 164, 276, 330f, 333, 335, 352, 363
Potato Marketing Board 99
Pottenger, F.M. 193
Potts, Dr G.R. 315
POWELL, L.B. 36, 276
PRACTICAL SELF-SUFFICIENCY 270–272
Pratt, Simon 123
Precautionary Principle 30
Preventive medicine (see Medicine)
Price, Prof. H.H. 171
PRICE, WESTON A. 180, 193, 203
Priestley, Dr 174
Prince Philip 146, 401
Private Eye 338
Produce Bulletin 225
PRODUCER CONSUMER WHOLE FOOD SOCIETY LTD. 224–226
Progressive Farming Trust 302, 319

Pye Charitable Trust 301
PYE, JACK 300, 302, 321
Pye, Mary 300
Pye Research Centre 301f
Pye-Smith, Charlie 285f
Pyke, Magnus 59–61, 66, 104, 276

Queen Mary 170
Quinny, John 116

Radical Right politics 25, 37, 289, 328–330, 412
Randal, Derek 100, 199, 224f
Rank, Joseph 64
Rank Hovis McDougall 64
Raven, Charles 171
Raven, Hugh 115, 250f
Ray, Joanna 191
Rayner, M.C. 38
Read, Steven 123
Reading University 27, 114, 127, 180, 277, 317
Reading Vegetarian and Food Reform Society 204
Real Foods, Edinburgh 252
Redgrave, Michael 228
Reductionism 94, 122, 130, 151, 171, 276, 325f, 364, 367
Reed, D.B.C. 280
Reed, Matthew 20
Reich, Charles 154
Reid, Tony 27
Rennie, Alastair 157
Research Institute of Organic Agriculture (FiBL, Switzerland) 89
RESURGENCE 92, 269, 275, 279, 282, 284f, 338
Revolutionary Communist Party 290
Reynolds, Fiona 115
Ribbentrop, J. von 330
Rickard, Sean 47
Ridley, Mr 97
Ridley, Nicholas 159
Ridsdill-Smith, Marcus 242
Riverford Organics 257
ROBB, PROFESSOR R. LINDSAY 73, 82, 151, 263, 266, 276, 298f, 352, 368
Robertson, Fyfe 100
Robertson, Noel 55f
Robinson, G.W. 310
Robinson, John 365
Roche, Laurence 279
Rodale, Jerome 97, 178

Rodale Press 71
Roddick, Sam 439
Rollinson, F. 29, 31
Rolt, L.T.C. 267, 362f
Rose, Julian 248, 250, 252, 285
Rose, Kenneth 106, 182, 193–195
Rose, Walter 284
Rosko, 'Emperor' 236
Roszak, Theodore 373
Roth, Mike 119f
Roth, Peter 81
Rothamsted Experimental Station 103, 298, 308, 315, 323
—, Broadbalk field 298
Rowlands, Gareth 285, 312
Rowlands, Rachel 134
Rowntree, B.S. 42, 51, 96
Roy, Harcourt 104
Royal Agricultural College, Cirencester 150, 156, 315
Royal Agricultural Show 105, 309
Royal Agricultural Society of England 159, 309, 344
Royal Army Veterinary Corps 217
Royal Bank of Canada 261
Royal College of General Practitioners 190
Royal College of Physicians 187
Royal Dutch Shell 109
Royal Horticultural Society 146, 309
Royal Institute of Public Health and Hygiene 187
Royal Society 178
Royal Society for the Prevention of Cruelty to Animals 217
Royal Society of Arts 140, 178
Royal Society of Medicine 182f
Rudd, Geoffrey 205
Rudel, Joan 86
Rudel, Siegfried 86, 93, 300
Rule of Return 35f, 71f, 96, 105, 118, 122, 122, 133, 232
RURAL ECONOMY 33, 95, 122, 330, 343
Rural life 30–33, 114, 269, 281, 285, 338, 358, 375
RURAL RECONSTRUCTION ASSOCIATION 33, 37–39, 330
Rusch, Dr Hans P. 323
Ruskin, John 353
Rust, Deidre 86, 242
Rust, Michael 78, 86, 106, 193, 205, 239, 242, 320

Ryton, HDRA trial ground and gardens 115, 144–147, 160, 247, 363

Sackville-West, Vita 137
SAFE Alliance 115, 250
Safeway supermarkets 247
Sahara University Expedition 259
Sainsbury's supermarkets 256
Saint Christopher's School, Letchworth 86
Sale, Kirkpatrick 77
SAMS, CRAIG 21f, 152, 197, 202, 206, 230–237, 252f
—, Green and Black's 230, 235
—, Whole Earth 230, 240
SAMS, GREGORY 152, 197, 206, 230–236, 350
—, Real Eat company 230
—, VegeBurgers 230
Sanderson-Wells, Dr T.H. 164
—, Lecture 192, 293
Sandy, Dr Clive 334, 353
Saunders, Robert 333
Savage, Winifred 106
SAXON, EDGAR 32f, 39, 106, 165f, 168, 196, 202, 204, 212, 331, 352, 355f
Scharff, Dr J.W. 106, 188, 304
Schofield, Alan 156, 161
Schofield, Debra 156
SCHOFIELD, LILIAN 227–231, 240
School of Economic Science 314, 323, 371
Schulberg, Budd 265
SCHUMACHER, E.F. 24, 54, 56, 88, 102, 106f, 109, 122, 124f, 132, 148, 153, 182, 202, 237f, 247, 266–269, 274f, 288, 308, 314, 318f, 337, 339f, 352, 359–365, 370, 377
—, religious philosophy of 359–362
Schweitzer, Dr Albert 357
Science/scientific method 21, 30, 35, 51f, 55, 59, 62–64, 69, 71, 74, 76, 83, 87–90, 94f, 98, 104, 119f, 125, 133, 138, 140, 143, 146, 148, 151, 157, 170f, 180–182, 184–187, 193–197, 201, 205, 214, 217f, 220, 259–264, 273, 276–278, Ch.7 passim, 338, 356–360, 364, 367f, 370, 372, 374, 376
—, ruination of true farming 216
—, vested interests and 41, 52, 69, 135, 201, 216, 262, 290
Scofield, Dr Tony 72, 314–318
Scopes, Nigel 157

Scott, Sir Leslie 28
Scott, Peter 277
Scott Report 28
Scott-Samuel, Alex 173
Scottish Academic Press 171f
SCOTTISH SOIL AND HEALTH GROUP 95, 98f, 188, 355
Scullion, J. 312
Seaweed 99, 133, 140, 142, 148, 336, 434
Secret Highway, The (film) 105, 300
Secrett, F.A. 134, 139f
Secularism 25, 84, 278, 358, 370
Seddon, Quentin 52
SEED 152, 197, 206f, 212, 230–236, 275, 279, 339f, 368f
Seed Restaurant 230
Seeds, patenting of 282
SEGGER, PETER 24, 90, 100, 109f, 112f, 153f, 157f, 160f, 196, 229, 244–246, 249, 251, 274, 311, 320, 341, 344, 371
Seifert, Alwin 77, 431
Self-sufficiency 36, 60, 99, 109, 118, 124, 143, 149, 152f, 203, 269–276, 280, 329, 341–344
Sempill, Lord 330, 333
Sensby, Malcolm 160
Sense About Science 289, 324
Seton, Lady Frances 137
SEVENTIES GENERATION 21f, 25, 87, 89f, 92, 102, 110, 115, 122, 150, 153, 236, 238–240, 244, 247, 251, 278, 329, 340–342, 345, 350–352, 362, 371, 373
Sex Pistols 58
SEYMOUR, JOHN 36, 109, 152–154, 202f, 232, 269f, 272–274, 280, 284f, 349, 352, 362–365, 377
Seymour, Sally 152, 270
Shears, Dr Curtis 206, 435
Shelley, Percy Bysshe 193
Shephard, Gillian 287
Sheppard, Rev Dick 99
Shepperd, Graham 155, 243–245
Sherman, Bob 145
Shewell-Cooper, Ramsay 139
SHEWELL-COOPER, DR W.E. 98, 138f, 232, 271, 352
Shiva, Dr Vandana 160, 282
Shoard, Marion 48f, 132, 281
Short, Rendle 186
Shuttleworth College 128
SINCLAIR, HUGH M. 139, 165, 179–182, 229, 290

Sisson, C.H. 356
Skinner, B.F. 88
Small farmers 46, 255, 283, 286, 349, 362
Small Farmers Assistance Scheme 46
Small Farmers Association 271
Smallholders' Training Centre 274
Smarden (poisoned land) 131
Smith, Dr Michael 346
Smuts, J.C. 76, 373–375
Snell, Peter 283
Snow, C.P. 88f
Soames, Nicholas 146
Social Credit 33, 37, 280, 331
Social Movement Theory 20
Socialism 235, 330, 332f, 337
—, Fabian 332
—, Guild 332f, 349
—, Liberal 332f
—, Monopoly 332
—, National 333
—, 'Real' 332
—, Regional 327, 333, 349
—, Socialist Party of Great Britain 339
—, State 37, 331–334, 337, 349
Socialist Environment and Resources Association 343
Society of Friends (Quakers) 99, 313
Society of Herbalists 209f
Society of Medical Officers of Health 307
Soil erosion 33, 99, 109, 131, 155, 258, 260, 272, 283, 285, 310, 314, 318
SOIL AND HEALTH 34f, 95f, 98, 189
SOIL ASSOCIATION (main references only. See also: BALFOUR, EVE; BRYNGWYN PROJECT; HAUGHLEY EXPERIMENT; *LIVING EARTH; MOTHER EARTH; SOIL ASSOCIATION QUARTERLY REVIEW; SPAN;* WHOLEFOOD) 101–113, 124f, 226–230, 286–288
Soil Association, The (journal) 108
SOIL ASSOCIATION QUARTERLY REVIEW 108, 111, 149, 188, 198, 200, 208, 213, 239, 246, 265, 274, 312f, 321, 324
Soil Magazine 95, 165, 331
Soil Survey 346
Somerset Organic Producers 159, 254–256
SOPER, JOHN 79f, 83, 367

INDEX

Soper, Marjorie 83
Sowden, H.W. 339
SPAN 106f, 109, 126, 205, 212, 238, 264, 269, 337–339
Spearing, Anne 249
Spedding, Sir Colin 130, 241, 246, 317, 322, 374
Spiritualist Association of Great Britain 367
Spowers, Rory 352, 359
Springhead/Springhead Trust 319
St. Ignatius of Loyola 360
St. Paul 375
St. Thomas Aquinas 361
Stable society 278
Stallibrass, Alison 171f
Stamp, Terence 235f
Standards 74f, 87, 92f, 110, 112, 130, 157–159, 161, 201, 208, 220, 222, 226, 231, 236, 238, 240–248, 286–288, 290, 319, 322, 324, 344, 347f, 350
—, Conservation Grade 92, 241
Standen, Anthony 294
STAPLEDON, SIR R. GEORGE 35, 38, 79f, 129, 149, 237, 258, 263, 274, 286, 308, 364, 373
STAR AND FURROW 79f, 83, 90, 212, 356, 367
Starling, Bill 92, 202, 208, 344
Steadman, Ralph 287
Steele, Judy 160
STEINER, RUDOLF 33, 37, 72, 77, 79, 83, 85–90, 92–94, 124, 127, 244, 289f, 317, 323, 352, 356, 366–368, 371, 376, 379
Stephenson, ,W.A. (Tony) 336
Stevens, John 157
Stevenson, G.D. 29, 31
STEWART, DR VICTOR I. 127, 129, 157, 310–312, 317
STICKLAND, DAVID 72, 78, 91f, 107f, 157, 231, 236–245, 251, 266, 309, 314, 316f, 319, 339, 347, 363f, 379
Stickland, Sue 145f, 271
Stitt, Sean 184
Stone, Prof. Dan 329
Stonehouse, Bernard 315
Stott, Catharine 327
Stott, Martin 342f
Stout, Ruth 108
Strange, Penny 116, 119, 272, 283
Stuart, Mick 423
STUART, ROBERT L. 84, 98f, 108, 177, 189, 306, 365

Stuckey, D.R. 333
Sufism 370
Sugar 57, 62f, 65, 186f, 189, 195, 230, 235, 281
Sun, The 151
Supermarkets 23, 67, 112, 158, 234, 240, 244, 246–252, 254f
Surveyor, The 306
Swanley Horticultural College 138
Sweeny, Sedley 155, 182, 273–275, 282
SYKES, FRIEND 34, 72, 80, 84, 101, 133, 177, 199, 243, 274, 280, 297, 308
Szekely, Prof. Edmond 215

Talbot, Kathleen 255
Tancock family 312
Taoism 120
Tati, Jacques 280
Taverne, Lord 289f, 323f, 367
Tawney, R.H. 330
Taylor, Elizabeth 255
Taylor, Dr Geoffrey 282
Taylor, Bishop John V. 373
Technology/technics 21, 25, 29f, 41, 43, 51, 54–56, 58–63, 66, 69f, 88, 114, 116, 120, 130, 148, 183, 185, 210, 216, 232–235, 249, 267, 272f, 278, 280f, 294, 317, 319, 337, 340, 342, 349, 357–359, 361, 370, 373, 377–379
Temple, Jack 148f
Temple, Archbishop William 353
Templegarth Trust 176
Tennessee Valley Authority 273, 277
Tesco supermarkets 247
TEVIOT, LORD 37, 200, 330
Thames Water Authority 303
Thatcher, Margaret 110, 250, 333f, 350
THEAR, KATIE and DAVID 270–272, 285
Theophilus, Linda 24
Theosophy 366
Third World Organic Support Group 146
Thompson, Michael 242
Thompson, Richard 242
Thorburn, James 104
Thorpe, W.H. 263
Tibet Society 274
Tibetan Ecoforestry Training Partnership 274
Tibetan Farm School 274
Tillich, Paul 171
Time Out 126

Tinker, Jon 265
Tolhurst, Iain 155–158, 160–162, 202, 254, 342
Tolhurst, Lyn 155f
Tolstoy, Leo 32, 349
Tomlinson, H.W. 29
Topping, Jane 224
Torrey Canyon disaster 337
Townley, J.S. 306
Townsend, Prof. Peter 172
Trading Standards Office 246
Transition Towns 23, 123, 349
Travers, Pamela L. 285
Trees 33
Trotter, Douglas 171f, 174, 177, 352
Trotter, Henrietta 172, 177
Trowell, Dr Hugh 182, 184f, 352
Truscott family 340
Turner, Everard 106, 181f, 193–195
TURNER, FRANK NEWMAN 35f, 72, 96f, 99–101, 133, 141, 147, 199f, 209, 213f, 218, 220, 223–226, 243, 251, 254, 327, 352, 355
Tustian, Brian 255

UK REGISTER OF ORGANIC FOOD STANDARDS 112, 130, 241, 246–248, 287f, 322, 347
Ulbricht, Dr Tilo 315
United States Department of Agriculture 75
Urwin, David 249f

Vaihinger, H. 368
Van Allen, Judith 234
Veblen, Thorstein 361
Vegan Society 207
Veganism 202, 204f, 207–209
Vegetarianism 20, 85, 87, 91, 152, 155, 185, 191, 202–209, 211–213, 223, 228, 230–232, 344
Venables, Bernard 306
Vernon-Dier, Dick 254
Vesey-Fitzgerald, Brian 215
Veterinary treatment 141, 181, 214f, 217–220, 243
Vickers, Brigadier A.W. 107, 144, 151, 192, 230, 237, 265, 324, 339
Vickery, Dr Kenneth 181f, 187, 195, 352
Village Produce Association 255
Vine, Anne 73f
Vitrition Ltd 222
Voelcker, J.A. 282

VOGTMANN, DR HARTMUT 89, 129, 157, 184, 283, 317, 319f, 322
Voisin, André 129
VOLE 232, 275, 279–281, 342

Wacher, Carolyn 89f, 153f, 161, 341
Wacher, Charles 114, 147, 153–155, 157, 160, 371
Wad, Y.D. 305
Waddington, Prof. C.H. 277
Wade, Julian 316
Waitrose supermarkets 252, 255
Waksman, S.A. 262
Wale, Michael 430
WALES, HRH THE PRINCE OF 77, 112, 146, 321
WALLER, ROBERT 24, 78, 80, 89, 100, 102f, 107, 111, 122f, 129, 140, 153, 175, 193, 237, 244, 263, 266f, 269, 274, 276, 279, 282, 285f, 335–337, 339, 344, 352, 363–365
Walston, H.D. 31f
Walston, Oliver 31, 115, 286, 430
Walters, A. Harry 182, 197, 266, 282
Walters, Kate 420
War Agricultural Executive Committees 29, 40
Ward, Colin 170f, 280, 350
Warren-Davis, Ann 212f
Watchom, Margaret 123
Water, pollution of 54, 96, 118, 120, 143, 196, 264, 281, 283, 295, 306f, 309, 344
Waterman, Charles (see Davy, Charles)
Watson, E.F. 96
Watson, Guy 257
Watson, Prof. J.A. Scott 291
Watson, J.D. 88
Watson, Lyall 340
Webb, James 444
Weeds 53, 75, 132, 134, 137, 152, 160, 213, 218
—, value of 134, 213, 218, 220
Welfare State, totalitarian nature of 170
Wellcome Trust 170, 179
Welsh Agricultural College 127
West Wales Organic Growers 254
WESTLAKE, DR AUBREY 79, 96, 168f, 176, 182, 215, 352
Westlake, Jean 215
Westminster Bank 41
Wheals, John 139
Wheel of Life 36, 77, 134

White, Col. H.F. 192
White, Lynn, Jr 352
Whitefield, Patrick 118, 121f
Whiteway Colony 280
Whitley, Andrew 202
Whittaker, Richard 103, 109
Whitwell, Betty 343, 345
Whole Food (newsletter) 225f
Whole Food Mark 224f
WHOLEFOOD (Soil Association shop) 91, 107, 190, 227–230, 240
Wholefood Trust 147, 229
Wholeness 21–23, 30, 66, 94, 98, 107, 122, 151, 157, 165–168, 172, 181, 222, 224, 245, 296, 303, 370, 372–376
Whyte, R.O. 123, 131, 258, 272, 283
Wicker Man, The (film) 443f
Widdowson, Elsie 64
Widdowson, R.W. 74, 129, 157, 222, 374
Wigens, Anthony 118, 344
Wightman, Ralph 36, 292
Williams, Dr Christine 180
WILLIAMS, DINAH 100, 125, 134, 155, 274, 311f
Williams, Stanley 100, 224
Williams, Tom 31, 37, 41f, 49, 52
WILLIAMSON, GEORGE SCOTT 34f, 38, 80f, 101f, 166f, 169–173, 175–178, 191, 214, 217, 294, 296f, 332f
Williamson, Henry 280, 285
Willson, Richard 279
WILSON, C. DONALD 82, 84, 98, 223, 226f, 293, 307, 335, 352
Wilson, Harold, MP 337
Wilson, Roy 80, 133
Windmill Hill City Farm 123
Winward, Mr 49
Wood, Barbara 267
Wood, Dave 432
WOOD, MAURICE 199
Woodley, Helen 119, 344
WOODWARD, LAWRENCE 21, 75f, 112f, 115, 129, 134, 229, 245, 248, 252f, 302f, 318–323, 328, 340–342, 374

Wookey, Barry 76, 114, 134, 219, 321, 341
Worcester College of Agriculture 126
WORKING WEEKENDS ON ORGANIC FARMS 109, 125f, 154, 212, 344
World Health Authority 166
World Health Organization 185
Worldwatch Institute 283
Worldwide Fund for Nature 112
Worthington, Jim 104, 276
Wrench, Dr G.T. 164f, 167f
Wright, Alec 160f
Wright, Iain 252
Wright, Mandy 160
Wye College, Kent 151, 269, 298, 312–315
Wylie, J.C. 305
Wynne-Tyson, Jon 208

Yates, P. Lamartine 332
Yellowlees, David 80
YELLOWLEES, DR WALTER 37, 80, 178, 181, 183, 185–187, 189–191, 200, 229, 232, 327f, 352
Yeo Valley brand 112
YEOMANS, P. A. 122f, 128f, 149, 192, 283, 311
Young, Mary 219
YOUNG, RICHARD 115, 130, 134, 216, 219–221, 285
YOUNG, ROSAMUND 134, 216, 219–221
Yudkin, John 63, 152

Zipperlen, Helen (see Murray, Helen)
Zuckerman Report 306f

Also by Philip Conford

The Origins of the Organic Movement

With recent media hysteria and public concern about BSE and genetically-modified crops, we could be forgiven for thinking that the upsurge of interest in organic farming is a reflection of modern debate, and a vindication of what left-wing alternative groups have been advocating for years.

However, in this first and authoritative history of twentieth century "green" culture, Philip Conford reveals that the early exponents of the organic movement actually belonged more to extreme right-wing, conservative groups, which were reacting to industrialization and the increasing threat to traditional country life, closely associated with socialist politics.

Drawing on a wealth of contemporary sources, Conford chronicles the origins of the organic movement in Britain and America between the 1920s and 1960s, and offers a perceptive portrayal of an organization which believes implicitly in the positive acceptance of the natural order and its laws. The author demonstrates convincingly that organic farming is not a recent issue, and traces the evolution of this now thriving movement.

With the recent EU directive banning animal products in animal feed, together with the public backlash against genetically-modified crops, this book provides powerful reinforcement to a debate that has raged for over a century, and which affects us all.

Praise for *The Origins of the Organic Movement*

'This is an excellent book ... the underlying issues are the most crucial of our time ... fine and valuable read.'
– *The Independent*

'At last, a thoroughly researched history of the organic movement! Jonathan Dimbleby's introduction sets the tone ... this book is authoritatively researched, with extensive glossaries and appendices. This book is alive with inspiration.'
– Ysanna Spevack, Organicfood.co.uk, Spring 2003

'For people like me and for many others within the organic movement, this book is a long-awaited and much needed exposition. I've been particularly excited to read about the earliest origins from around 1926. By detailing the contributions of each of the key players, it sets out, in possibly the best way ever done, what organic farming is and what it hopes to achieve.'
– Richard Young, Soil Association

'An invaluable guide to the hard facts behind the Organic Movement written in a clear and concise style.'
– *Worcester Evening News*

'Fluent, humane, elegant and beautifully balanced.'
– *Rural History*

'Conford has written an excellent and very readable book. A fine reference work. If you want a good example of "straight" environmental history writing, read this book.'
– *Organization and Environment*, March 2003

'This delightful volume contains a treasure trove of unusual information linked to inspired and determined people.'
– *The Food Magazine*, July/Septmber 2007